VIRTUAL LM

A PICTORIAL ESSAY OF THE ENGINEERING AND CONSTRUCTION OF THE APOLLO LUNAR MODULE

The historic spacecraft that landed man on the Moon

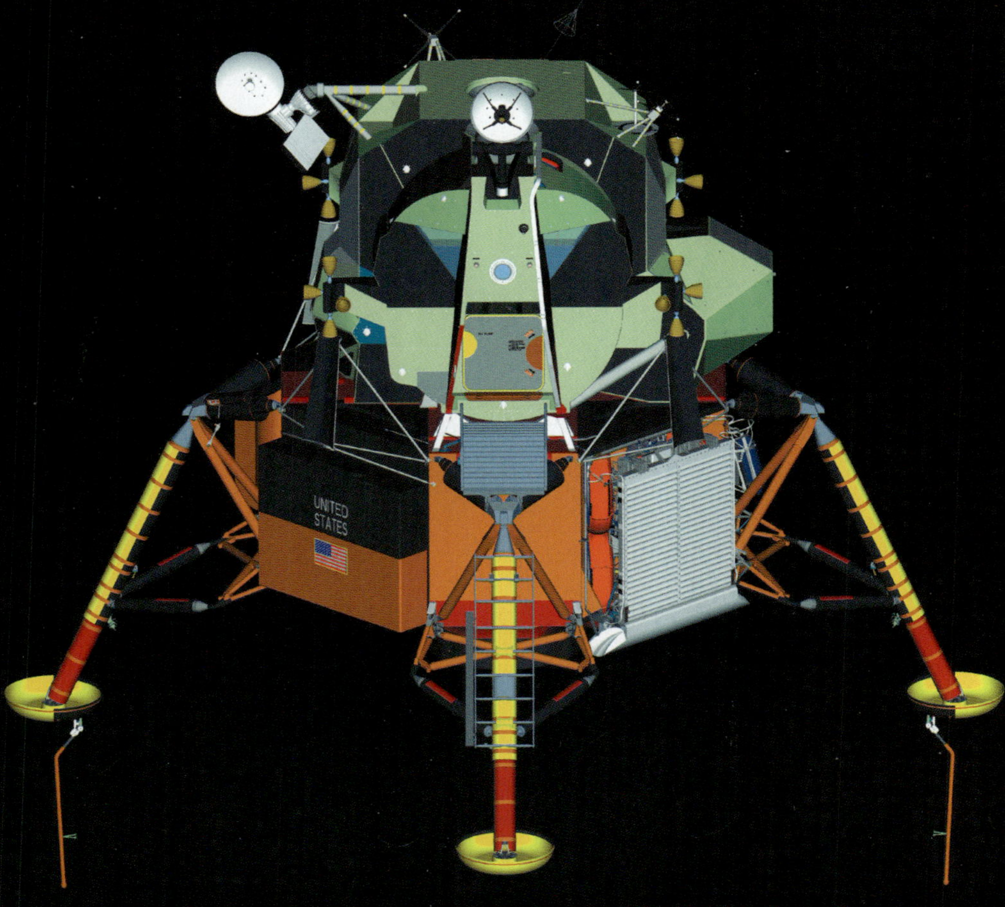

Also inside: The Lunar Rover and PLSS backpack

Scott P. Sullivan

We acknowledge the financial support of the Government of Canada through the Book Publishing Industry Development Program for our publishing activities.
Published by Collector's Guide Publishing Inc., Box 62034,
Burlington, Ontario, Canada, L7R 4K2 www.cgpublishing.com
Printed and bound in Canada
VIRTUAL LM
by Scott P. Sullivan
ISBN 1--894595-14-0
ISSN 1496-6921
Apogee Books Space Series

"First, I believe that this nation should commit itself to achieving the goal, before this decade is out, of landing a man on the Moon and returning him safely to the Earth. No single space project in this period will be more impressive to mankind, or more important for the long-range exploration of space; and none will be so difficult or expensive to accomplish."

From President John F. Kennedy's Special Message to the Congress, May 25, 1961

The author invites your comments and feedback.

Scott P. Sullivan
virtuallm2004@yahoo.com

4
LUNAR MODULE
Grumman
NASA
SCOTT SCHWEICKART
Mc DIVITT
USA
APOLLO IX
MANNED LUNAR SPACECRAFT
LM-5
APOLLO
STAFFORD YOUNG CERNAN
LM-6
MANNED LUNAR SPACECRAFT
APOLLO 11
MANNED LUNAR SPACECRAFT
APOLLO XII
Conrad · Gordon · Bean
Grumman
MANNED LUNAR SPACECRAFT
APOLLO XIII
EX LUNA, SCIENTIA
MANNED LUNAR SPACECRAFT
APOLLO 14
Shepard
Roosa
Mitchell
GRUMMAN
LM 10
EXTENDED STAY MANNED LUNAR MODULE
APOLLO 15
SCOTT WORDEN IRWIN
LM 11
EXTENDED MISSION
LUNAR SPACECRAFT
APOLLO 16
YOUNG · MATTINGLY · DUKE
LM 12
GRUMMAN
EXTENDED STAY
MANNED LUNAR MODULE
APOLLO · XVII
CERNAN · EVANS · SCHMITT

VIRTUAL LM

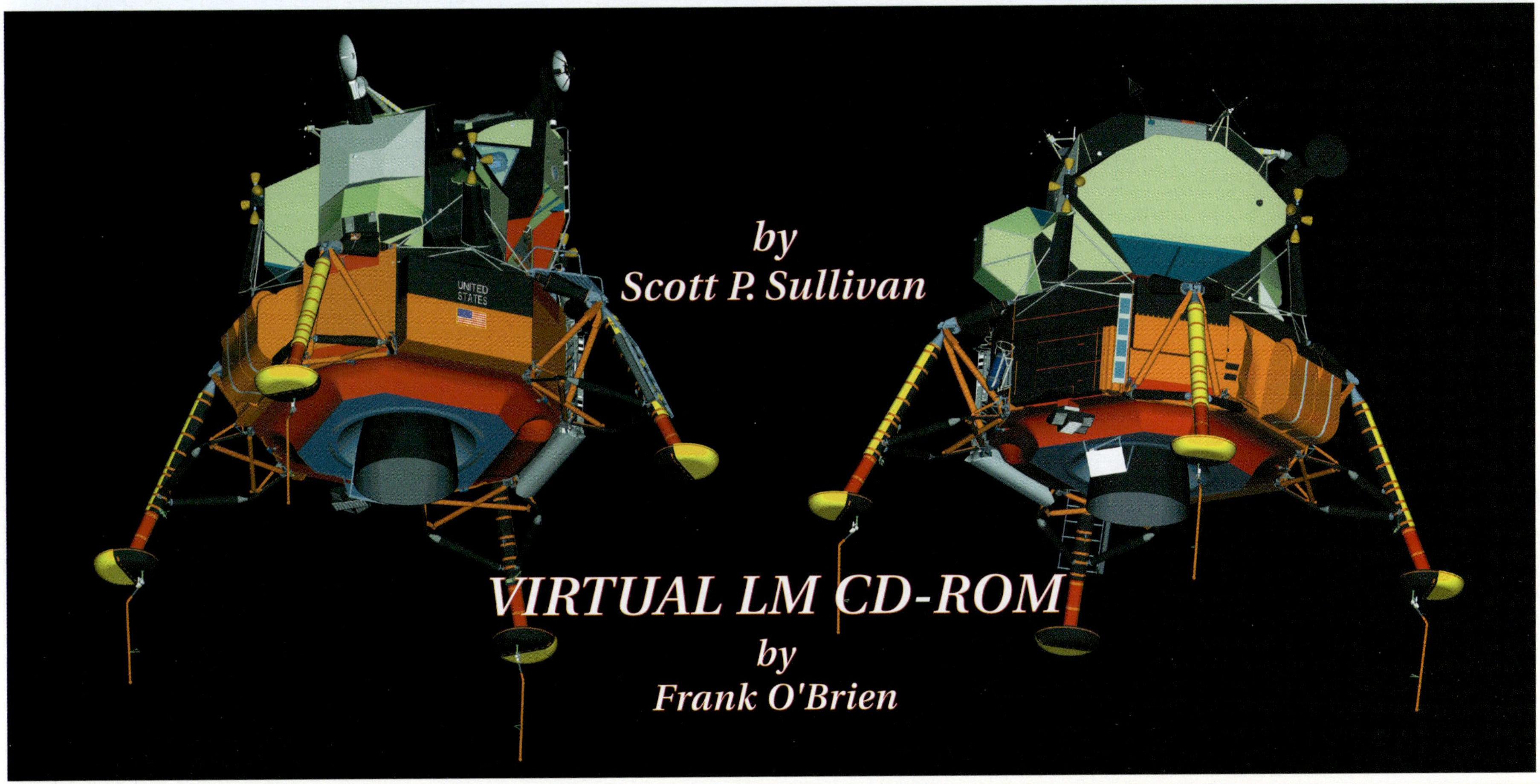

by

Scott P. Sullivan

VIRTUAL LM CD-ROM

by

Frank O'Brien

CONTENTS

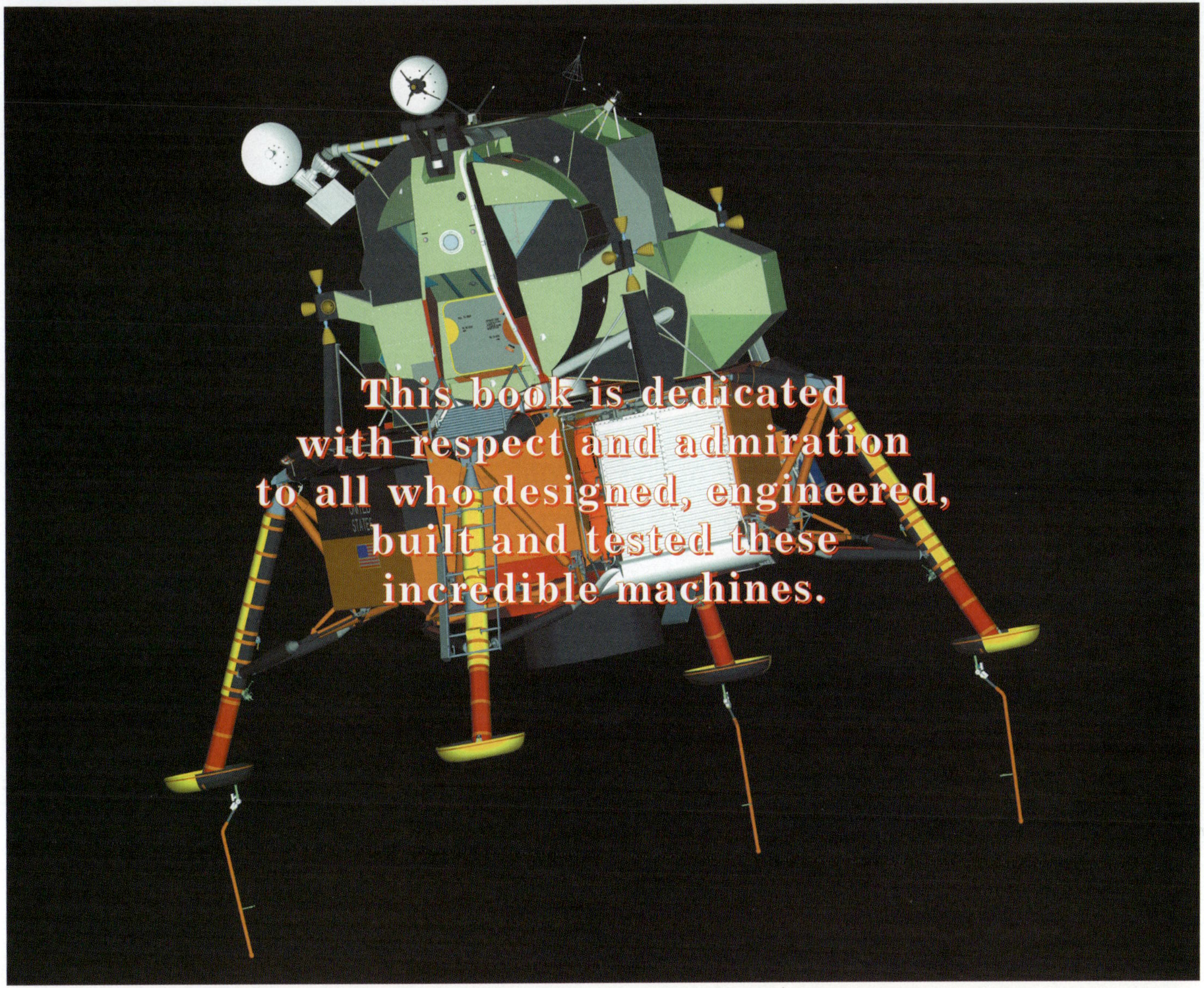

This book is dedicated
with respect and admiration
to all who designed, engineered,
built and tested these
incredible machines.

I would like to thank my Father and my Mother, George and Edie, for always being there for me and for being a constant source of inspiration.

I would like to thank Mr. Frank O'Brien, for generously supplying me with an invaluable wealth of information on the Lunar Module, and for doing a superb job creating the CD that accompanies this book. I asked Frank to include some biographical, and contact information on the CD, he's the guy to go to if you have any questions not answered in the book.

I would like to thank Mr. Kipp Teague for his excellent web site entitled "The Project Apollo Archive." Many of the photographs in this book came from his web site.

I would like to thank Mr. John Duncan for his excellent web site entitled "The Apollo Saturn Reference page." John's web site was another valuable resource.

I would like to thank Mr. Frederic Artner, and Mr. Barry Shanko for supplying me with some great research material for the book.

I would like to thank Mr. Paul Fjeld renowned LM aficionado for sharing his priceless knowledge and reference material on the Lunar Module with me.

I would like to thank Mr. David Weeks, scale modeler extraordinaire, for providing me with this outstanding drawings of the LM. I used these drawings extensively in creating my solid models.

And special thanks to Mr. Jack Morehart, the most gifted engineer I have ever had the honor of knowing, and the best friend I ever had!

INTRODUCTION

On May 25, 1961, just 20 days after Alan Shepard's 15 minute and 28 second suborbital Mercury flight, president John F. Kennedy announced to congress his plan of landing a man on the Moon and returning him safely to the Earth before the end of the decade. The project would be called Apollo.

Six days before president Kennedy's historic announcement, a man named Dr. John Houbolt had sent a hurried letter to NASA headquarters. His letter was not overly passionate and to the point. He stated," I do not wish to argue whether the direct way or the rendezvous way is best because of the lag in launch vehicle developments." It seemed to him that, "the only way that will be available to us in the next few years is the rendezvous way. For this reason alone it seems mandatory that rendezvous be as much in future plans as any item, and that it be attacked vigorously. . . . If NASA continues to dismiss Lunar Orbit Rendezvous [LOR] totally, as it has, someday there are going to be sorry NASA engineers."

What was Dr. Houbolt talking about? The concept of Lunar Orbit Rendezvous was to fire an assembly of three spacecraft into Earth's orbit on top of a single powerful three-stage rocket, the Saturn V. This 50,000-pound assembly would include: a mother ship or Command Module; a Service Module containing the fuel cells consumables and main propulsion system; and a small Lunar Lander or Excursion Module. Once in Earth's orbit, the last stage of the Saturn V would fire and expend itself, boosting the spacecraft and its crew into its lunar trajectory to the Moon. The Command and Service Modules would then separate from the spacecraft Lunar Module adapter, swing around 180 degrees and dock with the Lunar Module inside the lunar adapter of the Saturn V, extract it, and go on its merry way to the Moon.

Dr. John Houbolt explaining his plan for LOR, and below comparison between direct ascent and LOR vehicles.

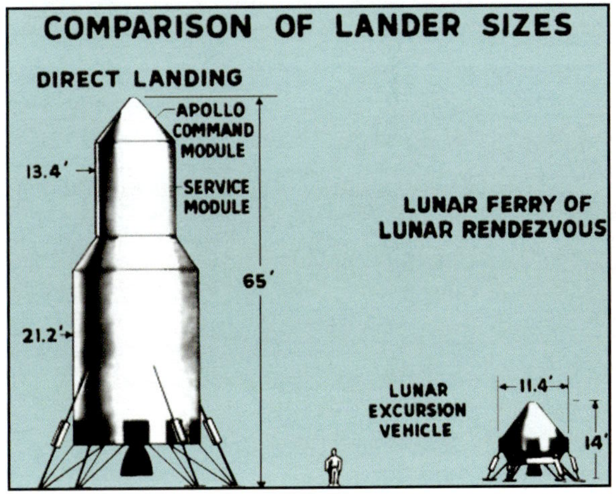

Sounds simple enough. But at the time that Dr. Houbolt wrote his note to NASA the concept of LOR was thought by many to be a risky and untried alternative to direct ascent.

There where three methods of reaching the Moon, known as flight modes, that where being hotly debated. The first was called Direct Ascent Mode, which would employee a single enormous rocket, called "Nova," to transport Apollo straight to the Moon with no rendezvous at all. The second, called Earth Orbital Rendezvous, called for two Saturn V boosters to launch and rendezvous in Earth orbit. In this mode, one rocket would carry a single Apollo spacecraft and its crew, and the other more fuel. The fuel would be transferred to the Apollo spacecraft while in orbit, and the spacecraft would then continue on to the Moon.

Dr. Houbolt clearly understood that everything was going to hinge on weight. The fuel-to-payload ratio was 400 to 1, meaning that for every pound of payload required for the trip to and from the Moon it would require some 400 pounds of fuel to get the job done. Opposition to the Houbolt concept continued to be strong. One member of the committee to select a flight mode said, "We thought it was too risky. Remember in 1961 we hadn't even orbited John Glenn yet. We certainly had done no rendezvous yet. And to put this poor bastard out there, separate him in a module, let him go down to the surface and then fire him back up and expect him to rendezvous . . . He didn't have a second chance; it had to be dead right the first time. I mean that just seemed like a bit much."

So, who was this man who persevered when some said he was probably in need of a long overdue vacation?

John Cornelius Houbolt had earned a B.S. and M.S. in civil engineering from the University of Illinois and was an aircraft structures expert who began working at Langley in 1942. In 1958 he completed his dissertation at the Swiss Federal Polytechnic Institute in Zurich on the heat related aeroelastic problems of aircraft structures in high speed flight to earn his Ph.D.

After returning from his graduate work in Switzerland, Houbolt and many other Langley researchers became increasingly curious about space flight.

Houbolt studied one particular problem related to rendezvous in space – the timing of the launch. NASA could not launch a mission at any arbitrary time and be assured of affecting a rendezvous with an orbiting spacecraft. To visualize the problem, Houbolt built a model with a globe for the Earth and a small ball on the end of a short piece of coat hanger wire, all connected to a variable-ratio gearbox. It simulated a satellite at different altitudes and in different orbital planes, enabling him to calculate the varying amounts of time it would take for the satellite to orbit the revolving Earth. From his consideration of orbital mechanics, he knew that a change in orbital plane at 25,000 feet per second without the help of any sort of aerodynamic lift would require an enormous amount of energy and realistically could not be made. With this simple ingenious model, Houbolt saw how long one might have to wait – a period of perhaps days – to launch a rendezvous mission. But he also found a way to circumvent the problem – if the orbital plane of the satellite could be made just one or two degrees larger than the latitude of the launch site, one could extend the launch "window" to four hours every day. Thus he began to understand how NASA could get around long waiting periods.

Houbolt was crushed to find out that his theory of LOR had been placed at the same level of disdain as lunar-surface rendezvous, an absurd concept that had an unmanned spacecraft land on the Moon full of supplies and fuel prior to the manned spacecraft's arrival. After the manned spacecraft had landed, having spent all of its fuel on the trip to the Moon, it would use a homing beacon on the unmanned supply ship to locate it and retrieve its cargo for the trip home.

Dr. Houbolt's second letter to NASA officials came on November 15th 1961. It was a nine-page summation of why LOR was the only rational decision to be made in reference to flight mode for the Apollo program. Houbolt knew that the letter might cost him his job; he was skipping normal channels by appealing directly to the Associate Administrator. He had framed his letter with lines such as, "Do we want to go to the Moon or not? And if so, why do we have to restrict our thinking to a certain narrow channel?" and " Why is Nova, with its ponderous size, simply just accepted, and why is a much less grandiose scheme involving rendezvous ostracized or put on the defensive?" Houbolt's biggest concern was against the bureaucratic guidelines that made it impossible to consider the merits of LOR. Houbolt warned, "The fact is completely ignored that they are dealing with a ponderous development that goes far beyond the state of the art."

Finally Houbolt began to gain allies in his push for LOR, including Wernher von Braun. On June 7th 1962 von Braun wrote, " We at the Marshall Space Flight Center readily admit that when first exposed to the proposal of Lunar Orbit Rendezvous Mode we were a bit skeptical – particularly the aspects of having the astronauts execute a complicated rendezvous maneuver at a distance of 240,000 miles from the Earth, where any rescue possibilities appeared remote. In the meantime, however, we have spent a great deal of time and effort studying the four modes (Earth-orbit rendezvous, LOR, and two Direct Ascent modes, one involving the Nova and the other a Saturn C-5), and we have come to the conclusion that this particular disadvantage is far outweighed by its advantages . . .

So, in the summer of 1962 NASA made the choice for LOR as the flight mode for the Apollo program. Thousands of factors contributed to the ultimate success of the Apollo lunar landings, but no single factor was more essential than the concept of LOR.

Then came the daunting task of designing and building the spacecraft that would land men on the Moon and do it in less than eight years.

The Grumman facility in Bethpage, New York where engineering work on the LM began, and below, Thomas J. Kelly of Grumman "the father of the Lunar Module."

In May of 1958, the Grumman Corporation began studies on manned space flight programs. Grumman, since 1930, had been in the business of designing and manufacturing military aircraft. They had submitted a proposal for Project Mercury, but because of prior Navy aircraft production commitments, their bid was unsuccessful.

The Grumman Corporation had competed for every major NASA contract and, except for the unmanned Orbiting Astronomical Observatory satellite, had always fallen short of winning with any of their proposals.

Grumman had a very active interest in pursuing the contract for the Lunar Module and refused to let it die, whether it was funded or not. In April of 1960 Tom Kelly began working on some in-house studies on manned lunar missions. Thomas J. Kelly, LM Assistant Program Director, was a mechanical engineer with a graduate degree from Columbia University. Kelly had been a propulsion engineer with Grumman on the Rigel missile program from 1951 to 1953 and F11F Aircraft Programs from 1953 to 1956, and with Lockheed in space propulsion system development from 1957 to 1958. Upon returning to Grumman, Kelly was Project Engineer for the Apollo Spacecraft Proposals and was Project Engineer for the Apollo Spacecraft effort. He was the Preliminary Design Engineer on the LM pre-proposal studies and the proposal to NASA. Mr. Kelly also served as the Engineering Manager and Vehicle Test Manager for the LM Program; he was responsible for all spacecraft engineering activity on the LM. Two other key people spearheading the Grumman team were Dr. Ralph H. Tripp, LM Program Director and Joseph G. Gavin, Jr., Vice President of Space Programs for Grumman Aircraft Engineering Corporation in Bethpage, New York. Dr. Ralph H. Tripp received his Ph.D. in Applied Mathematics and Theoretical and Applied Mechanics in 1942 from Iowa State College. Dr. Tripp had joined Grumman in 1942, working in the Structural Research Group, in charge of vibration and flutter. In 1948 he became the head of the Structural Research Group. Later that year he formed the Research Department and became the Department head. In 1949 he became head of the Instrumentation Department. In 1958 he became the Assistant Director of Flight Test and in 1962 he was appointed the Director of the Orbiting Astronomical Observatory (OAO). Joseph G. Gavin received his graduate degree in Aeronautical Engineering with honors from M.I.T. in 1942. After a tour in the Navy, Gavin began his career with Grumman as a Design Engineer. Gavin was assigned as Project Engineer on the F9F-6 program in 1950 and in 1952 as Project Engineer on the F11F Program. In 1956 he was assigned as Chief Experimental Project Engineer until 1957, when he was made Chief Missile and Space Engineer. In 1962 Mr. Gavin was promoted to Vice President, Director of the LM Program.

In 1960 Grumman submitted a preliminary study on the Apollo project to NASA. At that time, Earth Orbit Rendezvous was the most popular theory for lunar landing. Again, Grumman was unsuccessful in their bid to win the project.

The LM team had spent enough time researching the various aspects of designing a lunar lander from a conceptual standpoint that by 1961 they were convinced they had the ability to design and build a dependable spacecraft. In October of that year, NASA held yet another competition for the Apollo lunar spacecraft. Again Grumman proved unsuccessful in their attempt to win the contract.

Grumman engineers had resigned themselves to the possibility of winning the contract for the Lunar Module, although it was considered only a consolation prize. The Apollo Command and Service Modules were seen as being the long-term "workhorse" of the space program, having a practical use even after Apollo was over, with the applications for Skylab. The Lunar Excursion Module, or LEM, and eventually just LM for Lunar Module, would have practical uses for the lunar landings only, and although the historic significance and glory were obvious, its long-term production value fell somewhat short when compared to the CSM.

1960

1962

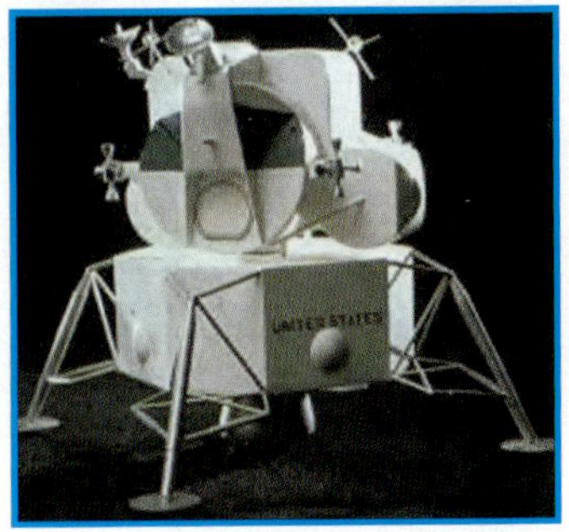

1965

In mid-1962, convinced that Lunar Orbit Rendezvous (LOR) was the method for making a lunar landing, Grumman launched a feasibility study on LOR. NASA then asked for proposals involving the use of the LOR concept with the lunar lander. Grumman submitted its proposal in September of 1962, with RCA as the principle subcontractor. On November 7, 1962, NASA announced that Grumman Aircraft Engineering Corporation, of Bethpage, New York, was selected to build the Apollo Lunar Excursion Module.

The initial design concept was an extremely simple thing, but it grew conceptually until finally, in April of 1962, the LM concept vehicles were being studied in a parametric form, that is: looking at all the different combinations, whether it should be a staged or unstaged vehicle; or what kind of engine compliment it should have; do you really need a throttleable engine, or could you use several discrete levels of engines; what kind of landing gear should be designed; etc. Another major design issue that was worked out early on was the weight trade-off factors, which were to serve well in the design process later on. The basic concept was a staged vehicle, which had a throttleable engine, and had the same overall mission requirements in very general terms – two men and approximately the same lunar surface stay-time as originally proposed by NASA.

LTA-2 above, and LTA-8 below.

The original concept allowed for docking at either hatch and had both astronauts seated in front of very large observation windows, but the most significant evolution in the design was the weight. This was limited to 22,000 pounds in the proposal and Grumman had said in its proposal that they thought they could make that weight requirement, if everything stayed the same as defined in NASA's design requirements. At 22,000 pounds it was still going to be a very tight squeeze. The LM was a severely weight limited vehicle. Grumman had pointed this out as the most significant technical challenge that they faced, and as it turned out, they couldn't have achieved the 22,000-pound weight limit even if nothing else changed – and there were plenty of design changes.

One of the first things the Grumman team did was to create two separate groups. One group worked on the design of a vehicle that had essentially a bare minimum pressurized cabin volume and as much of the equipment external as possible. The other group was assigned to work on a fully pressurized cabin, a long cylindrical-type thing with the crew having access to all equipment in-flight. A number of configurations were derived from those two starting points, with considerations for weight and operational advantages to both. Both teams came to conclusions that tended to meet in the middle. The advantageous configuration tended to be at neither extreme, but rather a combination of the best attributes of each.

Another thing the Grumman team did was to negotiate with NASA a contract for the job in its entirety. As Tom Kelly said, "there was an example of ignorance in action, if I ever saw one, at least on our part. I think NASA was a lot smarter than we were, at that time. But neither really understood the scope of what we were getting into, I don't think, at that point. When we negotiated, we thought we were negotiating essentially to build this vehicle, or as close as we could get to it, and that really wasn't what the NASA people had in mind. What they wanted to launch into, and subsequently directed us to launch into, was an extensive preliminary design phase, wherein every aspect of the design was gone over again, and not just from a conceptual standpoint." That phase of the project lasted for about 9 months, during which gross configuration changes were tried and studied, such as changing the tanks around, changing the number of landing gear, changing the shape of the cabin, and so on.

Near the end of 1964 it was painfully obvious that Grumman was at the end of its rope with the weight situation. They had to resize the fuel and oxidizer tanks again, and the way the vehicle was built the whole thing wrapped around the tanks, so if the size of the tanks changed, consequently so did everything else.

In November of 1964, Grumman negotiated with NASA a change in the control weights and simultaneously resized the fuel and oxidizer tanks. At that point it was a rapid progression from the end of the preliminary design phase. The end of 1964 essentially froze the concept and in 1965 they froze all the details on just about everything that could be frozen on paper, down to pin assignments and wire numbers.

The descent stage fell into place more rapidly than the ascent stage. Grumman went through more than twenty descent configurations. But the ascent stage proved to be far less clear-cut. The problems that the Grumman team were looking at were analogous to the difference between the Mercury and the Gemini spacecraft. The Mercury spacecraft had everything inside the pressure hull, and it proved to be difficult for the ground crew to get at, but was relatively easy for the astronaut. The Gemini spacecraft was just the opposite; it had everything accessible from the outside, with the crew packed into a minimum of pressurized space.

The final design was a bit of a compromise, with the majority of the components being relegated to the exterior of the spacecraft and only the components that would have an immediate life supporting function on the inside of the crew cabin. The environmental control system would reside inside the pressurized cabin and the PLSS backpacks and their assorted hardware would also be within the reach of the two astronauts, but all other consumables, such as water and oxygen, would be outside the crew cabin, either on the ascent stage or in one of the four quads in the cruciform descent stage.

In 1965 Grumman began test-firing the ascent stage engine at White Sands. At this time the engine firing program was behind schedule and this was proving to be a common, but unwanted theme which plagued Grumman throughout most of the program. Budget overruns by subcontractors and seemingly relentless design changes and modifications where also all too common to the Grumman team.

During 1965 Grumman built six lunar test articles (LTAs) to form the backbone of the LM test program. Bethpage shipped LTA-2 to Huntsville for vibration testing to see if it could withstand launch pressures, and LTA-10 to Tulsa, to check its fit in the adapter for the Saturn V, and LTA-1 was a "house" spacecraft, used to address problems during fabrication, assembly, and checkout. Three more LTAs were under construction: LTA-8 for thermal-vacuum testing in Houston, and LTA-3 and LTA-5 for combined structural shaking, vibrations, and engine firings.

Flight test plans for early production LMs were flexible to accommodate schedule differences with the Command Module. LEM-1 naturally received the most attention, since Grumman had to get it ready for an unmanned "LEM-alone" mission (Apollo-Saturn 206A). LEM-1 would have to be ready at least three months before the Block 2 Command Module, however, or its first mission would be part of a combined test of the two spacecraft.

But Grumman was moving slowly and NASA believed that LEM-1 would be nearly a year late. Many factors were contributing to the slower than anticipated progress, but ground testing topped the list.

In May of 1965 Grumman looked at possible launch dates for the first vehicles but could decide on nothing due to problems in the fiscal year 1966 funding. Hardware production had to be cut back in an attempt to absorb some of the loss. In July, Houston directed Bethpage to delete LTA-4, a vibration test article, and two flight test articles (FTAs). To replace the FTAs, two LTAs would be refurbished when they finished ground tests. After trials with full-scale models had been run at Langley and elsewhere, Houston also canceled a landing gear test model as an unnecessary expense.

Full scale training models, these two early versions of the LM featured a round front hatch and the one above incorporated a rope for ingress and egress.

Program spending began to equal schedules in importance. Just as the LM was on its way toward flight hardware production, it was caught in a budgetary squeeze by Congress. Again Grumman had to shoulder most of the financial burden. Expenditures had risen dramatically, from $135 million dollars in 1964 to an estimated $350 million in 1966.

In January of 1967, during a plugs-out test on the pad, fire broke out in the Apollo-Saturn 204 Command Module, killing the crew. In March of that year NASA announced that since the 204 launch vehicle was undamaged by the fire it would be used for the first unmanned launch of LM-1. Changes in the Command Module after the fire caused more delays and LM-1 did not arrive at Cape Kennedy until June 27 1967, three months behind schedule.

LM-1 was legless because it would burn up upon entry into the atmosphere and therefore required no landing gear. Just before dusk, at 5:48 on the evening of January 22, 1967 (after several hours of delay because of various equipment problems), *Apollo 5*, with LM-1 on board, lifted off from Cape Kennedy. Less than 45 minutes after launch its attitude control engines pulled LM-1 away from its berth in the Saturn S-1B adapter. After checking out the spacecraft for two orbits, ground control signaled the descent engine to fire for 38 seconds. Four seconds later, LM-1's guidance system sensed that the spacecraft was not going fast enough and stopped the engine. Ground control sent a switch-off signal to the guidance computer and cut in a mission programmer to command the LM's maneuvers. There were two more descent engine firings, two ascent engine firings, and one fire-in-the-hole (FITH) abort maneuver. All preliminary objectives had been achieved. LM-1 met a fiery end upon its entry into Earth's atmosphere on February 12, 1967, its remains falling into the Pacific Ocean several hundred miles southwest of Guam.

On February 8, 1968, NASA made the decision to forego the second unmanned Apollo LM flight using LM-2. The decision was reached after flight data on LM-1 had been reviewed. Now Grumman geared up for the first manned flight of the Lunar Module, LM-3, in Earth orbit with *Apollo 9*. On August 7, 1968 the Apollo review board certified the design of LM-3 as flight worthy. On March 3, 1969 *Apollo 9* lifted off from Cape Kennedy with LM-3 on board. The Grumman team held their collective breath while LM-3 was put through its paces hundreds of miles above the confines of Mission Control. LM-3 performed flawlessly, achieving all of its specified objectives, including transposition and docking with the CSM, intravehicular and extravehicular crew transfer, descent propulsion engine burn, Lunar Module active rendezvous and docking, and access to Lunar Module consumables.

On May 22, 1969 LM-4 (*Apollo 10*) descended to within 60 nautical miles of the lunar surface, and less than two months later *Apollo 11* astronaut Neil Armstrong stepped off of the landing gear pad of LM-5 to become the first human being to set foot on the Moon.

Within seven years of winning the contract to build the Lunar Module, the Grumman team had accomplished what many said was simply impossible. They persevered with relentless determination just to win the contract. They continued to strive towards their monumental goal while every imaginable setback was thrown at them, again and again. From endless budgetary overruns, to engines exploding during testing, it took a very strong constitution and above average commitment to excellence to be a member of the team that created the Lunar Module. Grumman had no benchmark to follow with regard to the design of the LM, as North American had had, to a large extent, when they developed the Command and Service Modules. A spacecraft designed specifically to fly in space, land on the Moon, and then return its crew safety to the CSM had never before even been imagined.

During a conversation with one of the Apollo astronauts I was surprised to learn that the astronauts who were not selected to be the first to land on the Moon had actually felt sorry for the crew of *Apollo 11*. They were certain the crew would return home safely, but without having landed on the Moon. None of their fellow astronauts actually thought we would be successful on our very first attempt.

I guess, to a large degree, we have people like Tom Kelly to thank for proving them wrong.

References:

Chariots for Apollo, by Courtney G. Brooks, James M. Grimwood and Loyd S. Swenson, Jr.
Enchanted Rendezvous: John Houbolt and the Genesis of the Lunar-Orbit Rendezvous Concept, by James R. Hansen

A note from the author:

The first time I saw the Lunar Module I thought to myself," that's not a spaceship; spaceships are shinny bullet-shaped things, with aerodynamic fins and huge engine nozzles at their tails." As a very young boy, the sum total of my familiarity with manned spacecraft was what I had seen on television of the Apollo Command and Service Modules, and Saturday morning reruns of old B movies, usually involving some kind of laughable space monster.

My layman's opinion of the Apollo Lunar Module was a common one, though, even with the astronauts who would pilot it to the Moon. They called it ungainly looking and far less attractive than it's more conventionally sleek counterparts, the Command and Service Modules. But in this case, the age-old saying that, "beauty is only skin deep," couldn't have been more applicable.

The Lunar Module proved to be a major milestone in human innovation and engineering excellence. Some say that, because it was designed to travel only in the vacuum of space, it was the first true spacecraft. This much is true: the Apollo Lunar Module was the first spacecraft to land humans on the Moon.

With this continuing series of books, I intend to share with you just what engineering marvels these machines were, that pioneered America's manned space program.

I am not a self-proclaimed expert on the manned spacecraft of the Apollo program, but simply an avid enthusiast. As such, I readily admit that there are countless learned experts out there, some who are hopefully reading this now, whose knowledge far surpasses mine. I received some criticism on my first book on the Command and Service Modules for not getting everything as accurate as some believe I could have. I must plead guilty to this charge. I approached these projects from the viewpoint of an artist who wanted to showcase these machines as works of art in their own right, recreating the intricacies of design and craftsmanship in a form that could be viewed as art by anyone who appreciates what human beings are capable of creating.

I had to take some artistic liberties with some of the finer points of the spacecraft's interior. With the structural support for the MESA deployment mechanism, for example, I knew what the mechanism looked like and how it functioned, but I could not find any reference material regarding the support structure that held it in place. In this instance, I had to draw upon my own experience and knowledge as a Mechanical Designer to come up with a reasonable estimation.

I also decided not to depict the external thermal coverings in their true state – appearing quite wrinkled due to the fact that they were made of layers of very thin laminated Mylar. I chose instead to show the true lines of the spacecraft. In my opinion, the wrinkled skin of the Lunar Module was a distraction from its actual structural appearance.

So, if you would keep that in mind while you look at these renderings of the imagination and technical expertise that went into designing and building these spacecraft, my intent of trying to take you inside the spacecraft, if only for a moment, will have been achieved.

They where designed at a time when computers where the size of an entire room, but had little more power than the average hand-held calculator of today. People who had grown up in a time before television or cell phones had designed them with nothing more than slide rules, pencils and paper, and an overwhelming desire to succeed. They achieved what most said was impossible, because they believed they could. This is the true embodiment of the human spirit, and is what I marvel at most when I look at these spacecraft. I hope you enjoy Virtual LM.

THE APOLLO LUNAR MODULE (LM)

The Apollo Lunar Module (LM) was the entire structure that sat beneath the Command and Service Modules, inside the spacecraft LM adapter, atop the Saturn V launch vehicle. The Lunar Module was a two-stage space vehicle consisting of an ascent stage and a descent stage. During descent to the lunar surface, both stages would act as a single unit. After the lunar mission, the descent stage would become a launch platform for the ascent stage. The ascent stage would separate and function independently on its journey back to the Command and Service Modules in orbit around the Moon.

The Lunar Module, or LM, was a fragile machine, designed solely for travel in the vacuum of space. It was so thin in some areas of the crew cabin that an astronaut could punch through the wall if so inclined. Each spacecraft took two and a half years to build. Each was fabricated almost entirely by hand. Its design life was a mere 72 hours on average, and after it had accomplished its mission it was completely expendable. Thirteen Lunar Modules were built, and six landed on the Moon between 1969 and 1972.

The Spacecraft Lunar Module Adapter (SLA)

THE J-SERIES LUNAR MODULE

THE H-SERIES LUNAR MODULE OVERVIEW

The Apollo Lunar Module stood 23 feet, one inch tall from the bottom of the landing pads to the top edge of the rendezvous antenna. The ascent stage was 14 feet across; 13 feet, 3 inches deep; and 9 feet, 3 and a half inches high. The descent stage was over 10 feet, 7 inches tall and 31 feet wide measured from the edges of the landing pads at their farthest points. The H-Series LMs weighed 33,205 pounds, fully loaded and fueled, and the J-Series LMs weighed about 36,244 pounds, fully loaded and fueled. The spacecraft's maximum speed was approximately 17,500 miles per hour.

The thermal coverings for the LM changed from one Lunar Module to the next to some degree. The thermal coverings and micrometeoroid shielding in this book depict LM 5 (Apollo 11) for the H-Series, and LM 11 (Apollo 17) for the J-Series.

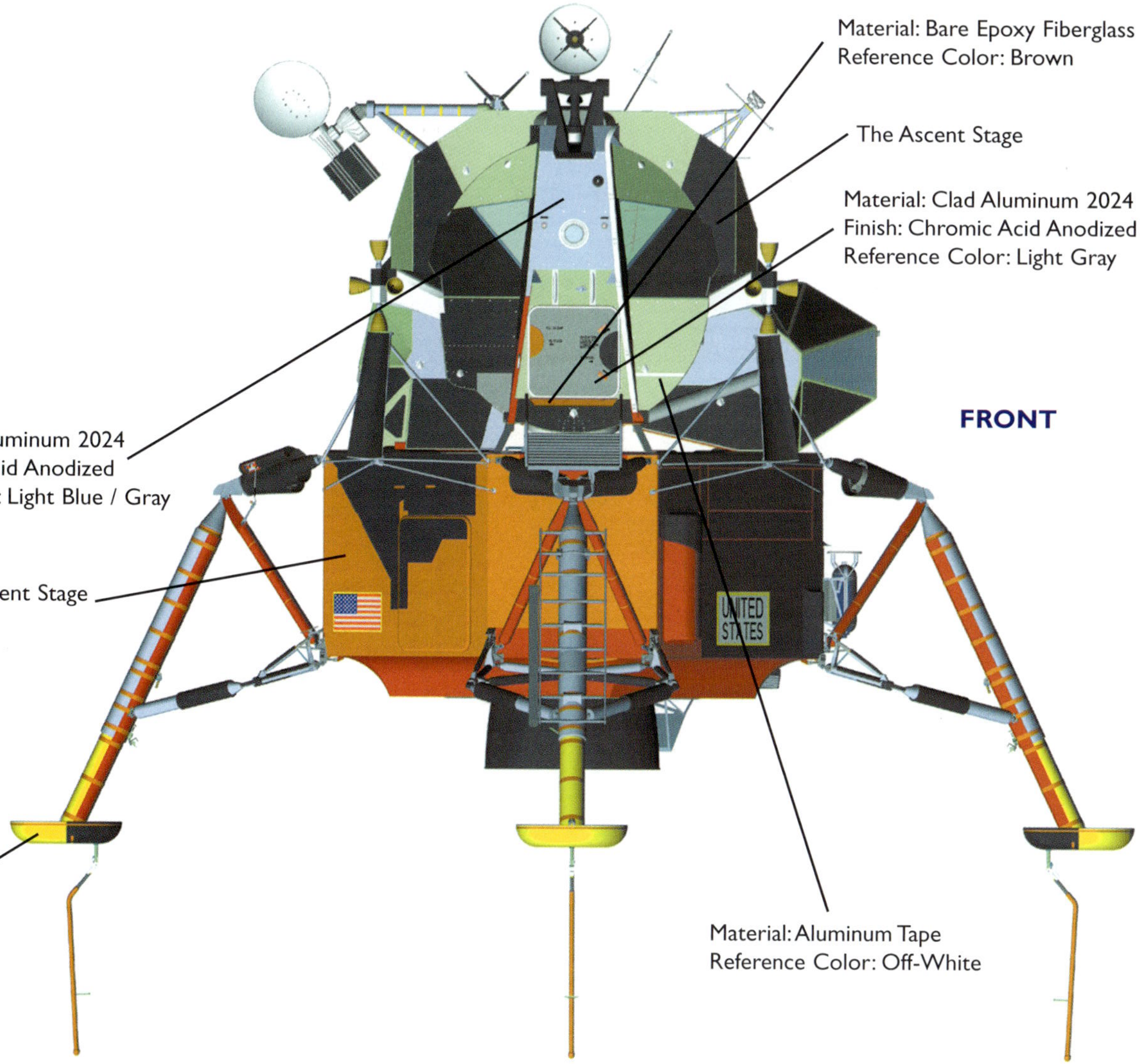

The Lunar Modules consisted of two design types, the H-Series and the J-Series. The H-Series LMs consisted of Lunar Modules 1 through 8 (*Apollo 1* through *14*) and the J-Series were LMs 9 through 11 (*Apollo 15* through *17*). The main difference between the two was that the J-Series was designed for extended stay and extended hardware.

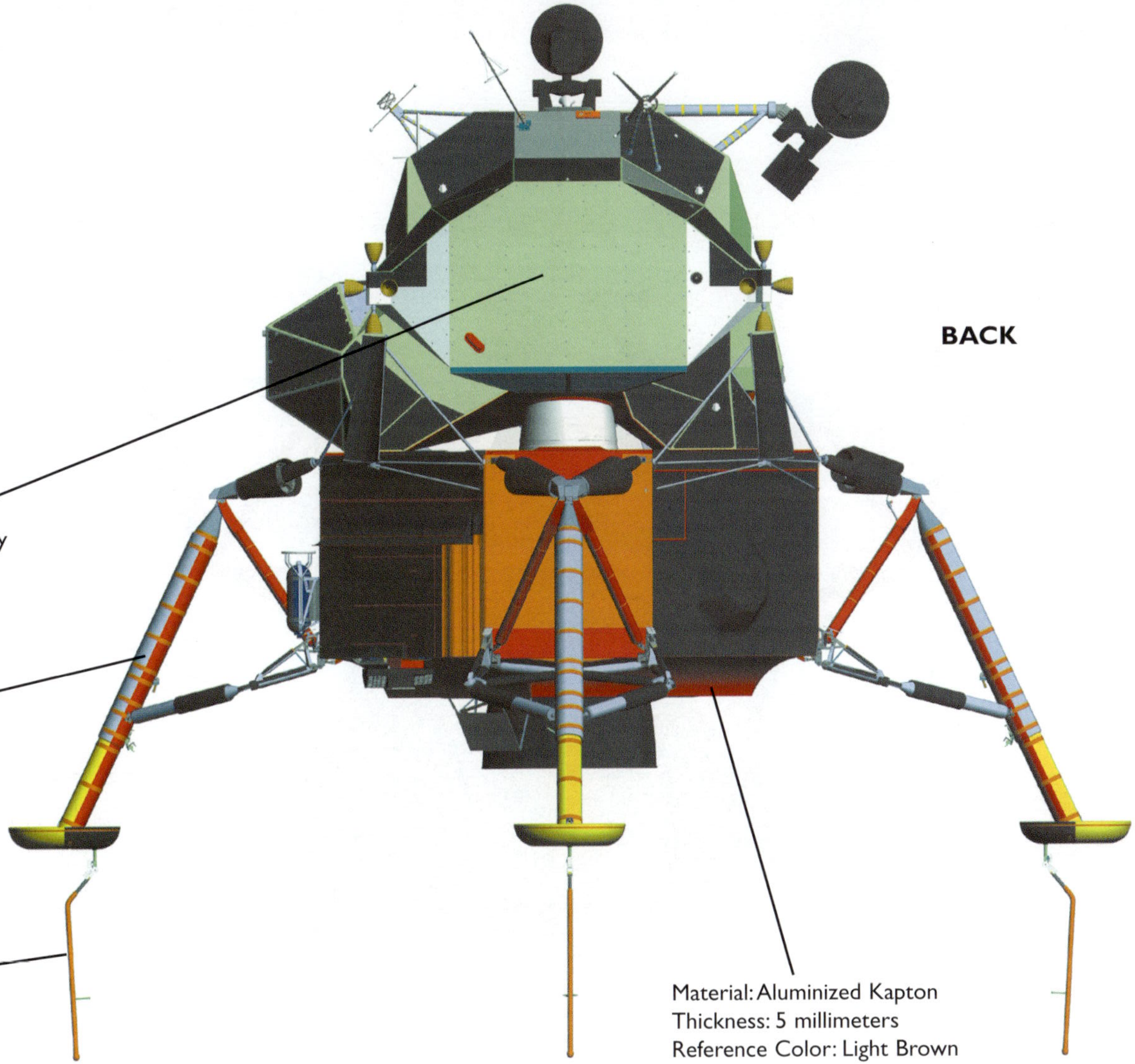

The unit price for a single LM in the 1960s was 50 million dollars.

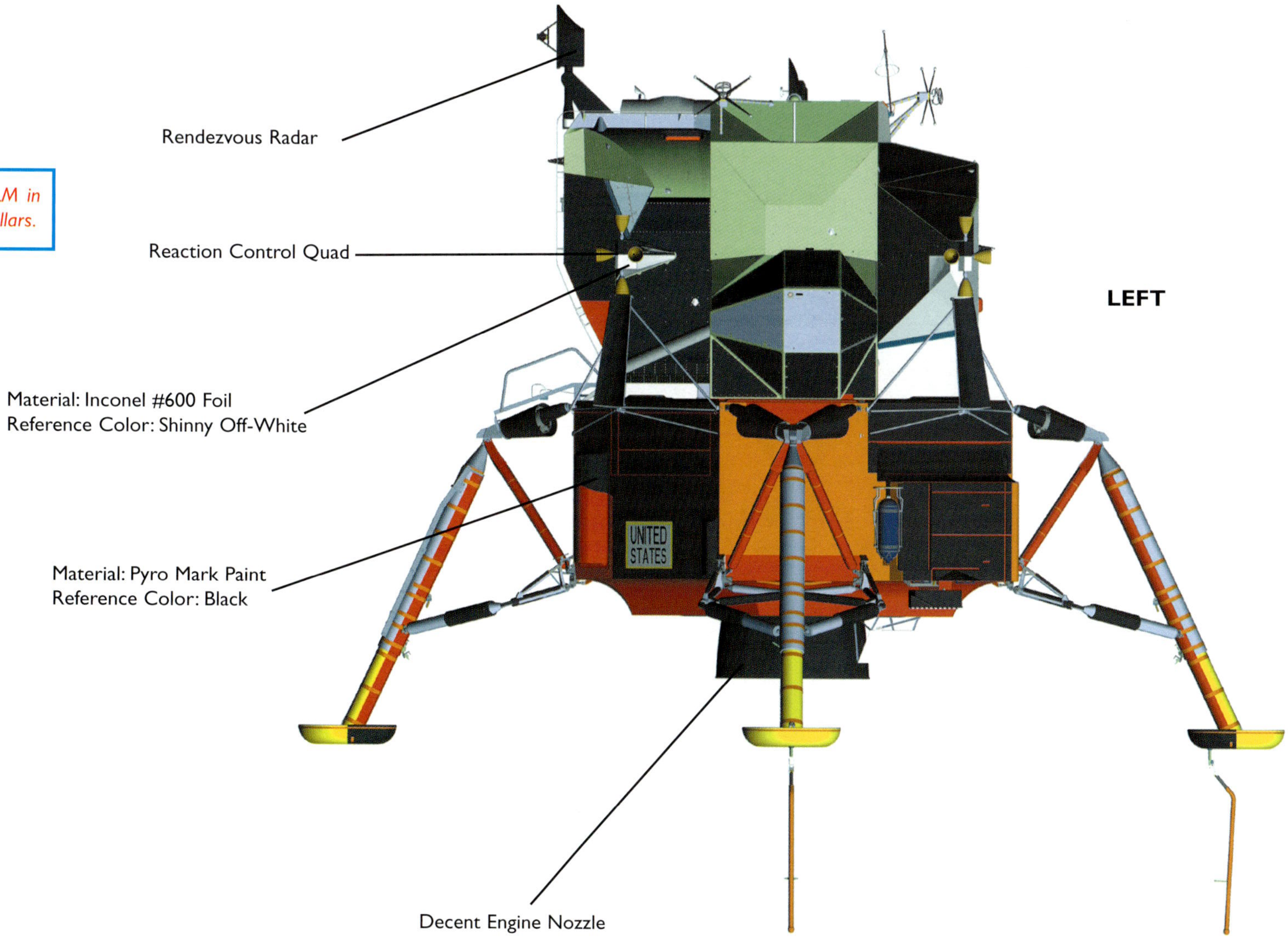

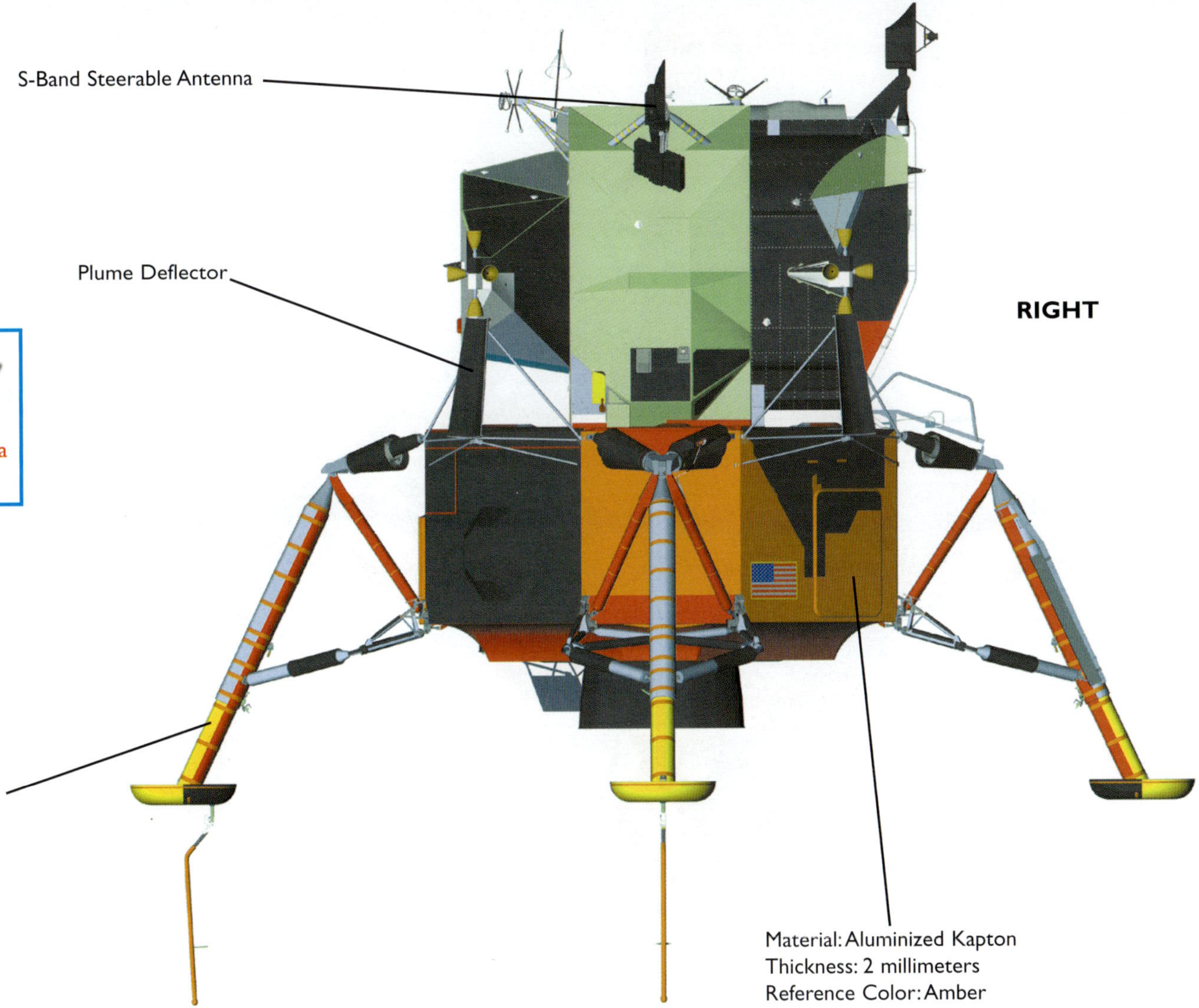

Lunar Module Chronology

December 8, 1959
Possible configurations are proposed for a manned lunar landing using direct ascent.

Lunar Module Chronology

January 30, 1960
Manned lunar landing and return study (MALLAR) indicates that Earth Orbit Rendezvous would be necessary.

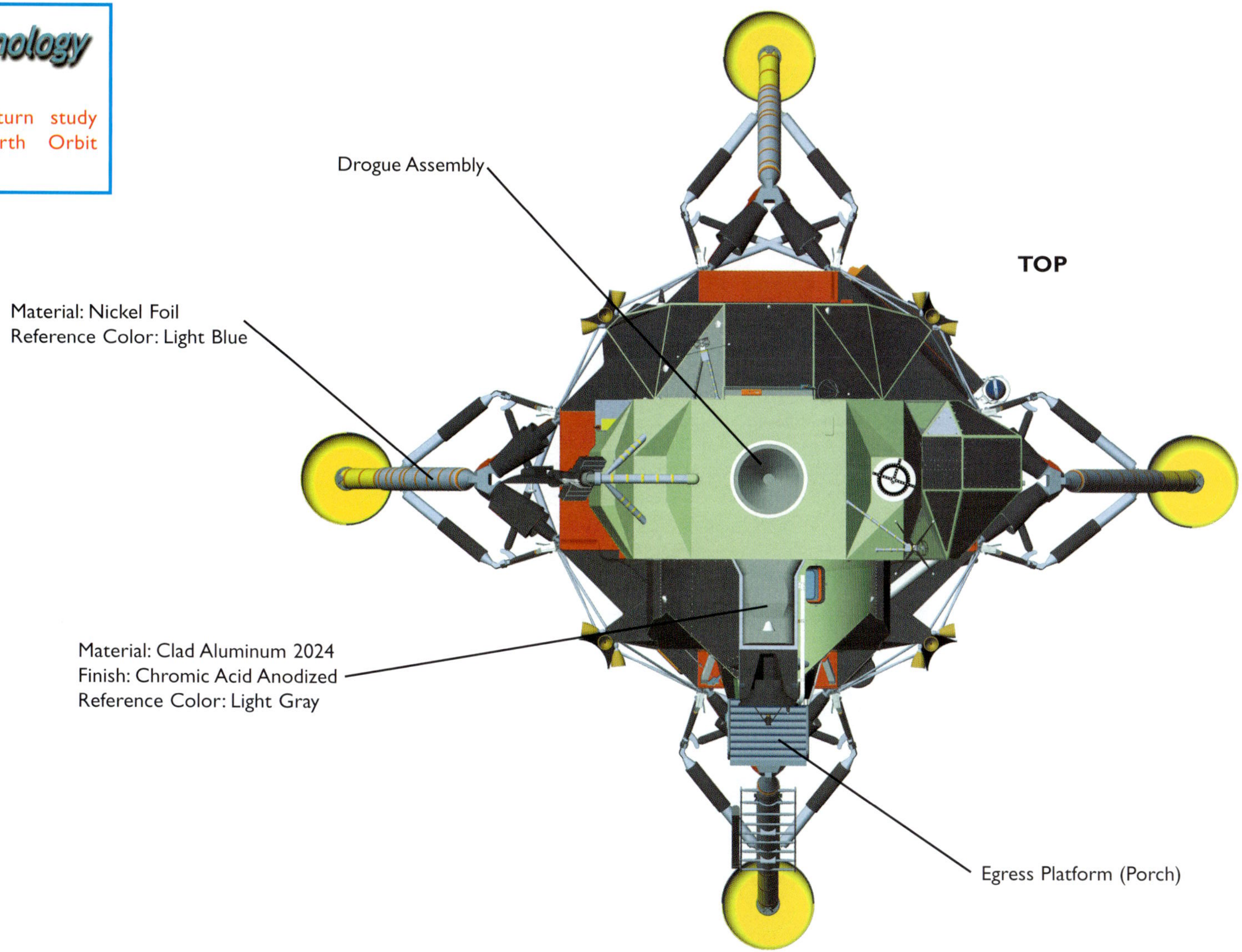

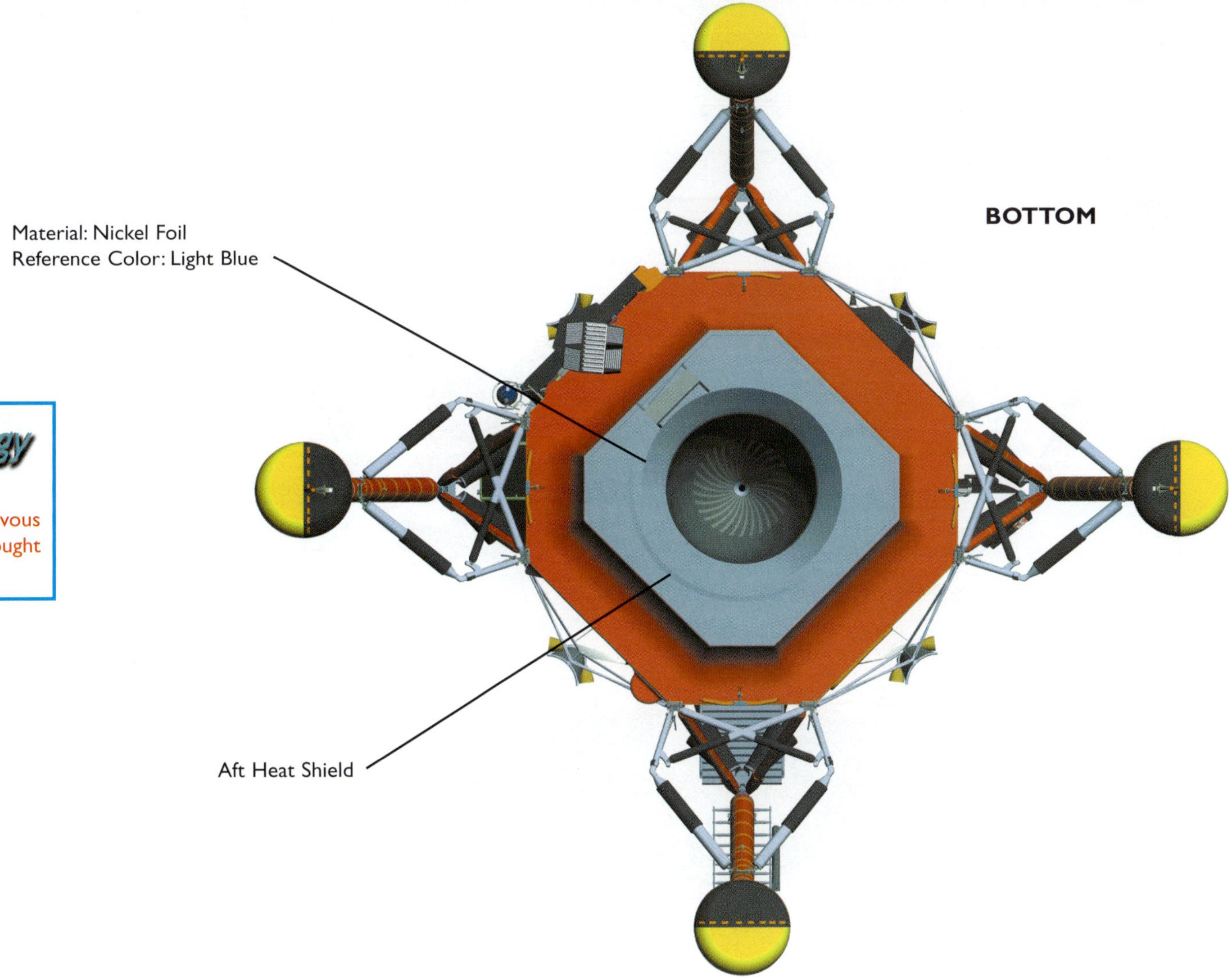

Lunar Module Chronology

March 21, 1960
A study on the Lunar Orbit Rendezvous method is conducted by the Chance Vought Corporation.

THE J-SERIES LUNAR MODULE OVERVIEW

The J-Series Lunar Module differed from the H-Series in that it was designed for extended stay missions, which required more equipment and supplies. The J-Series included the battery-operated Lunar Roving Vehicle (LRV) and the addition of extra equipment pallets in Quad 3 of the descent stage, as well as an extended stay MESA (Modular Equipment Stowage Assembly) in Quad 4.

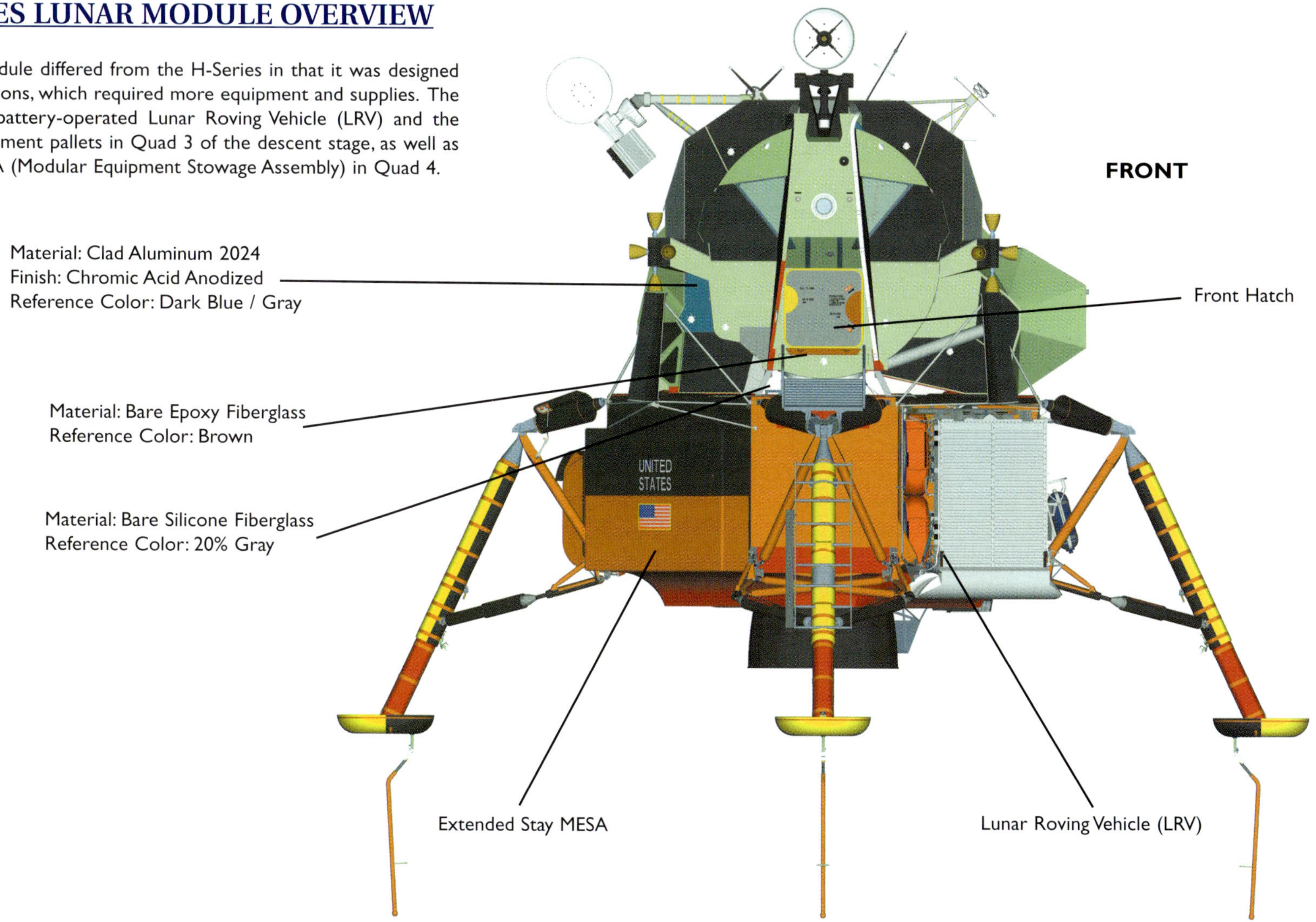

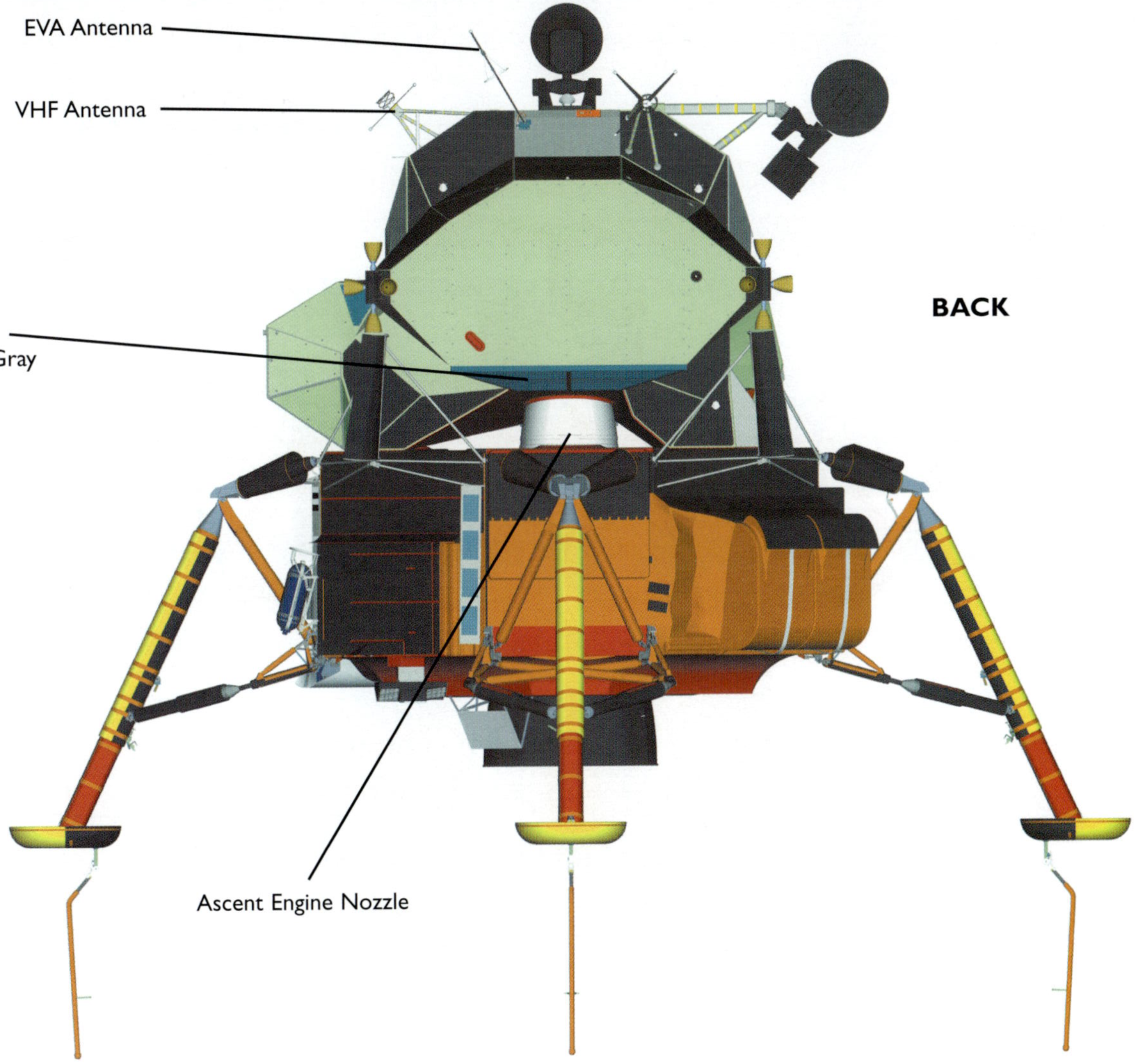

Lunar Module Chronology

April 5, 1960
John C. Houbolt presented a paper on rendezvous in space with a minimum expenditure of fuel.

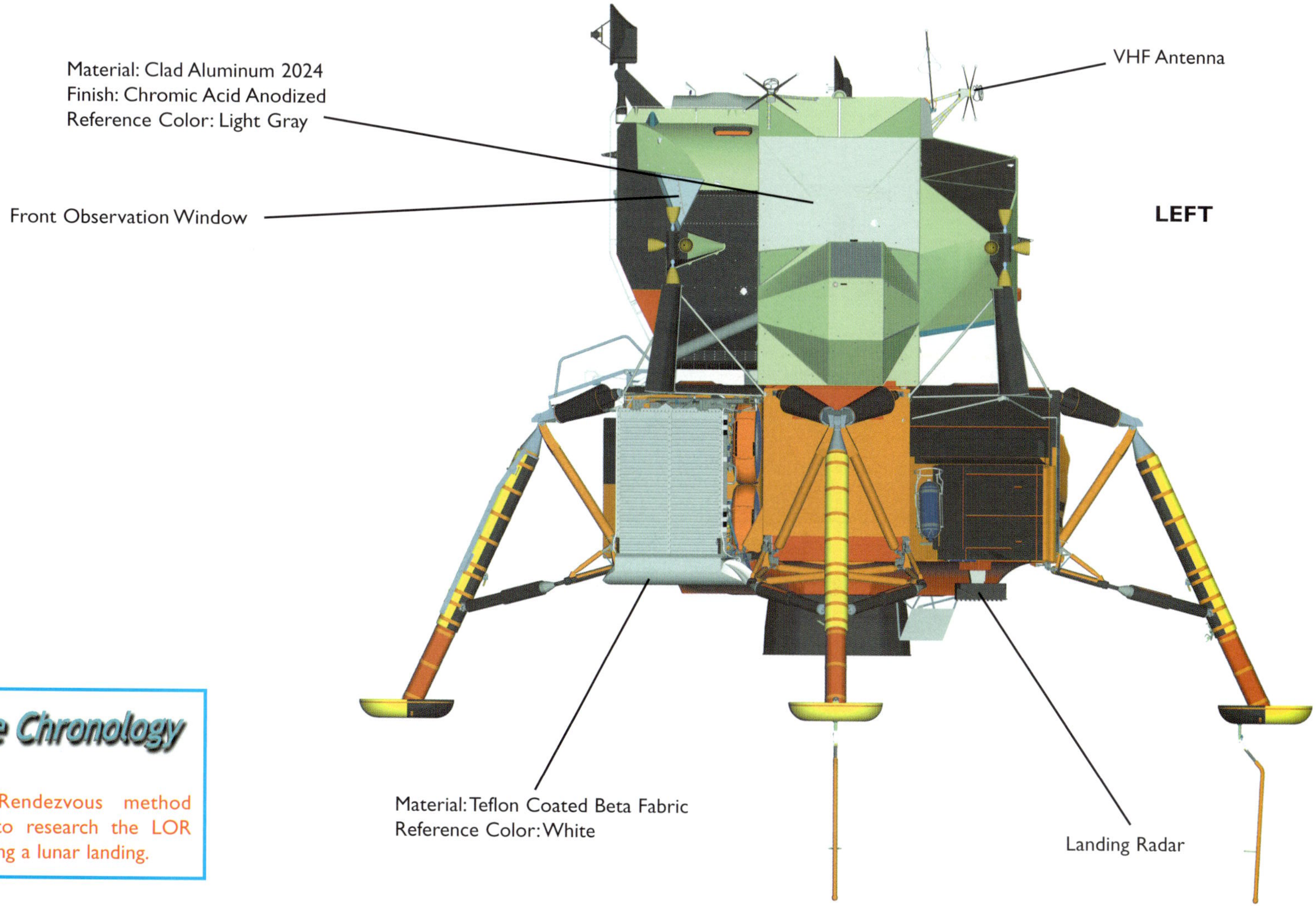

Lunar Module Chronology

December 10, 1960
The Lunar Orbit Rendezvous method program was begun to research the LOR method of accomplishing a lunar landing.

Lunar Module Chronology

December 14, 1960
Robert Seamans briefed on the Lunar Orbit Rendezvous method.

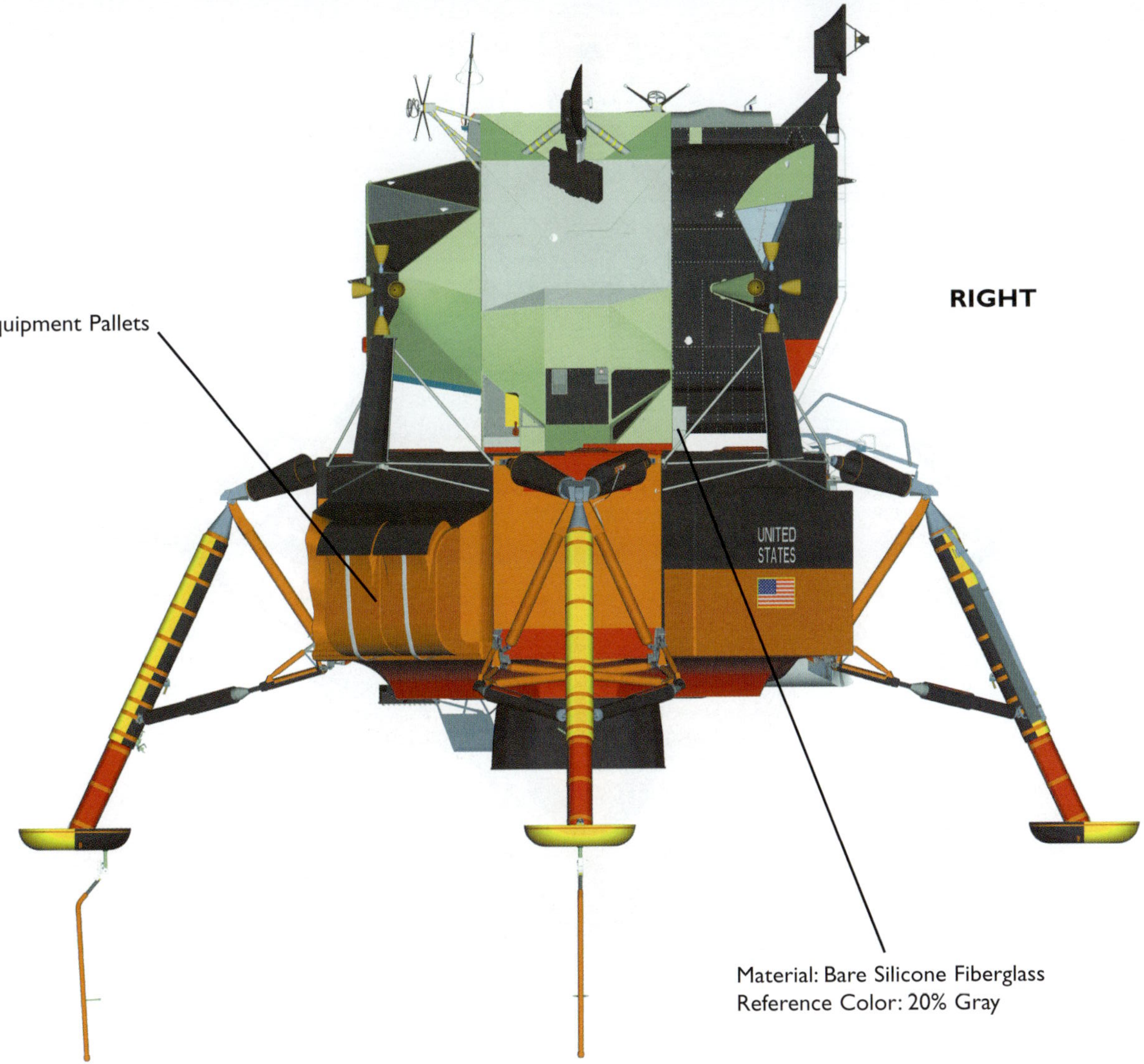

Lunar Module Chronology

December 29, 1960
The Grumman Corporation began work on a Lunar Orbit Rendezvous study.

TOP

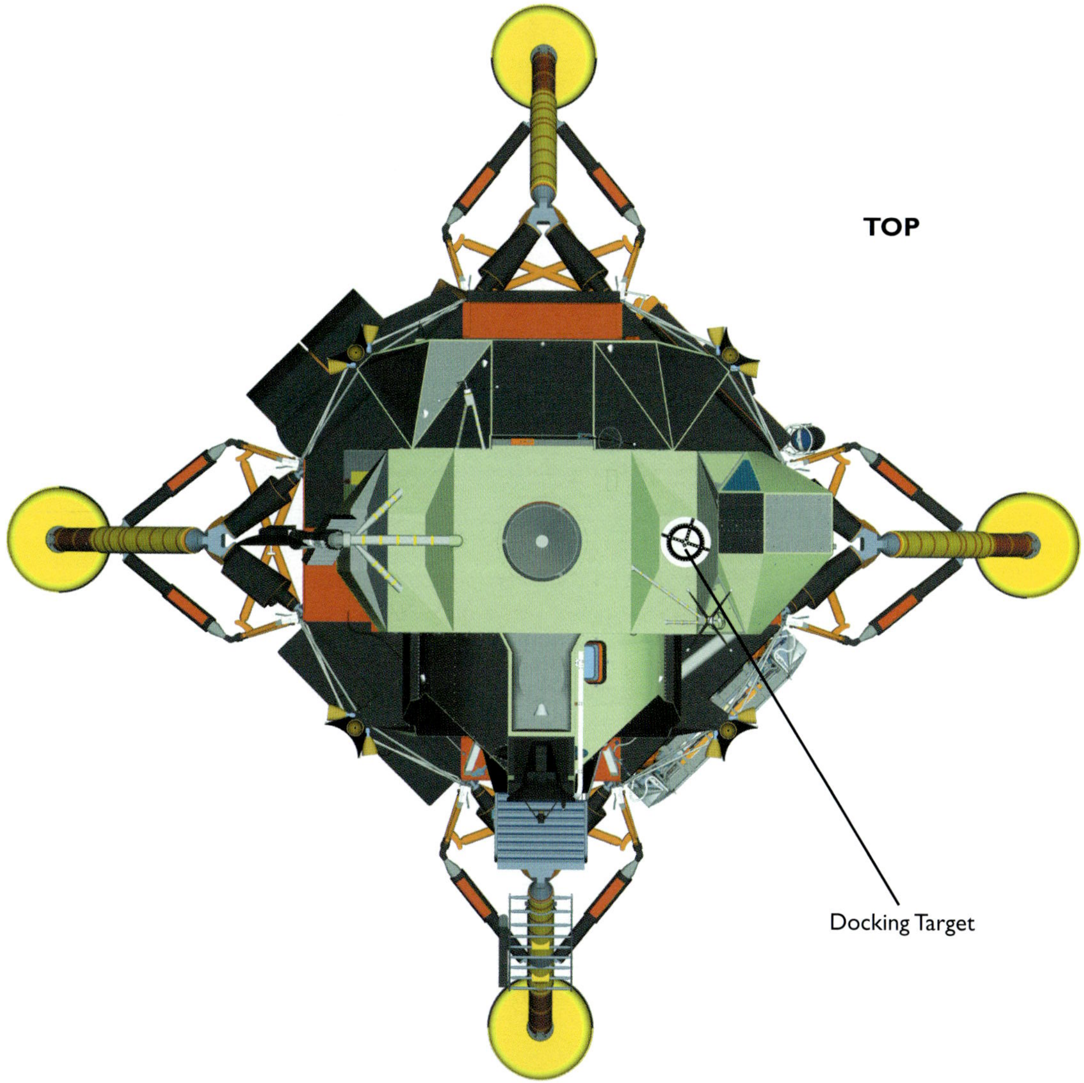

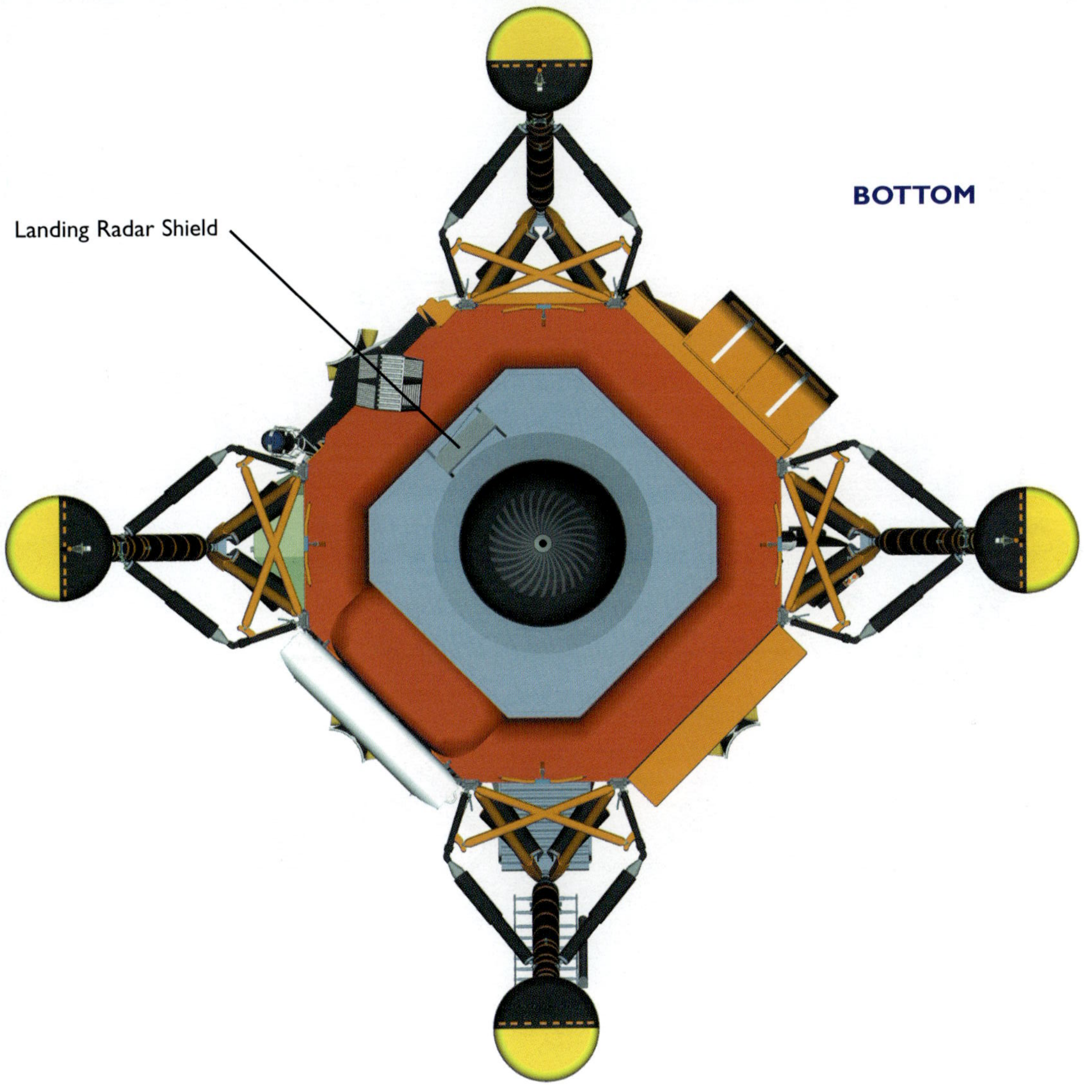

Lunar Module Chronology

May 25, 1961
NASA creates the Lundin Committee to assess lunar landing missions.

IN SPACE THE CONCEPTS OF BEING UP, DOWN, RIGHT SIDE UP OR BACKWARDS ARE COMPLETELY IRRELEVANT

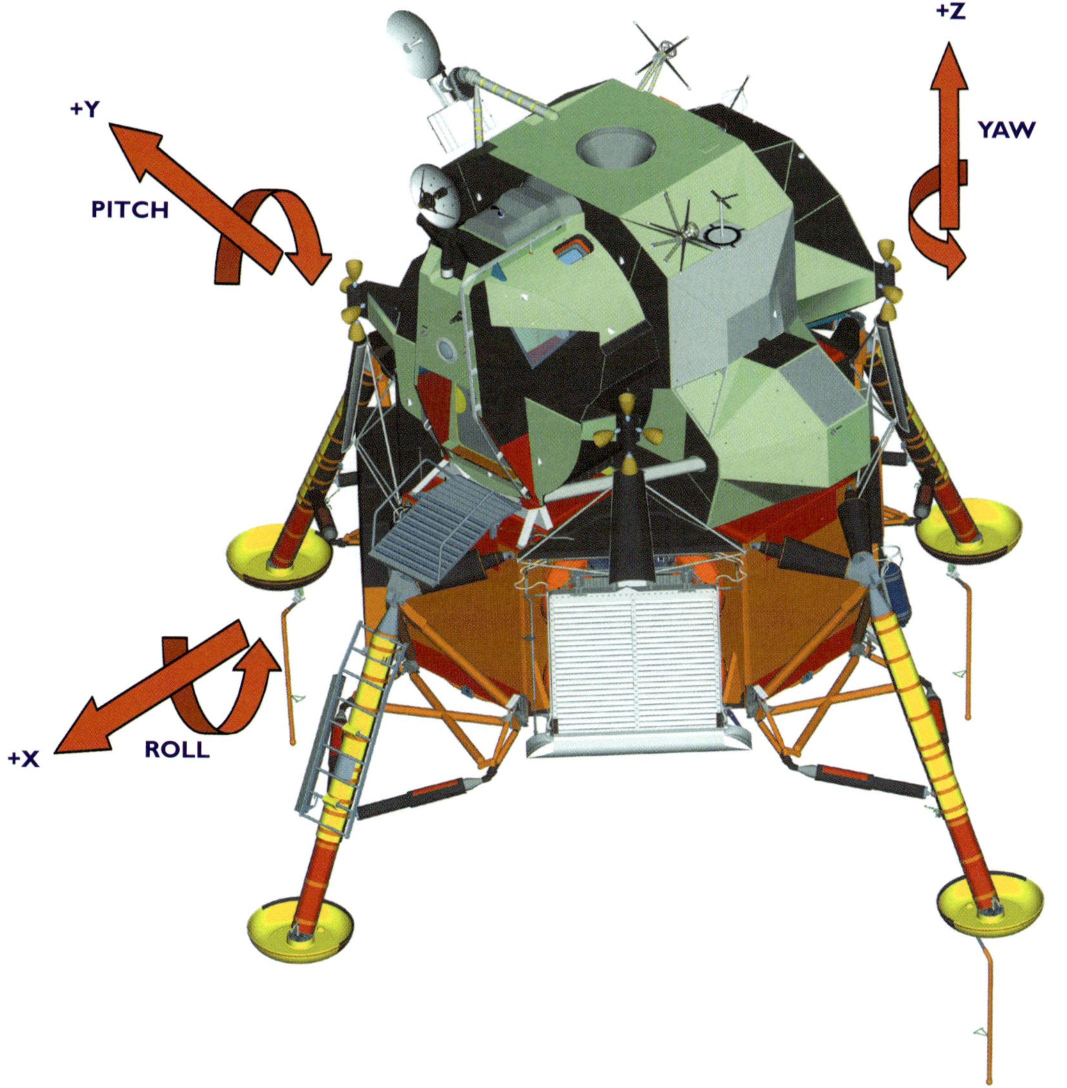

LUNAR MODULE AXES: PITCH, ROLL AND YAW

The axes of the spacecraft are three straight lines, each at a right angles to the other two. They are used for reference and to describe the spacecraft's movements. The X axis is the line running laterally from the front of the LM through to it's back; the Y axis is the line running laterally from side to side; the Z axis is the line running up and down, or from the head to the feet of the astronauts while standing at their flight stations. The CM's movement about the X axis is called roll, about the Y axis is called pitch, and about the Z axis is called yaw.

THE H-SERIES LUNAR MODULE STOWED (LAUNCH READY) CONFIGURATION

FRONT

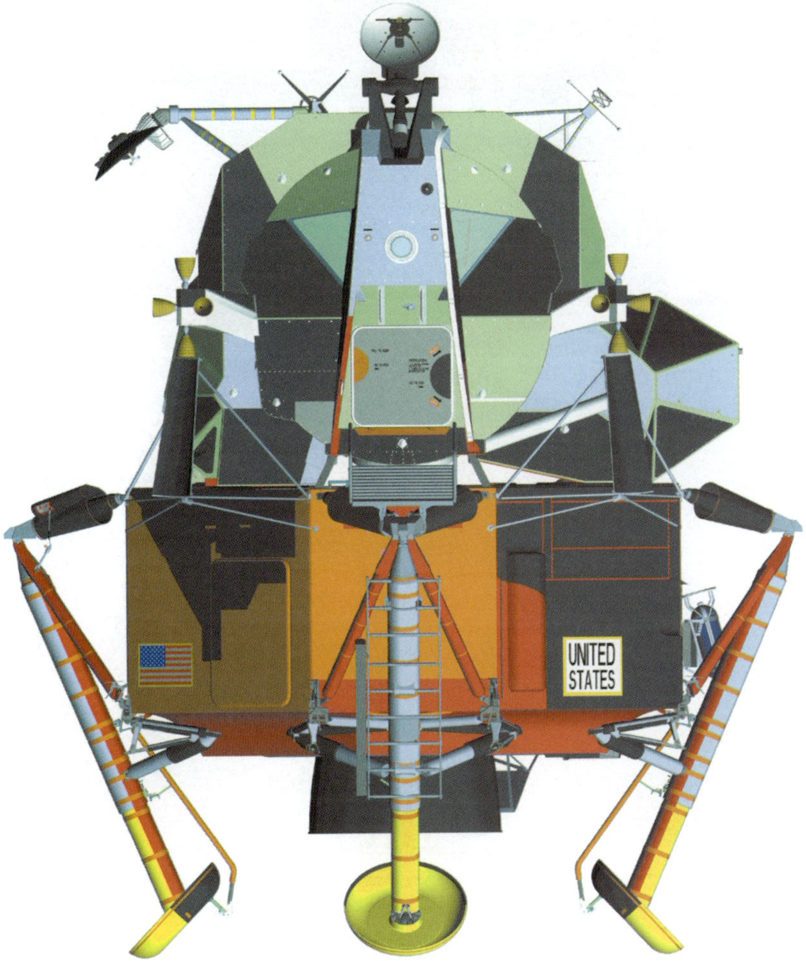

BACK

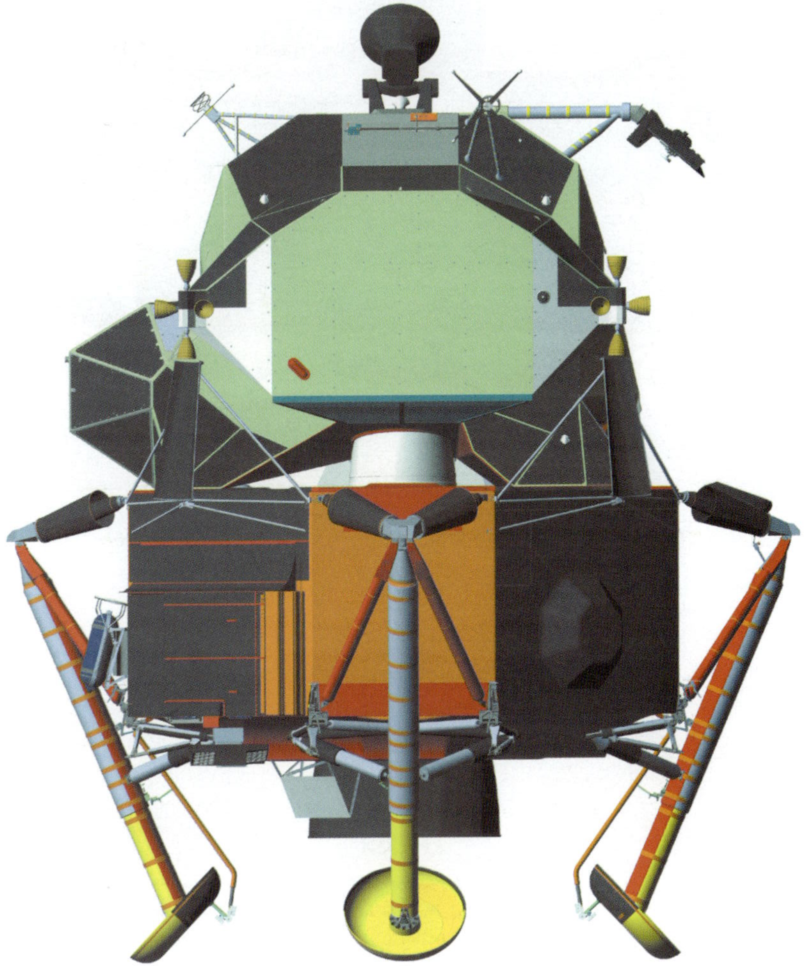

– Scott P. Sullivan –

The Lunar Module was cocooned within the protective shell of the Saturn Lunar Module adapter, atop the Saturn V launch, vehicle during the launch phase of the mission and until the spacecraft assembly was well on it's way to the Moon.

LEFT

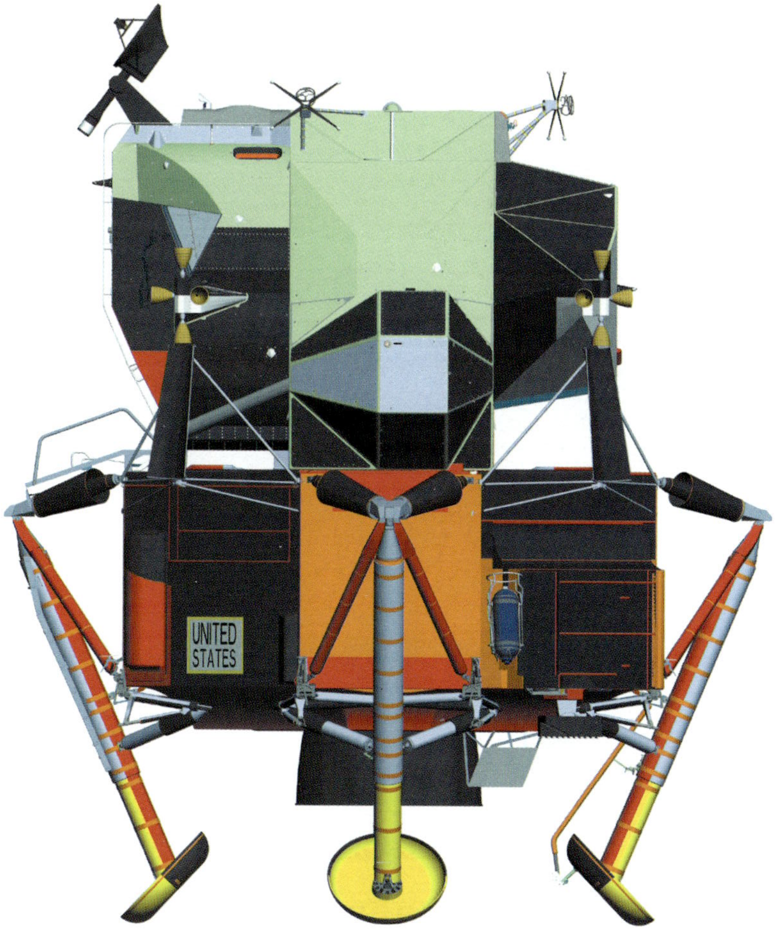

RIGHT

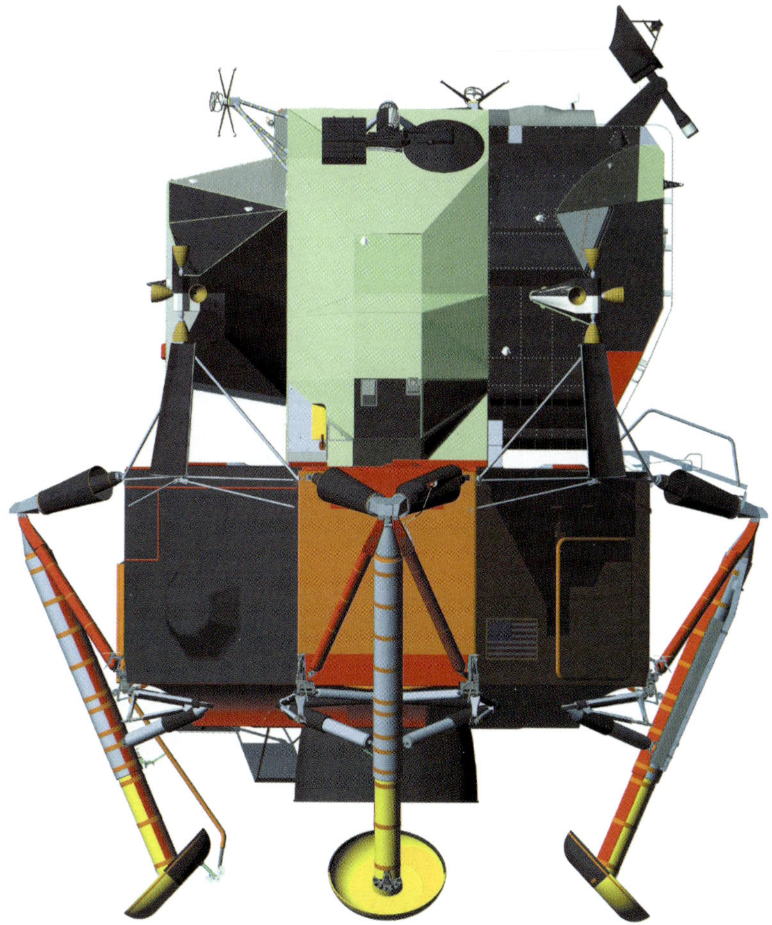

THE SATURN LUNAR MODULE ADAPTER (SLA) BEING LOWERED DURING ASSEMBLY WITH THE LAUNCH VEHICLE

H-SERIES LM WITH LANDING GEAR IN STOWED CONFIGURATION

LUNAR MODULE READY FOR ASSEMBLY OF SATURN LUNAR MODULE ADAPTER

The landing gear assembly was designed to fold in on itself for stowage, allowing for a larger overall structural footprint for the Lunar Module. The LM was attached to the SLA at four points on the LM descent stage.

TOP

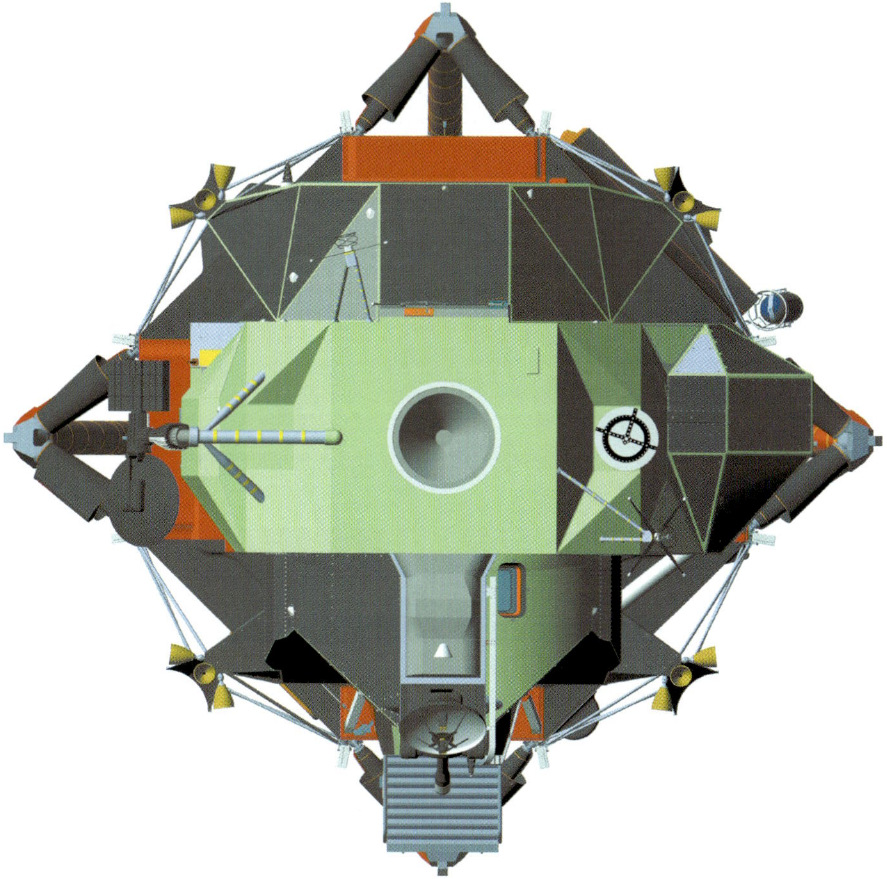

BOTTOM

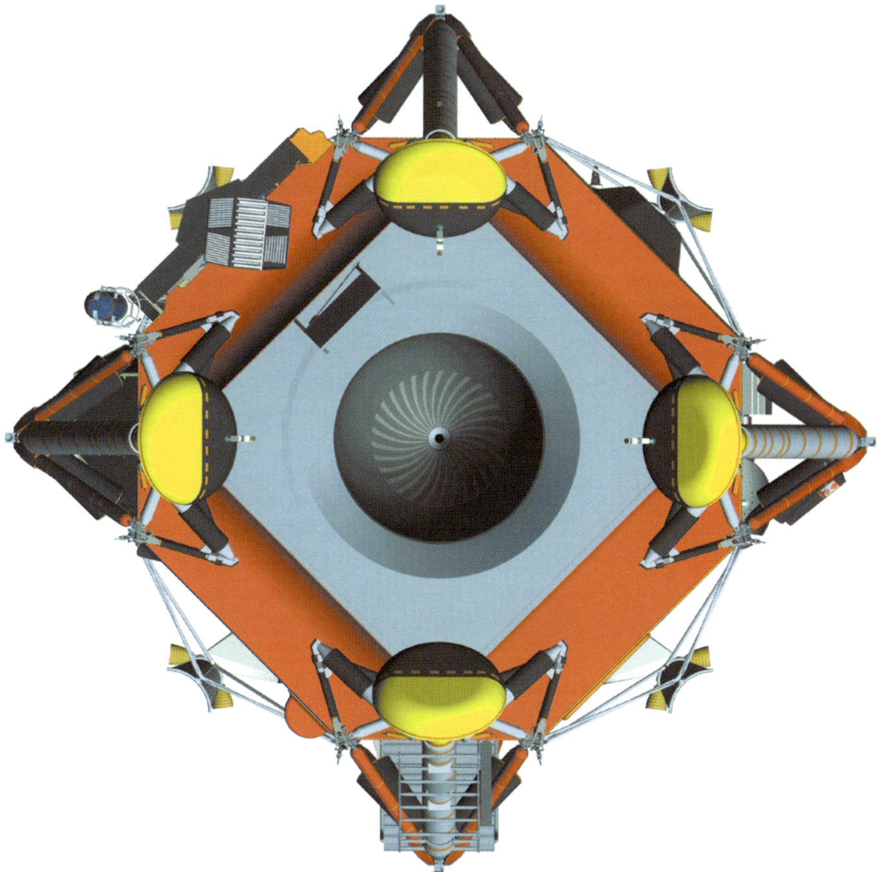

SLA / LM Attach Point (4)

THE H-SERIES DESCENT STAGE

The unmanned descent stage contained equipment and components used to deorbit and land the Lunar Module on the Moon. The descent stage also acted as a launch platform for the ascent stage after completion of the lunar mission. The descent stage housed a descent engine, as well as its fuel and oxidizer tanks and plumbing, the landing radar, electrical power and pyrotechnic components, the Apollo Lunar Surface Experiment Package (ALSEP), the MESA, and some of the consumables used by the astronauts.

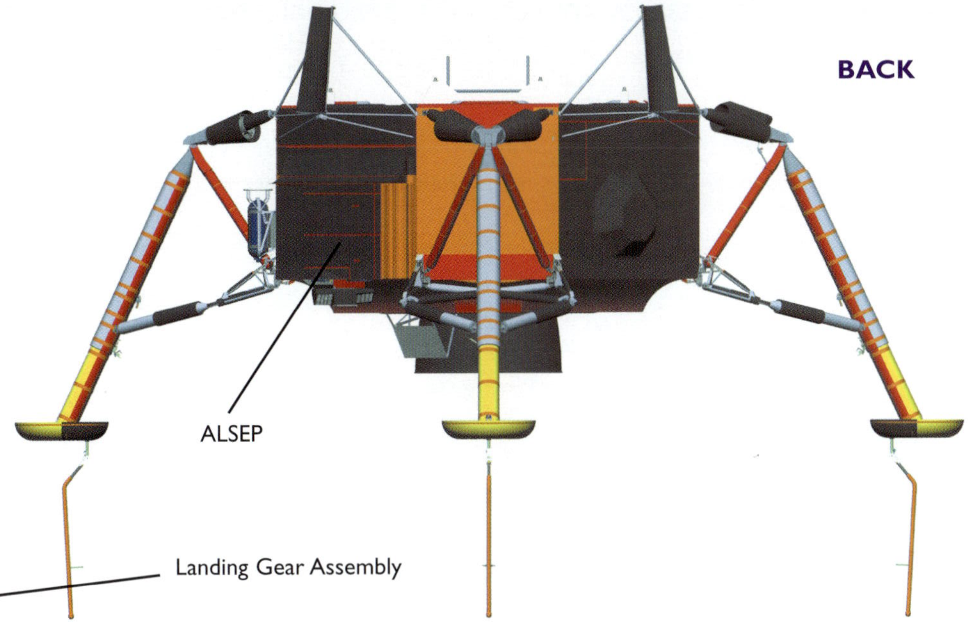

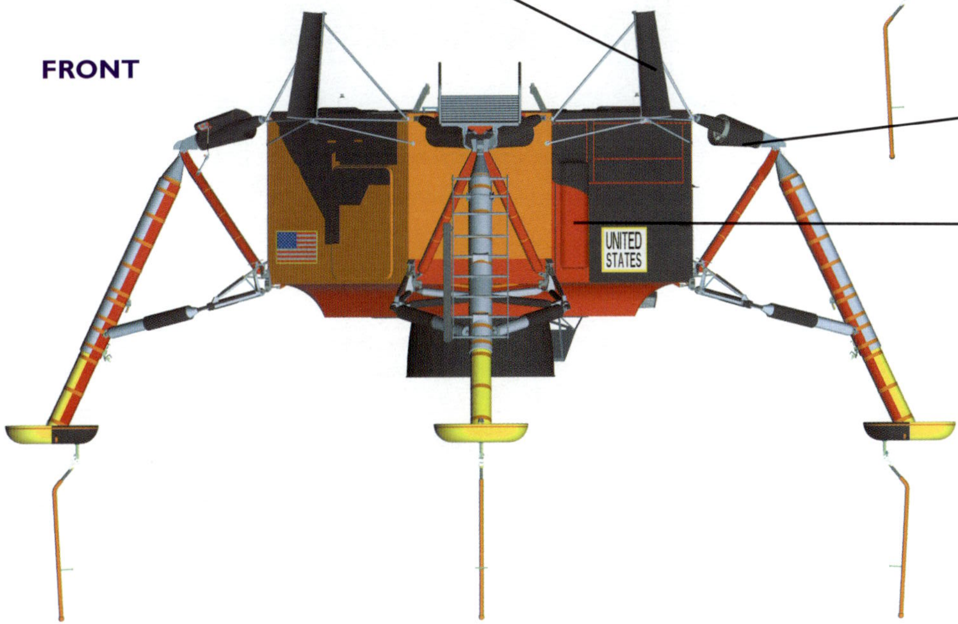

MAJOR SUBSYSTEMS:

(A) Provide Velocity Change for Lunar Deorbit and Lunar Landing (Throttleable)
(B) Protect Ascent Stage from Landing Damage
(C) Provide Ascent / Descent Stage Staging
(D) Provide Ascent Stage Launch Platform
(E) Stow Lunar Scientific Equipment
(F) Provide Attachment for Landing Gear
(G) Support LM in the SLA

– Scott P. Sullivan –

WEIGHT DRY: 4,260 POUNDS
WEIGHT FUELED: 17,944 POUNDS

Lunar Module Chronology

May 31, 1961
Basic concepts for Lunar Orbit Rendezvous were presented to the Lundin Committee by John C. Houbolt.

LEFT

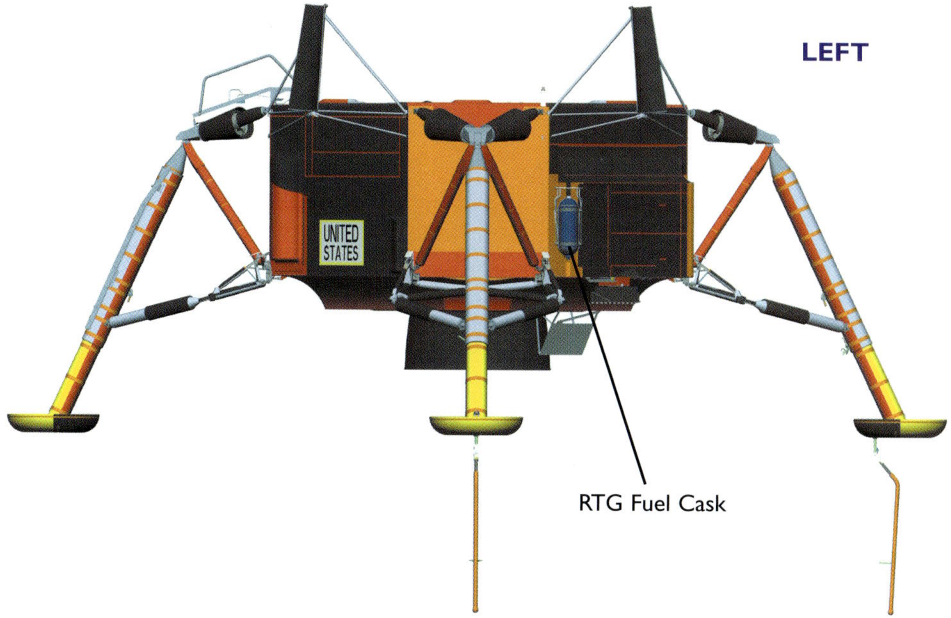

RIGHT

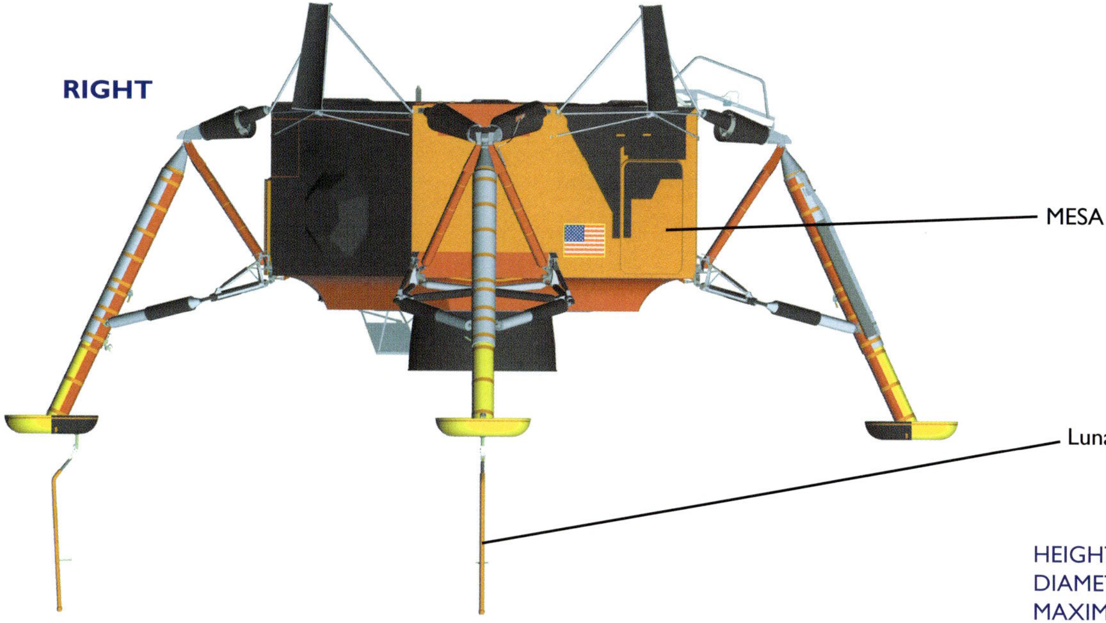

HEIGHT: 10 FEET 7 INCHES
DIAMETER: 14 FEET 1 INCH
MAXIMUM DIAMETER DIAGONALLY ACROSS LANDING GEAR: 31 FEET

TOP

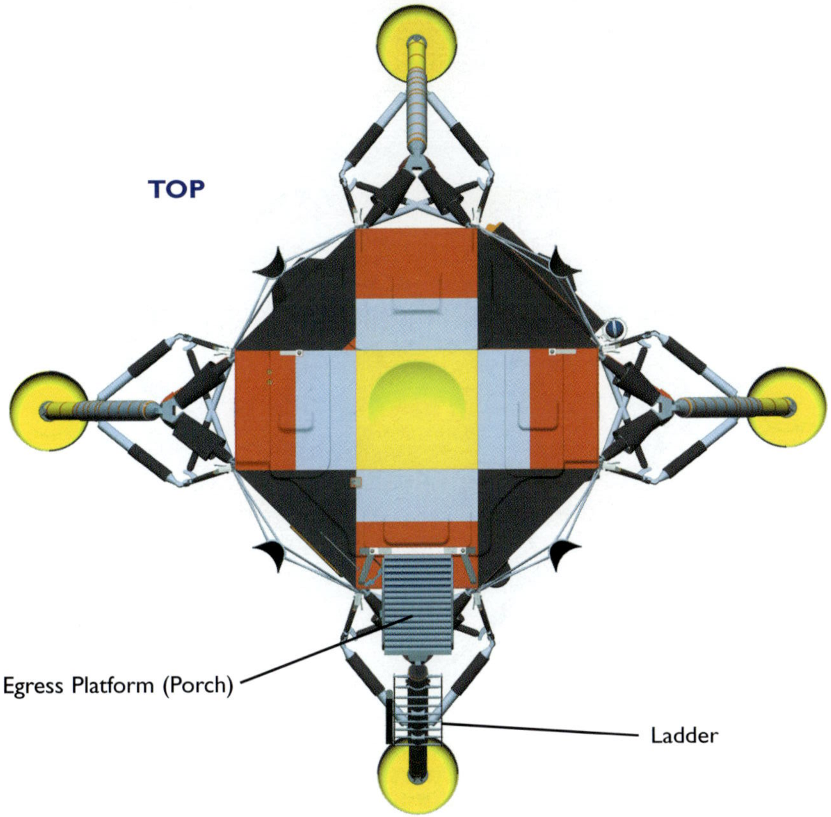

THE DESCENT STAGE COMPRISED APPROXIMATELY TWO THIRDS OF THE WEIGHT OF THE LM AT EARTH LAUNCH

BOTTOM

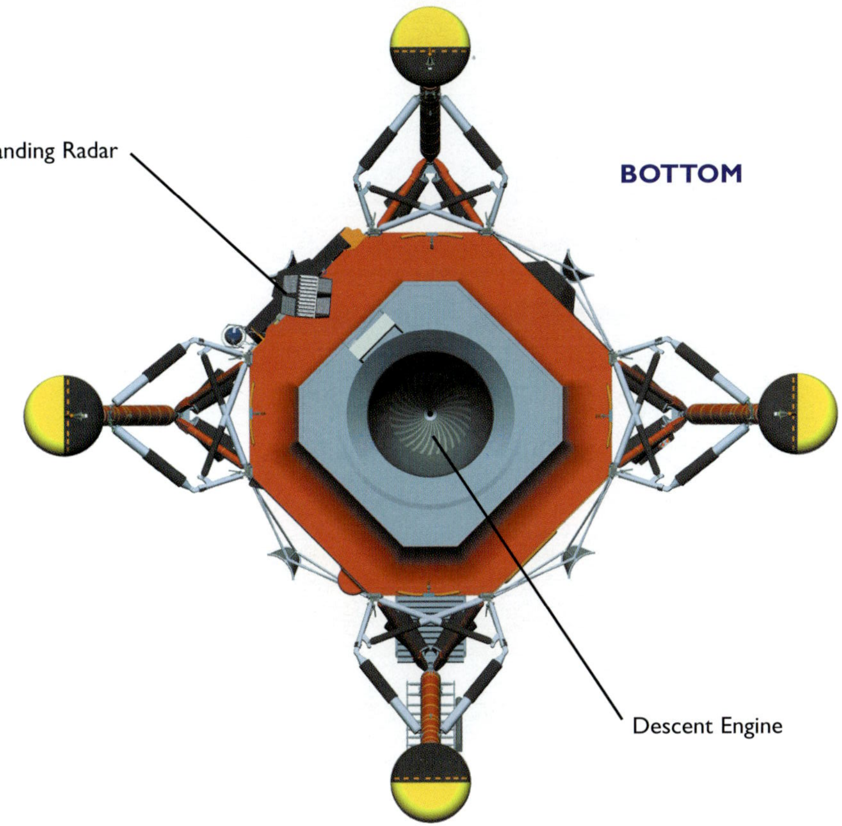

Lunar Module Chronology

August 23, 1961
A NASA committee evaluates three rendezvous methods for manned lunar landing. These methods were: Earth Orbit, Lunar Orbit, and Lunar Surface Rendezvous.

QUAD 1

The main structure of the descent stage consisted of two pairs of parallel beams arranged in a cruciform, with a deck on the upper and lower surfaces, approximately 65 inches apart. The ends of the beams, approximately 81 inches from the center, were closed off by aluminum beams to provide five equal-sized compartments – a center compartment, one forward and one aft of the center compartment, and one right and left of the center compartment. A four-legged truss (or outrigger) at the end of each pair of beams served as a support for the landing gear primary strut. Struts between the ends of all main beams formed triangular bays, or quadrants, to give the descent stage its octagon shape. The quadrants were designated 1 through 4, beginning at the left forward compartment and continuing counterclockwise as viewed from the top.

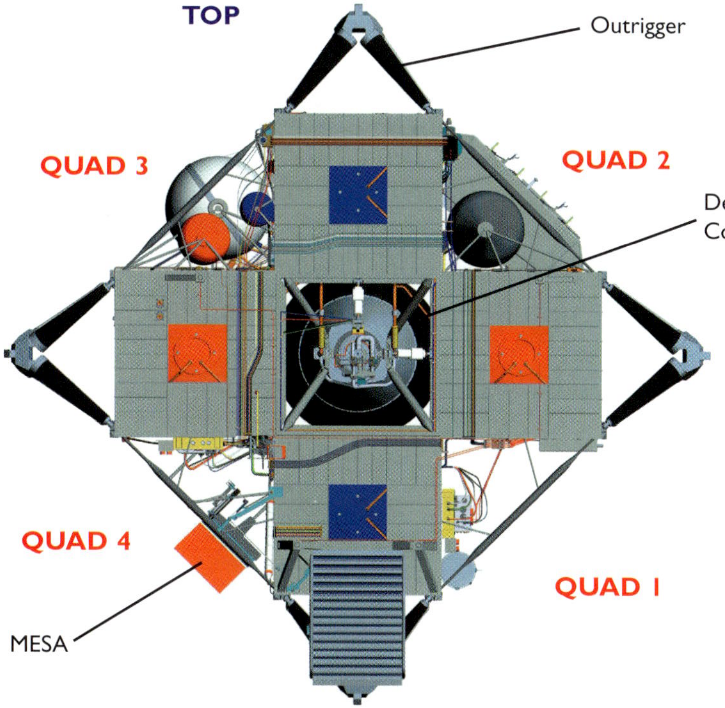

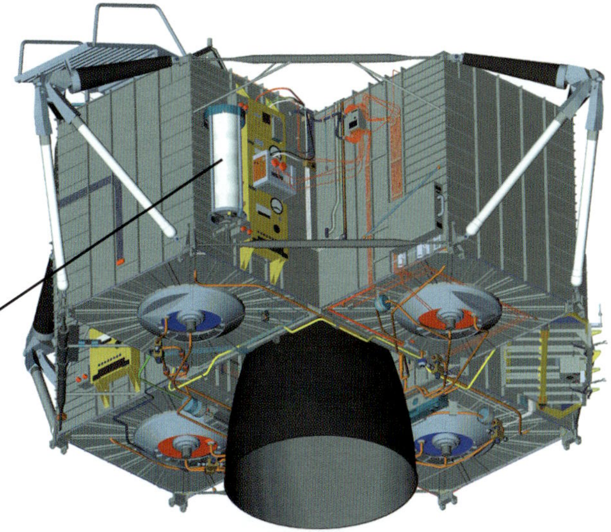

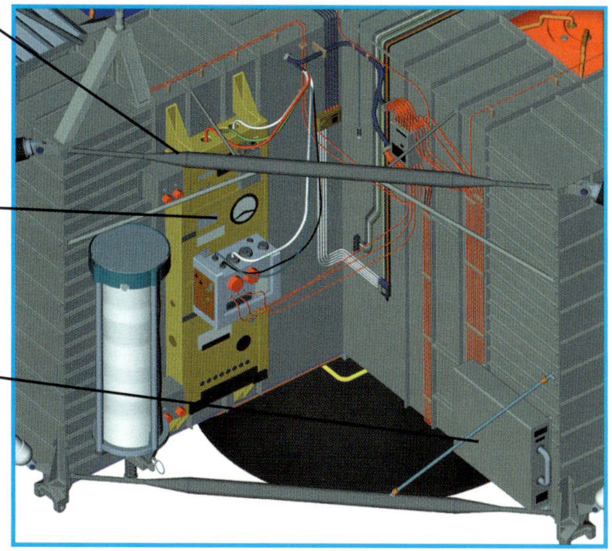

QUAD 2

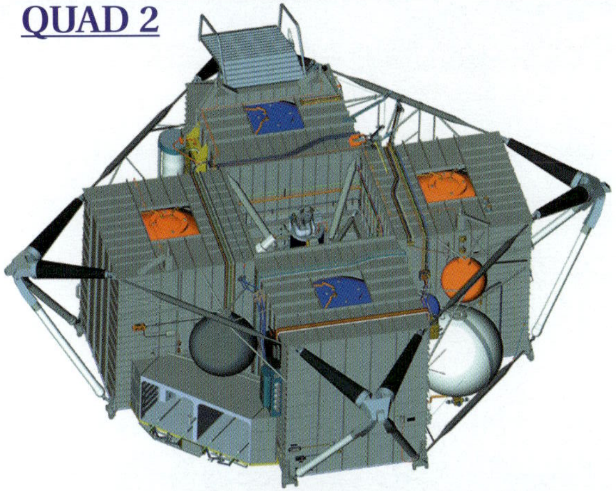

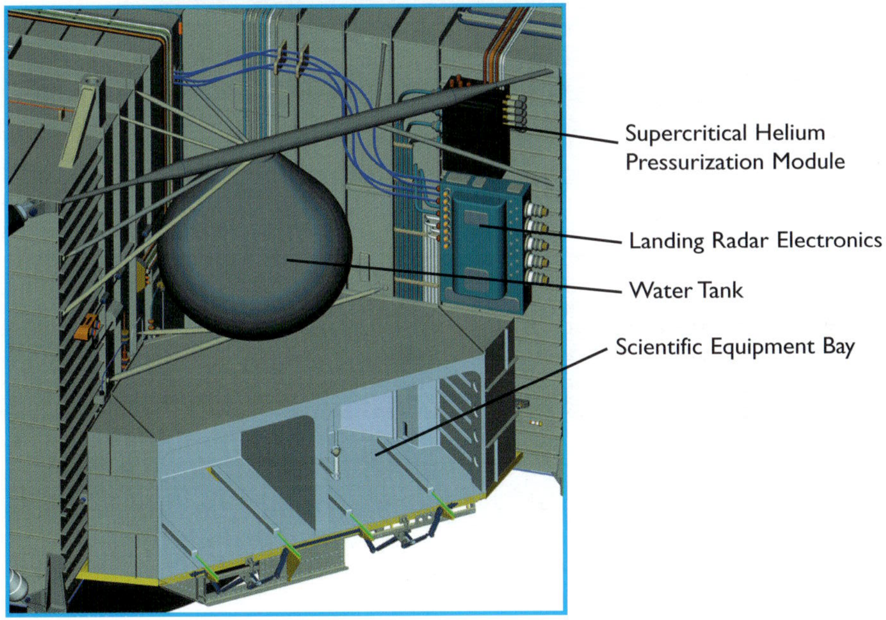

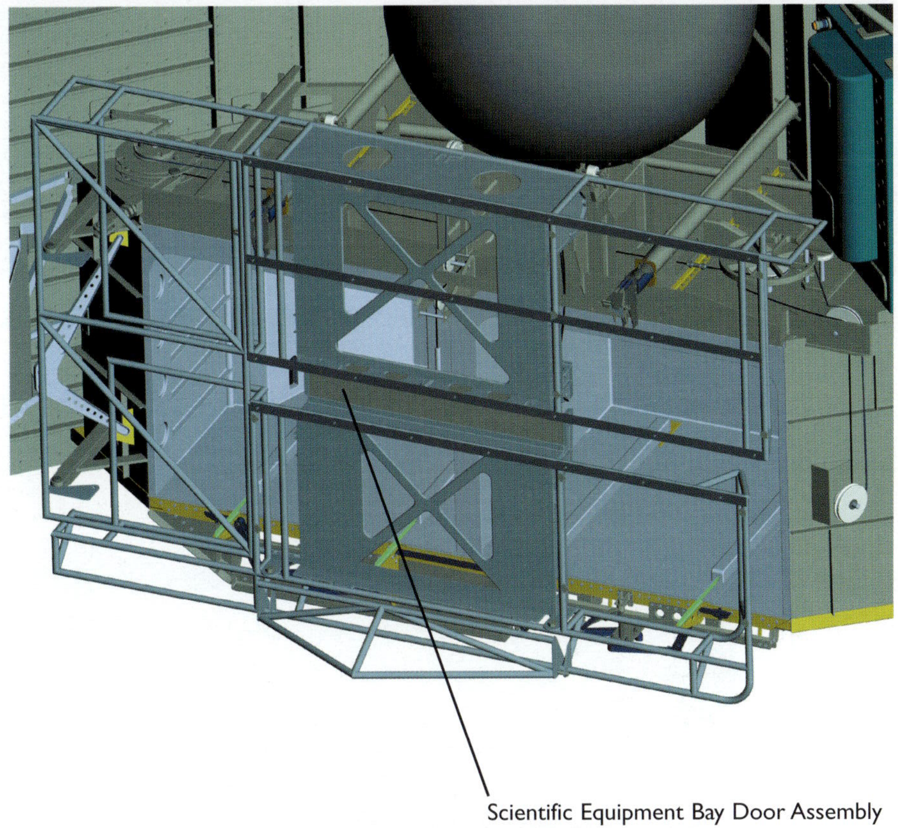

The quadrants housed components from the various subsystems. In addition, there was the Modularized Equipment Stowage Assembly (MESA) in Quad No. 4. Quad No. 2 housed the Apollo Lunar Scientific Experiments Package (ALSEP). The ALSEP was packed on two sub-pallets and consisted of various experiments. Energy for the ALSEP was provided by the nuclear-powered Radioisotope Thermoelectric Generator (RTG). The last ALSEP experiment left on the Moon ceased working in December of 1974.

The scientific equipment bay door was operated with a complex set of cables and pulleys acting on the accordion-like bay door assembly. The lanyards (not shown) extending from pulley A actuated the pulley system to either raise or lower the bay door. When the astronaut pulled the lanyard cable H towards himself, the pulley system raised the double hinged bay door, and when he pulled the lanyard cable G towards himself, the system reversed itself, causing the bay door to lower. As the cables pulled the double pulley B either clockwise or counter-clockwise the cables in turn actuated another set of pulleys, marked E (see details on page 44), which turned the door rollers marked C and D. The rollers acted against mating rollers on the bay door, causing it to go up or down. The roller set E also simultaneously pulled the double pulley marked F, opening the side door, which was designed to act as a protective cover for the RTG fuel cask, and contained the fuel cask lanyards on its inner face.

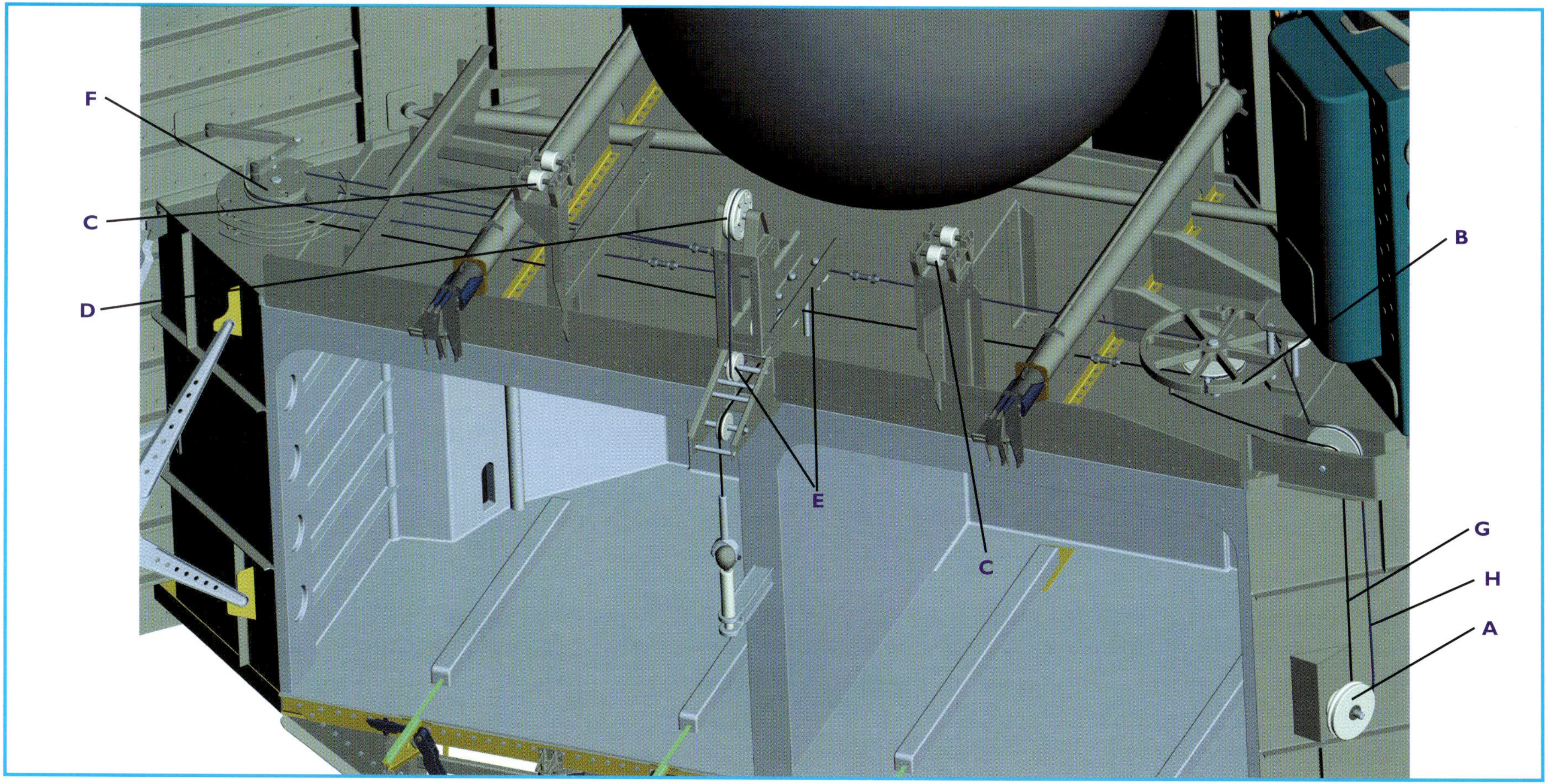

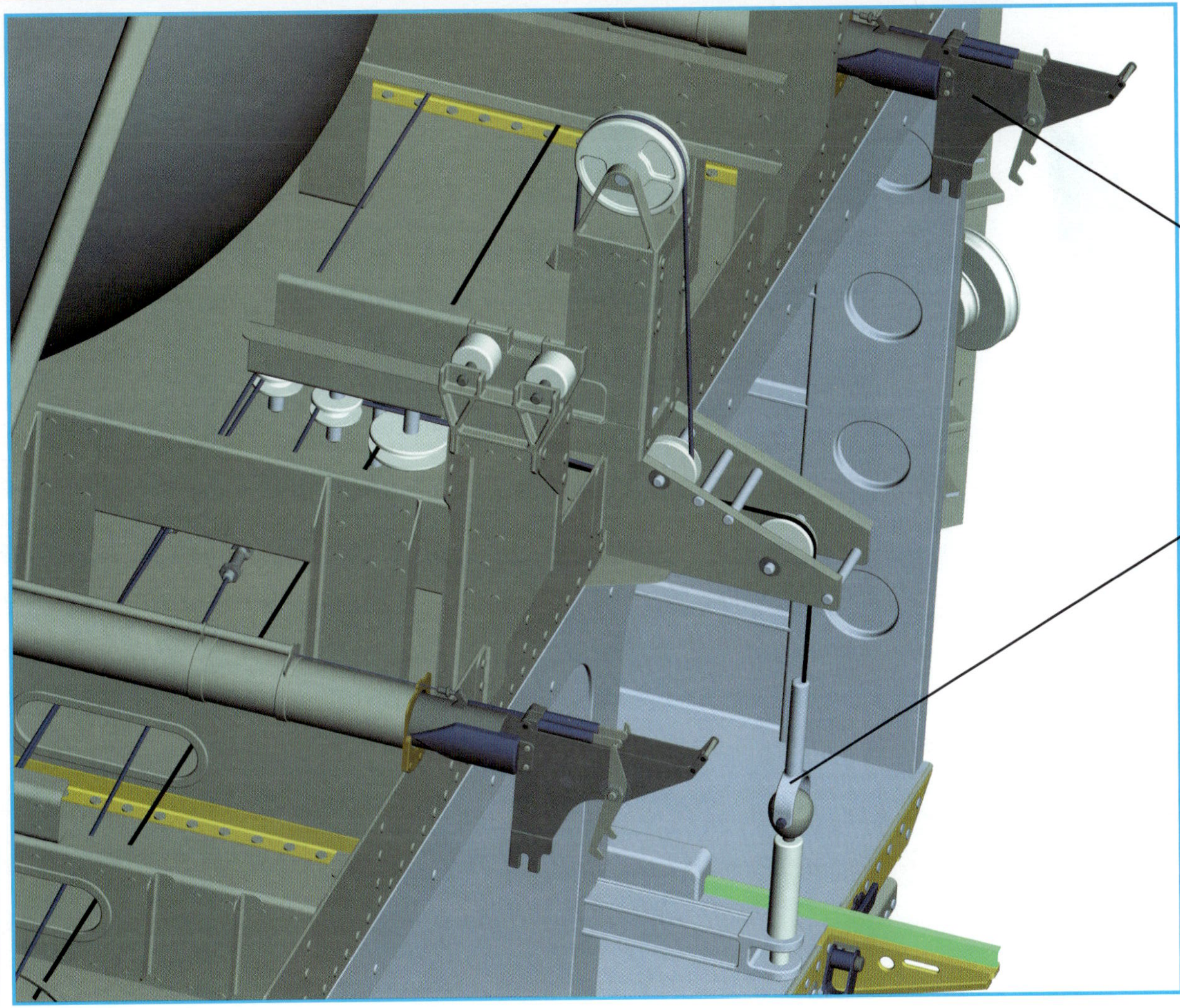

QUAD NUMBER 2 ALSEP BAY DOOR
DURING ASSEMBLY

DOOR ASSEMBLY IN OPEN POSITION

The descent stage water tank was an aluminum tank with an internally mounted standpipe and bladder. The bladder contained water. The space around the bladder was charged with nitrogen according to a schedule-dependent loading (48.2 psia maximum at +80 degrees F). The nitrogen forced the water out of the bladder through the standpipe and into the system.

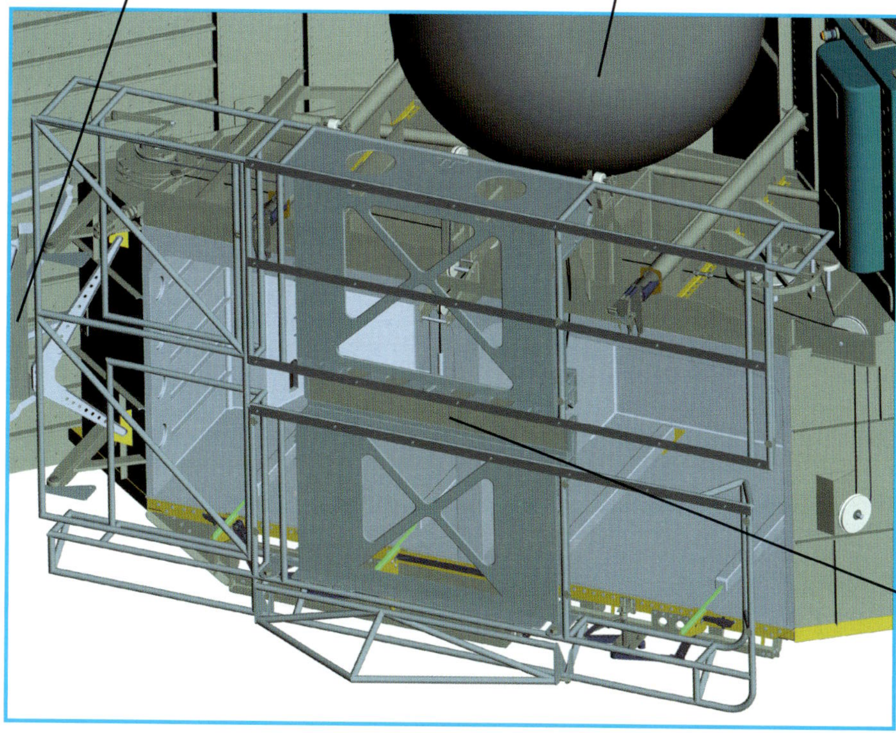

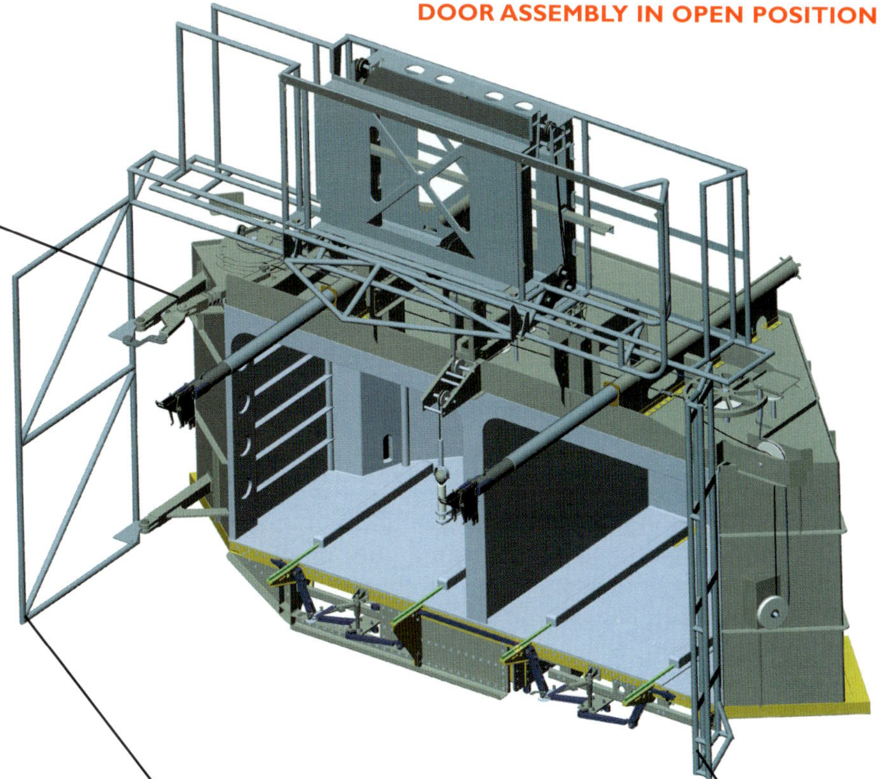

QUAD 2 ALSEP BAY DOOR ASSEMBLY

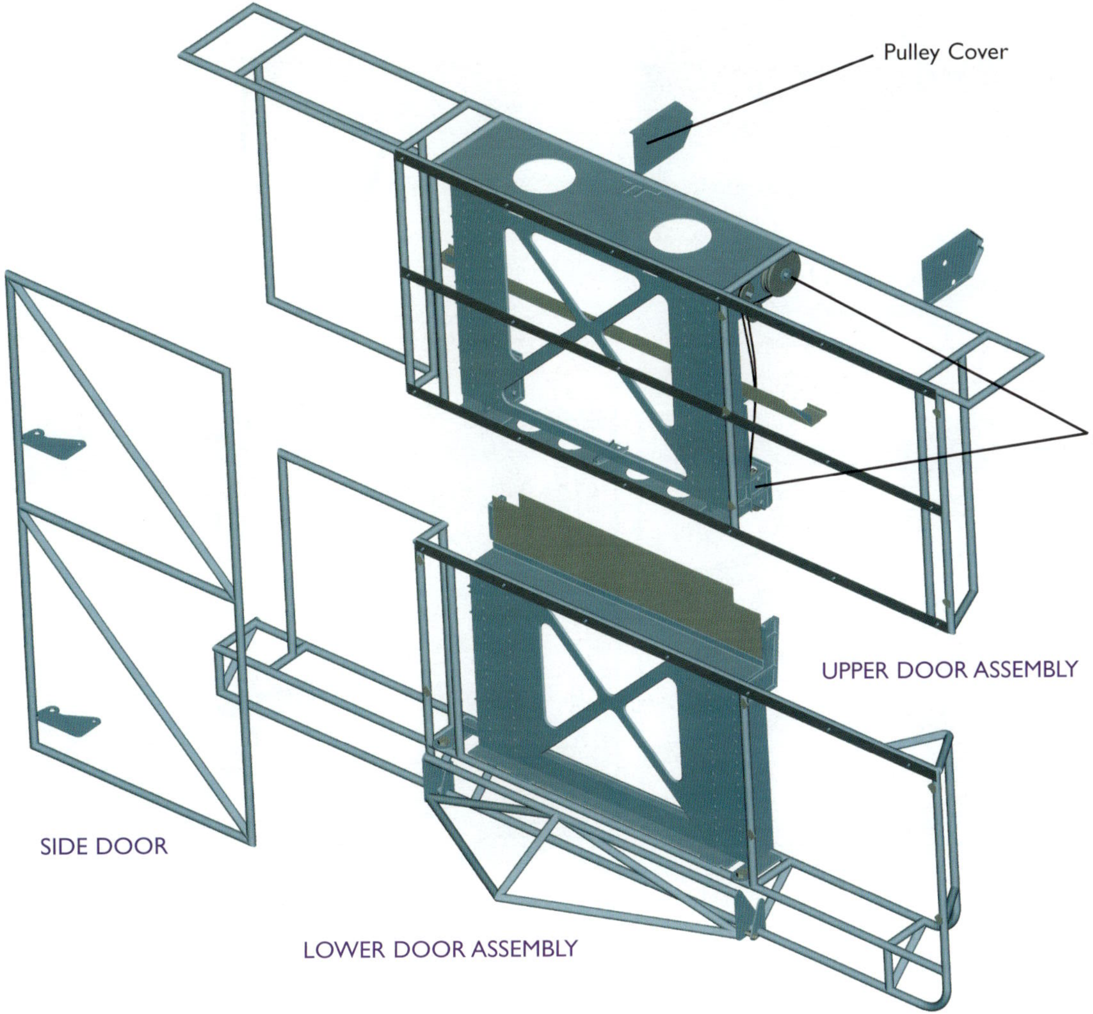

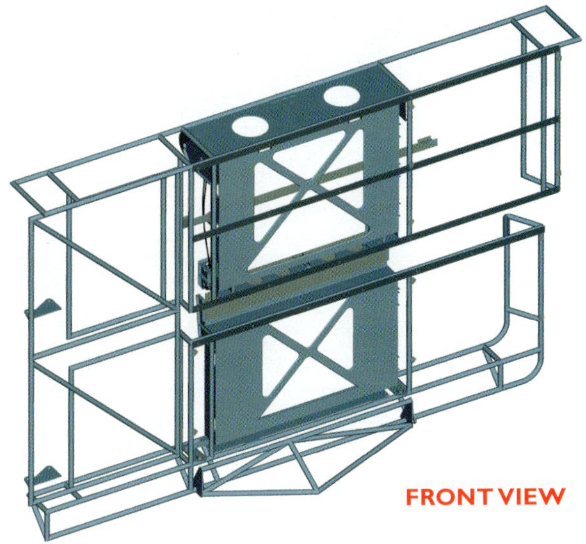

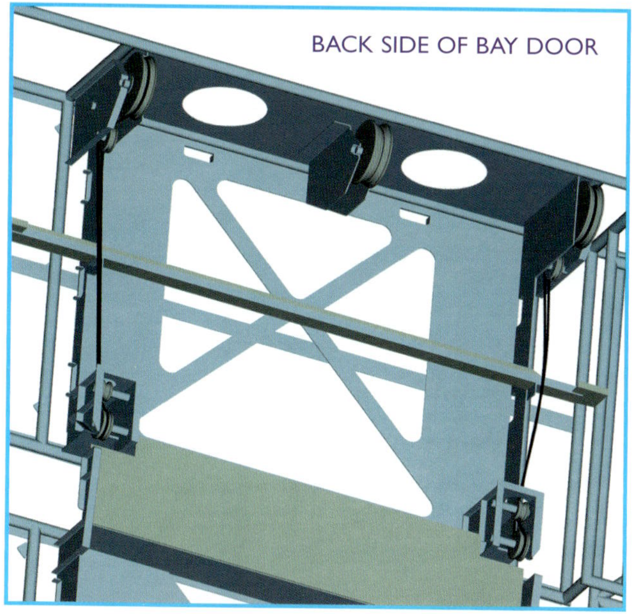

DETAIL VIEW SHOWING THE INTERACTION BETWEEN THE DRIVER ROLLERS C AND D, AND THE PULLEY / ROLLERS ON THE BAY DOOR

DETAIL VIEW OF PULLEY / ROLLER ARRANGEMENT ON INSIDE OF UPPER BAY DOOR

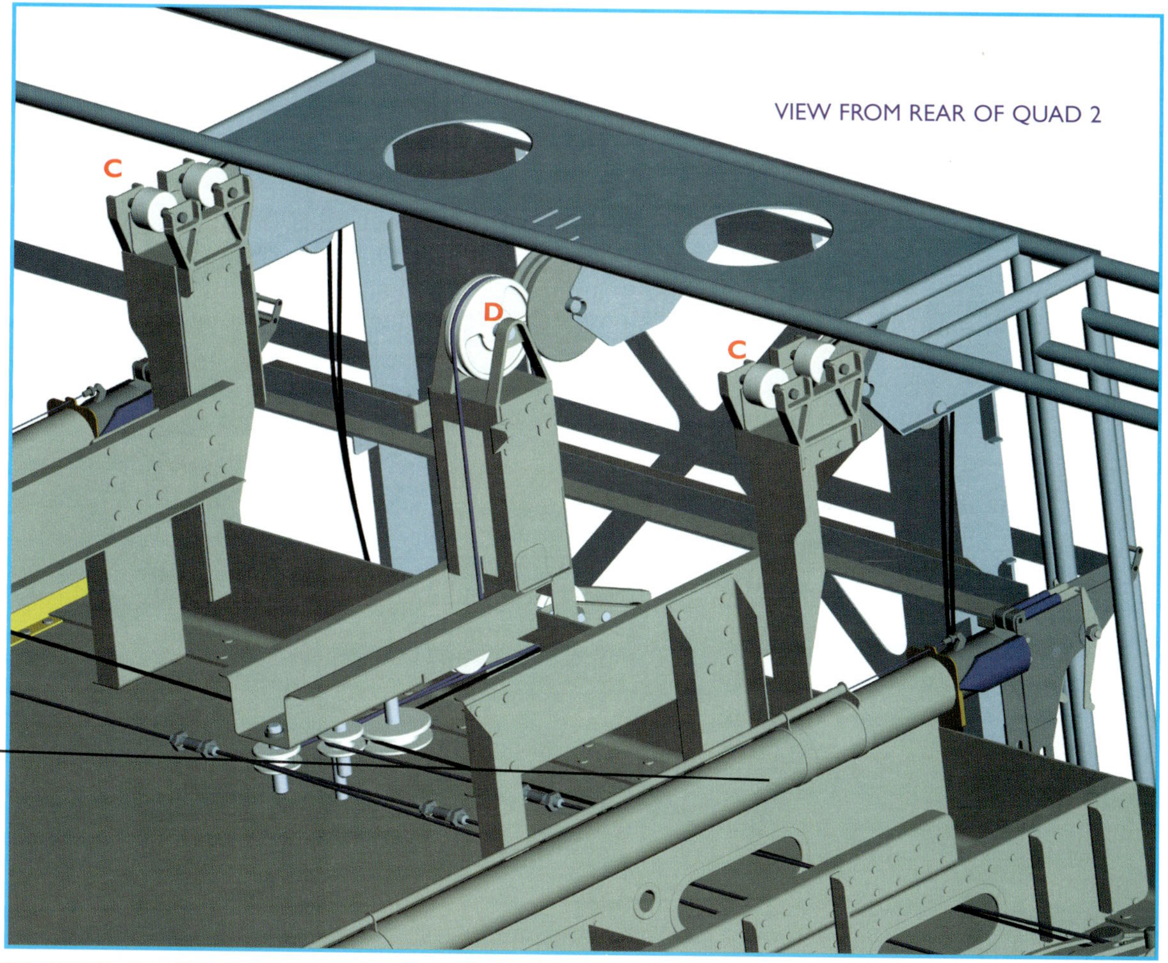

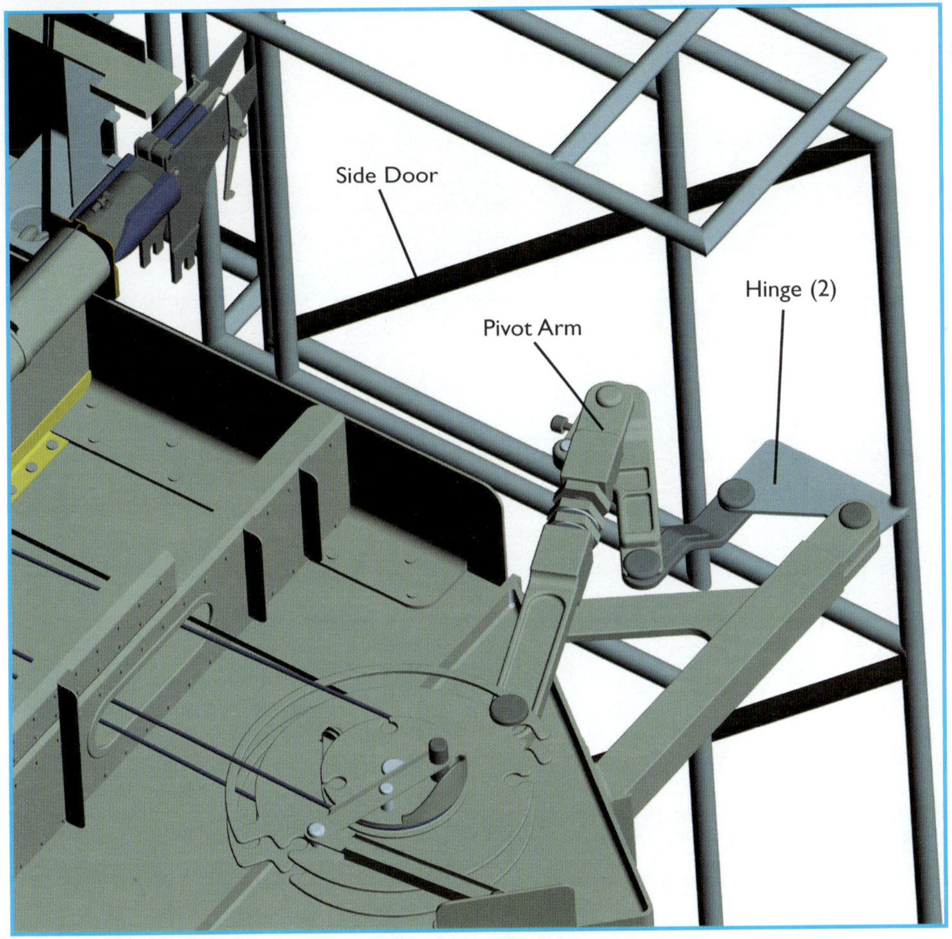

THE PIVOT ARM THAT OPENS AND CLOSES THE SIDE BAY DOOR

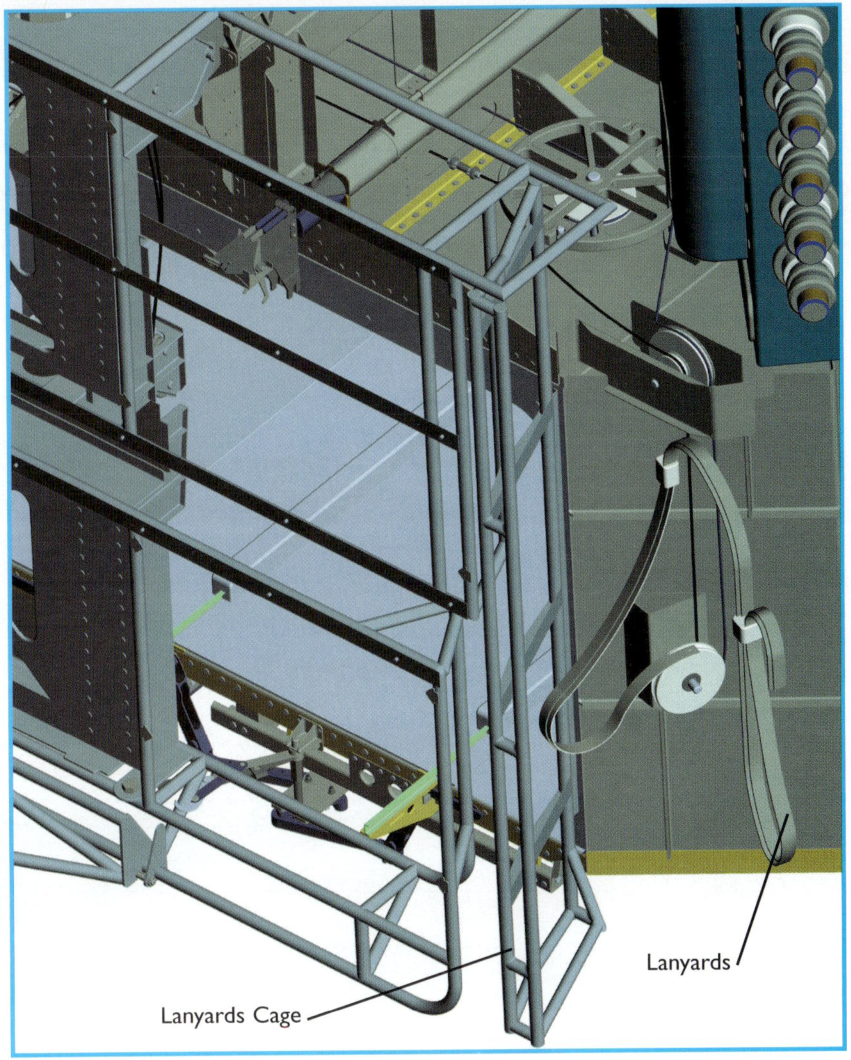

THE LANYARDS IN THEIR STOWED CONFIGURATION

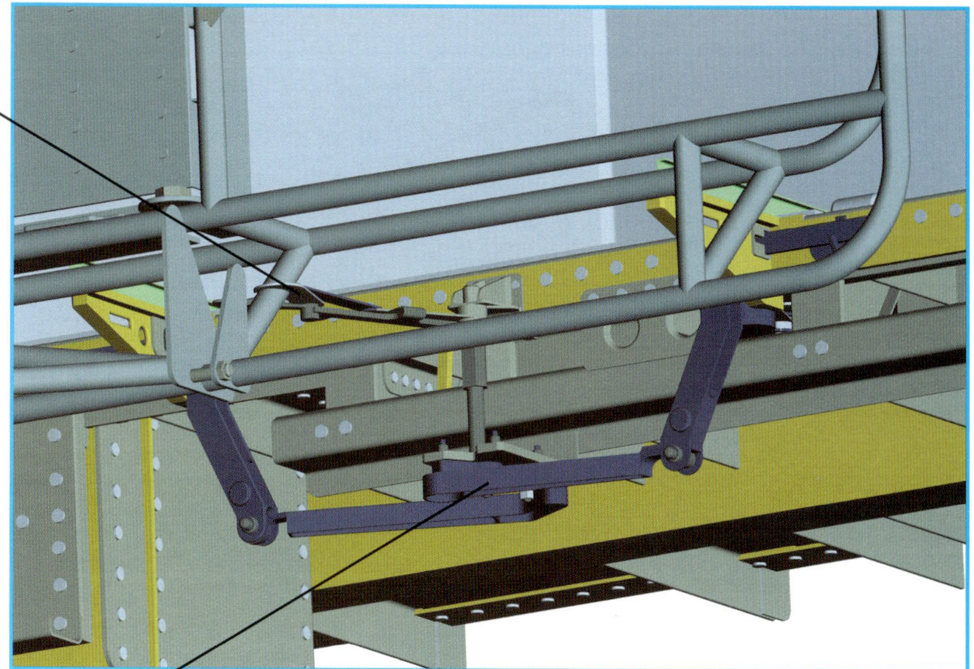

BOTTOM VIEW

The ALSEP bay door was locked at the bottom by two sets of interconnected locking mechanisms. The astronaut was required to manually release the locking mechanism by pulling a lever that would cause the linkages to work in tandem to disengage the two latching claws.

THE ALSEP BEING ACCESSED BY AN ASTRONAUT

The second flexible hinge was comprised of the thermal blanket that covered the exterior of the descent stage (not shown). In further chapters we will see how the thermal blanket, attached to the upper bay door, was responsible for helping to control its path as it was raised and lowered, and how it mounted the bay door assembly to the body of the LM.

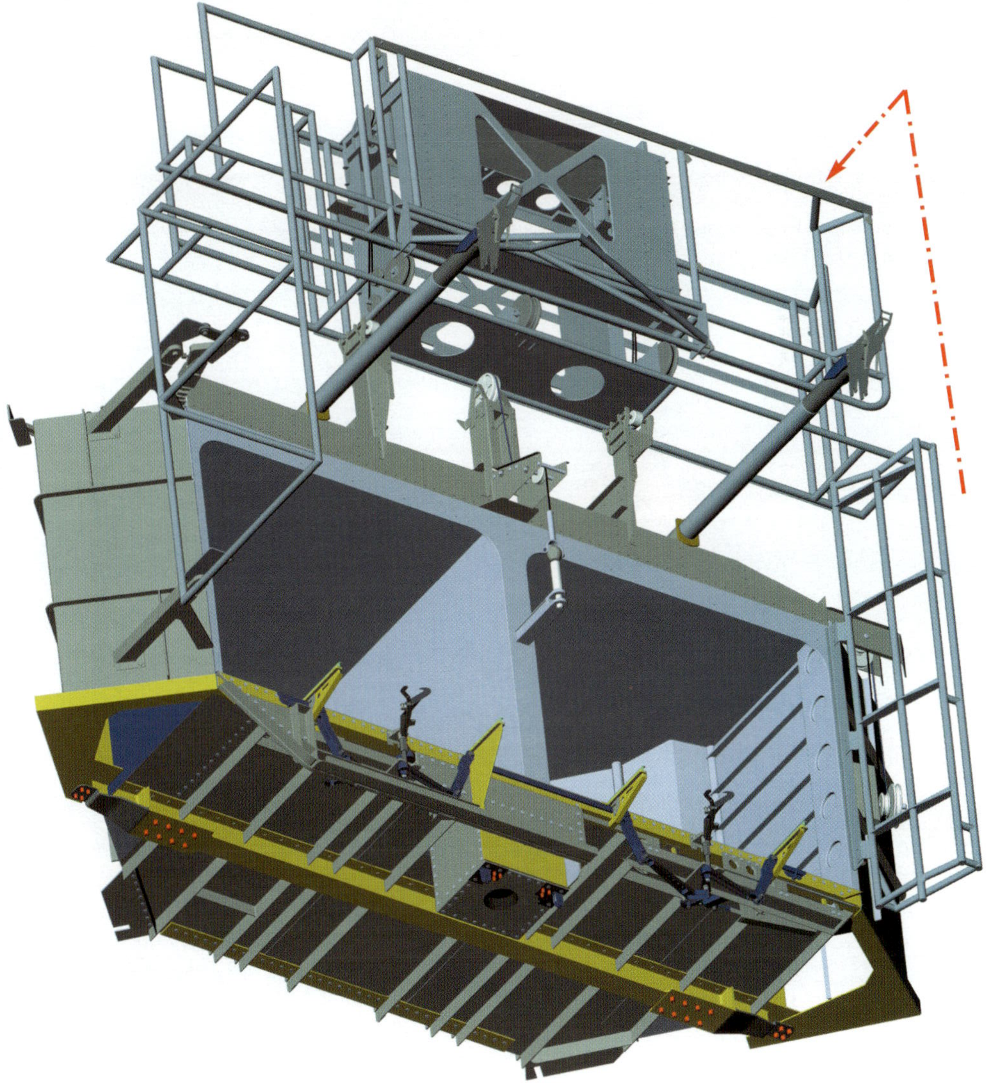

QUAD 3

Quad number 3 contained a supercritical helium tank, an ambient helium tank, and an oxygen tank. Quad 3 also contained descent engine electronics and wiring, the descent engine control assembly, and the high-pressure oxygen control assembly. The supercritical helium tank was double walled and consisted of an inner spherical tank and an outer jacket. The void between the two was filled with aluminized Mylar insulation and evacuated to minimize heat transfer to the tank. The vessel had fill and vent ports, a burst disk assembly, and an internal helium / helium heat exchanger. The high-pressure oxygen control assembly consisted of a high-pressure oxygen regulator, a bypass oxygen pressure relief valve, an overboard relief valve, and a burst diaphragm assembly.

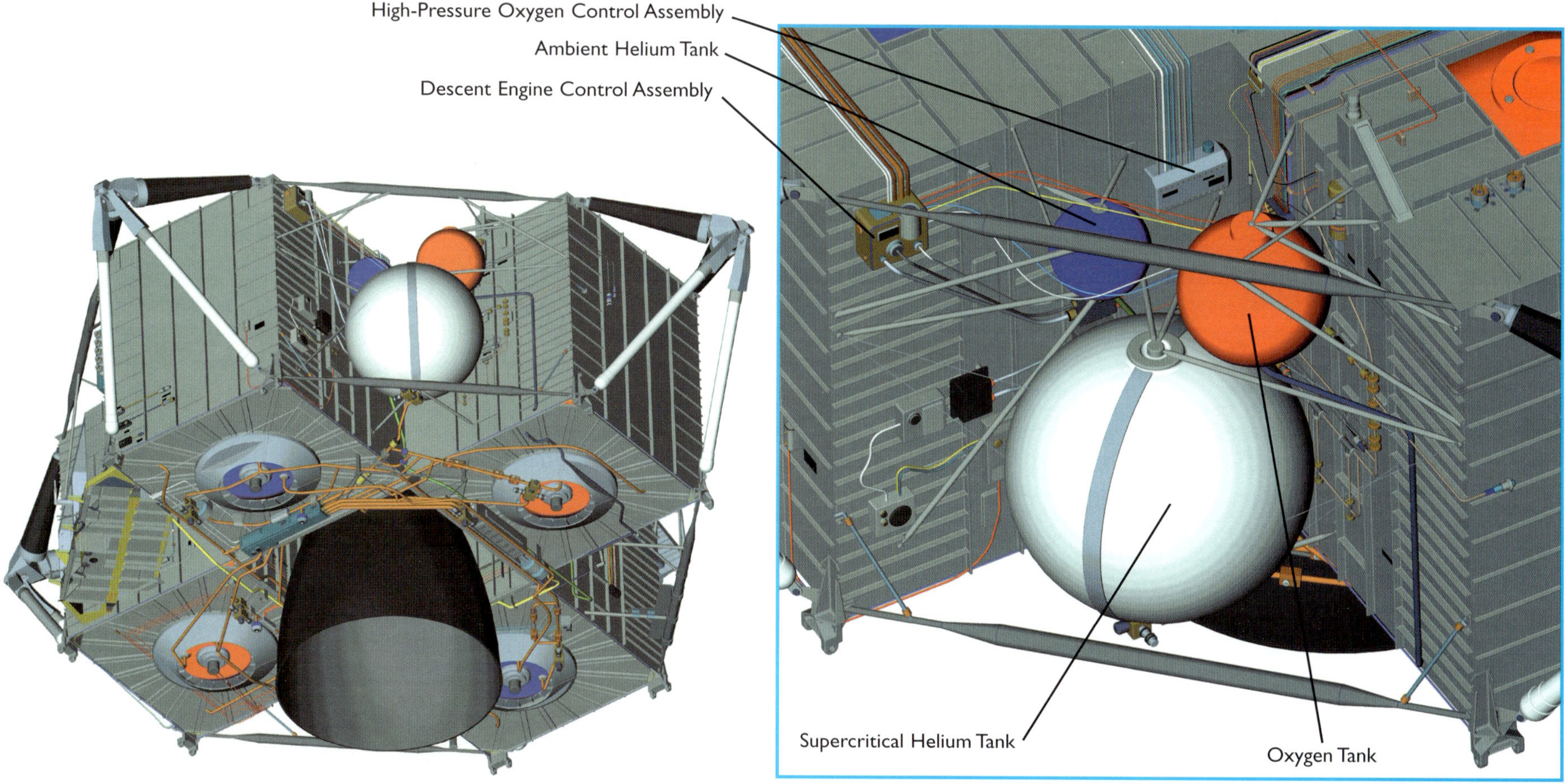

QUAD 4

Quad number 4 contained batteries, an electrical control assembly, a power outlet for scientific equipment, an explosive relay box, the interstage umbilical guillotine assembly, pyrotechnic batteries, the S-band antenna cable, and the Modular Equipment Stowage Assembly (MESA). The MESA contained fresh PLSS backpack batteries and LiOH cartridges, a TV camera and cable, a still camera, tools for obtaining lunar geological samples, food, film, and containers in which to store the samples. It also had a folding table used to place the sample return containers.

On the following pages the MESA deployment mechanism will be detailed. The mechanism that was responsible for deploying the MESA was very complex. To try and make my explanation of this device as simple to understand as I possible I have numbered all of the various components and shown as many different views as I thought necessary, some with the MESA in place and some without. The reader will more than likely need to cross-reference the diagrams a few times to grasp the mechanism in its entirety.

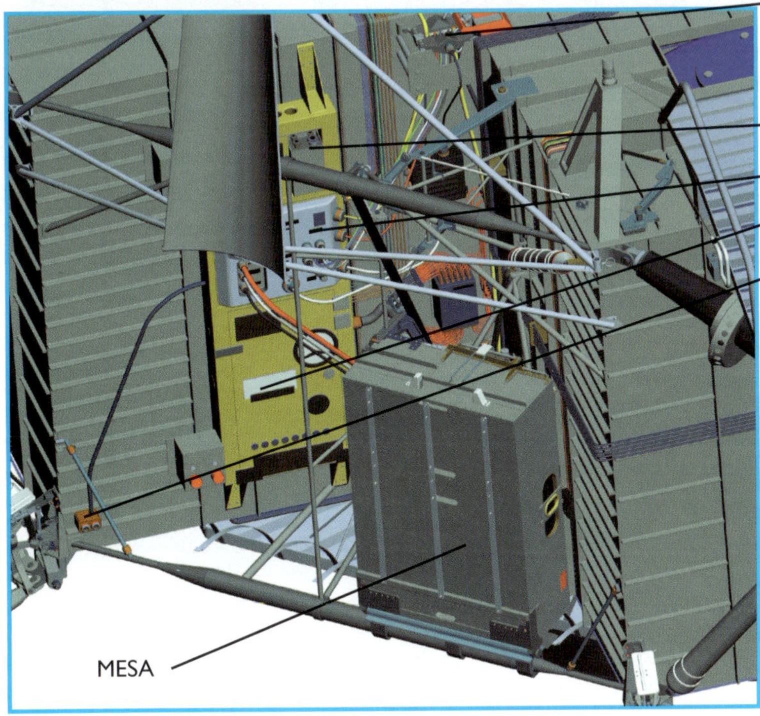

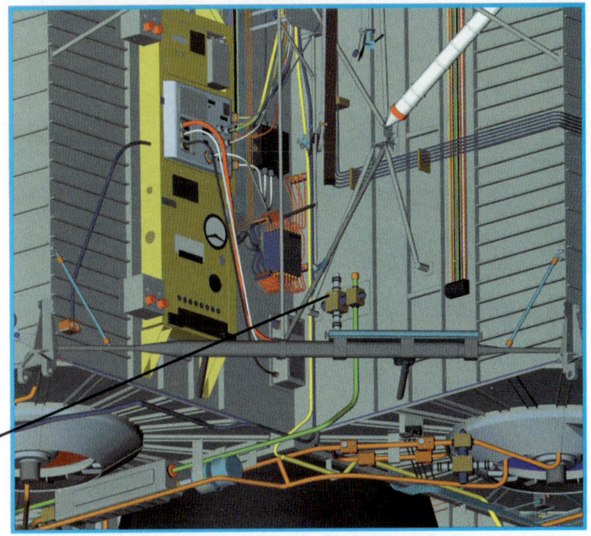

QUAD 4 SHOWN WITHOUT MESA

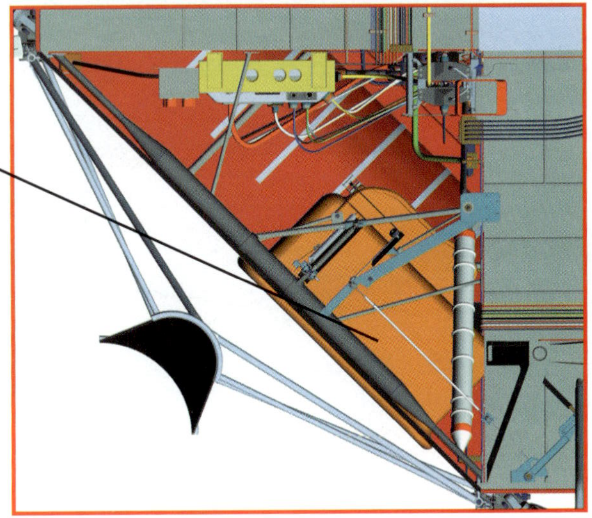

TOP VIEW OF MESA WITH THERMAL COVERING

DESCENT STAGE UPPER THERMAL COVERING NOT SHOWN

MESA DEPLOYMENT SYSTEM

The H-Series and J-Series MESA deployment mechanisms were essentially the same; the main difference was that the J-Series (extended stay) MESA was much larger.

The MESA deployment was initiated by an astronaut pulling the actuating handle marked A.

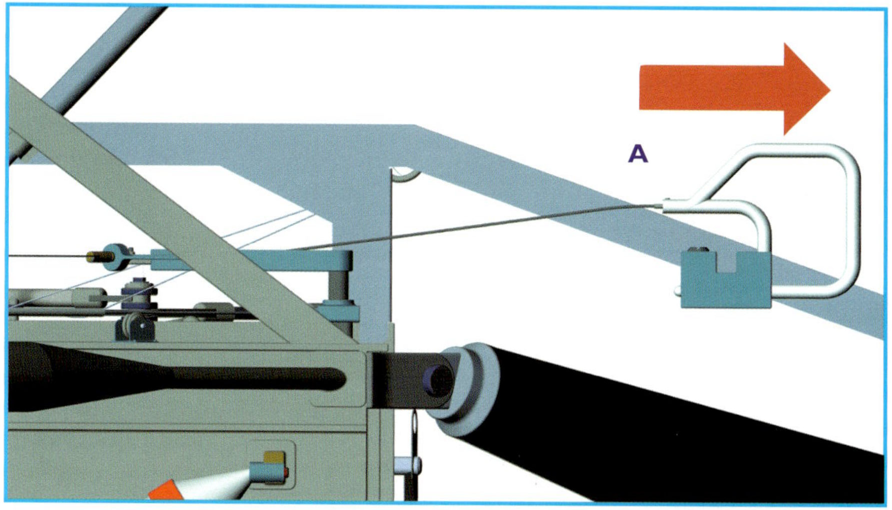

MESA DEPLOYMENT ACTUATING HANDLE

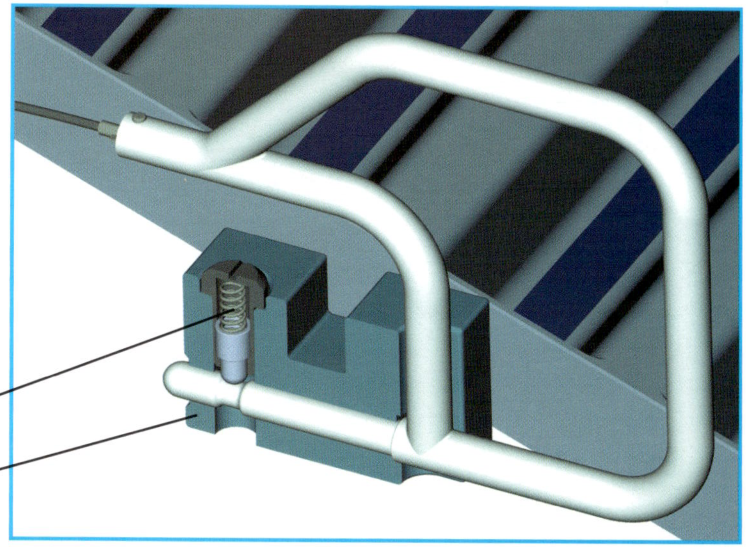

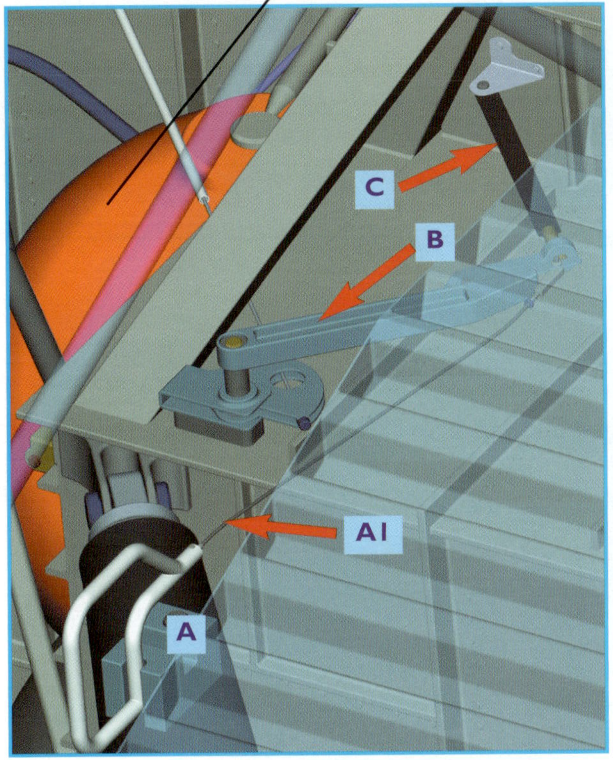

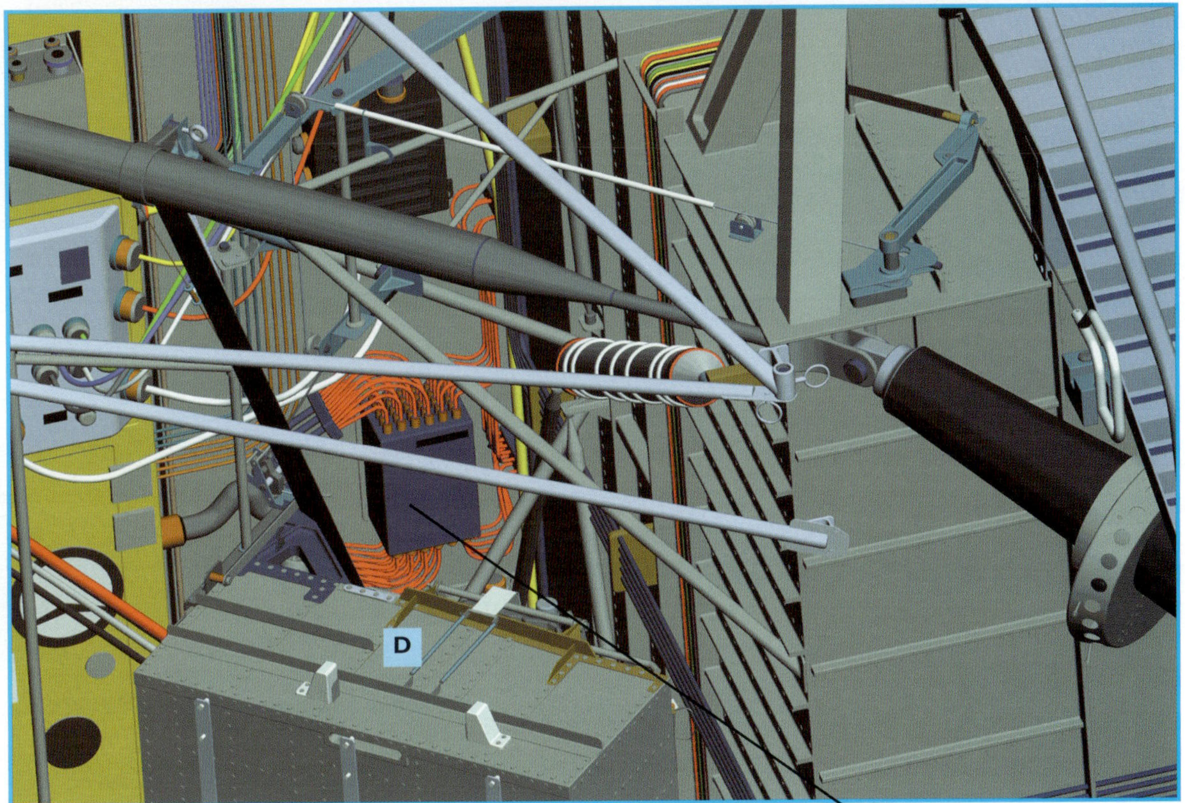

EGRESS PLATFORM SHOWN TRANSPARENT

As the handle was pulled, cable A1 caused cable arm B to rotate clockwise. As the cable arm rotated through its full range of motion it disengaged the MESA locking mechanism. The electrometric strap C acted as a spring to pull handle A back into its mounting block.

Cable G (see next page) made its way through pulley assembly H and down to locking mechanism assembly J. Locking mechanism assembly J mated with locking post F on the MESA. The MESA, marked D, pivoted about pivot point E next page).

As the cable pulled cam K, cam K disengaged lock pin M from pivot arm N (see page 53). Key L prevented lock pin M from interfering with its own disengagement from locking post F on the MESA as it slid into a slot on locking post F. Pivot arm N rotated about axis P and pulled the MESA over the center of its pivot axis using the pulling action of extension spring Q (see page 54). At this point gravity (what little there is on the Moon) took over and, lowering strap S mounted in yoke R, gently lowered the MESA to its resting position.

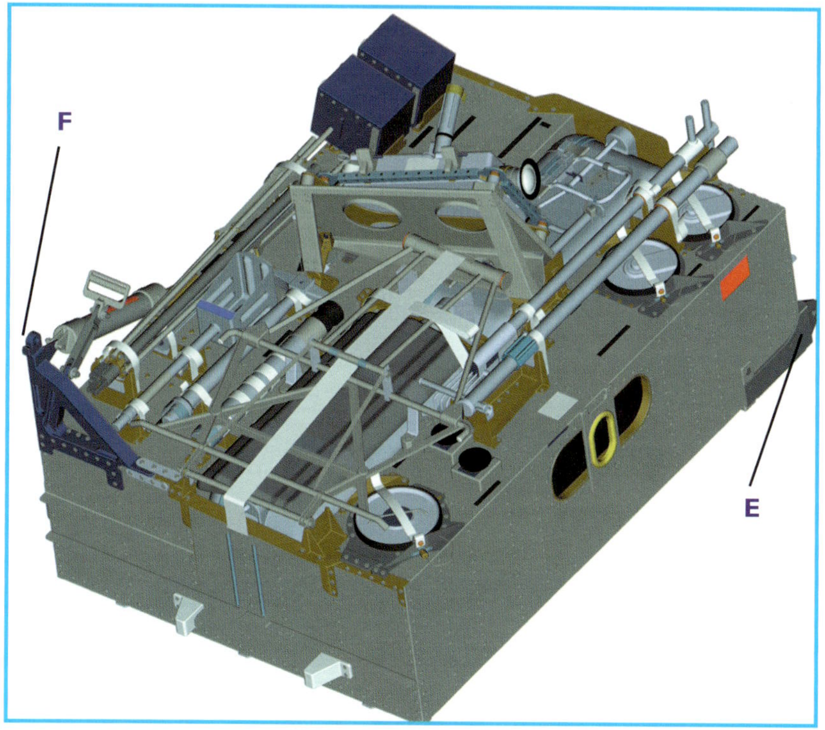

MODULAR EQUIPMENT STOWAGE ASSEMBLY

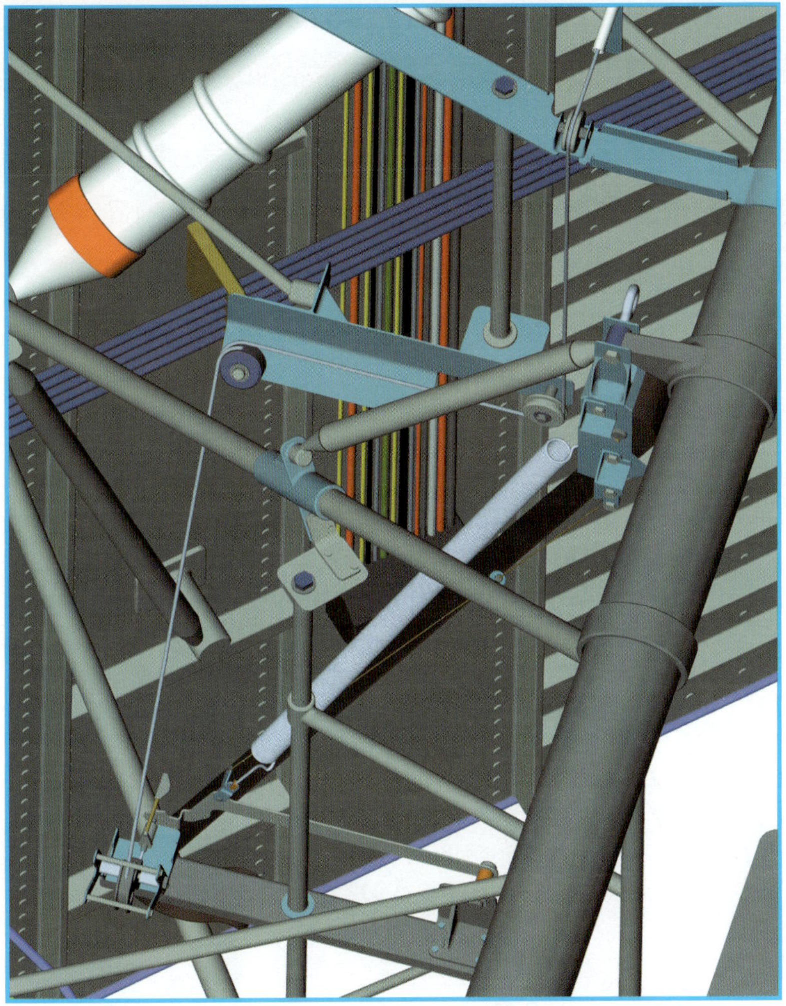

CABLE PATH FROM CABLE ARM ASSEMBLY
(MESA NOT SHOWN)

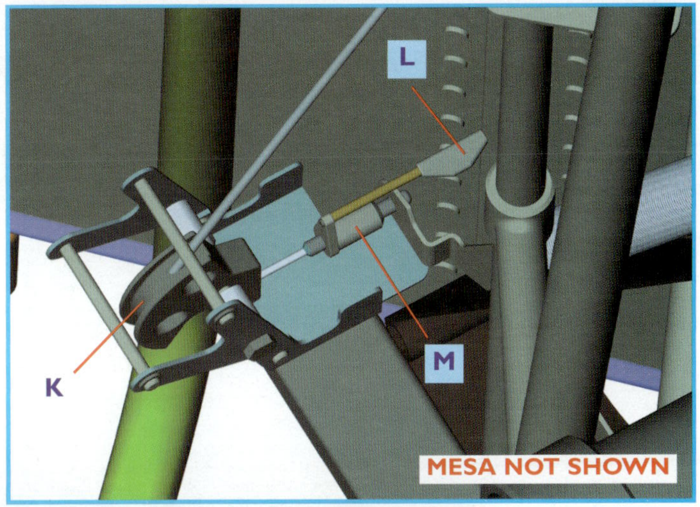

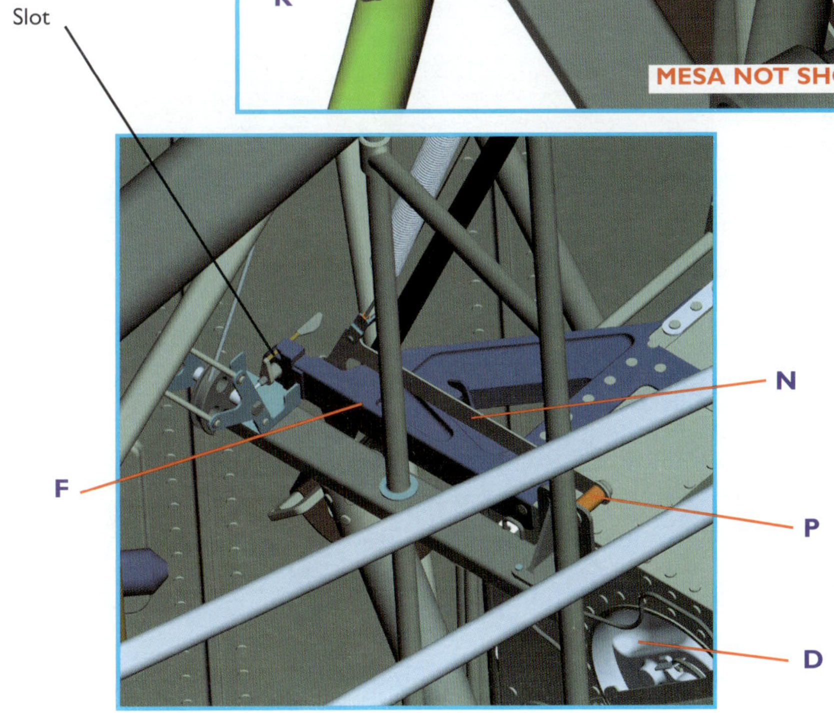

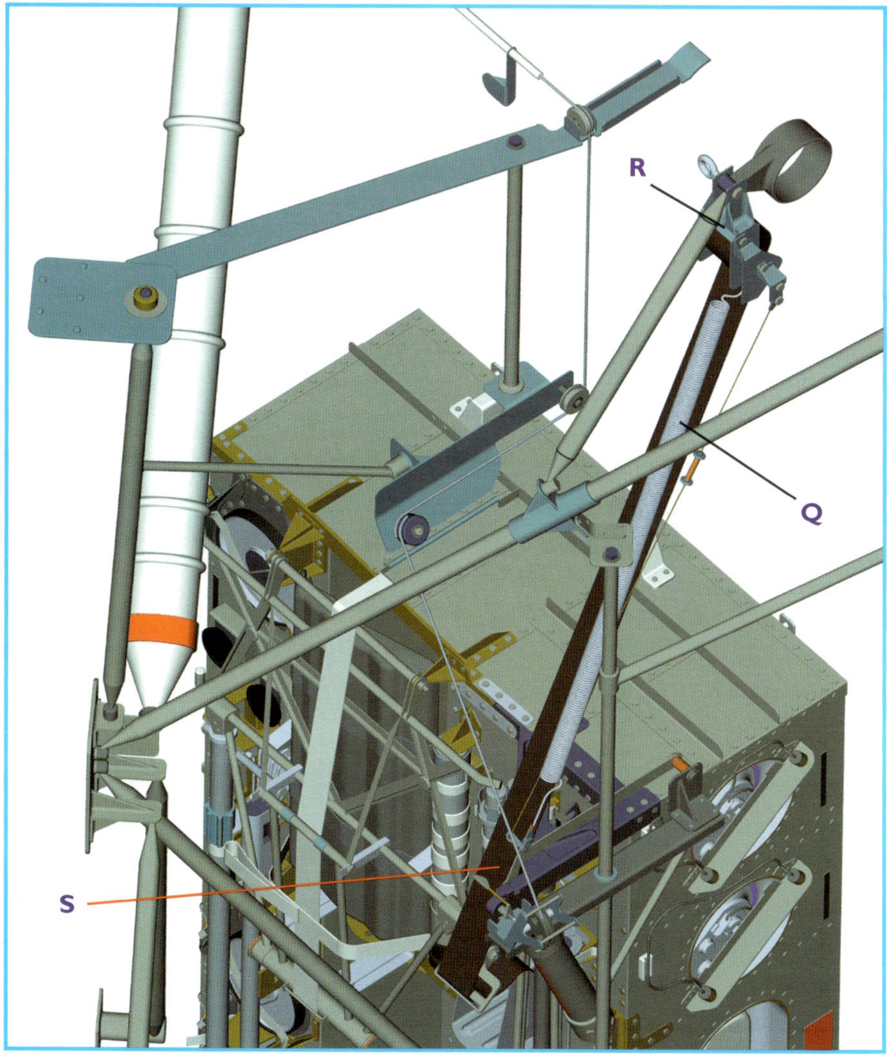

MESA SUPPORT STRUT (LM NOT SHOWN)

H-SERIES MODULAR EQUIPMENT STOWAGE ASSEMBLY (MESA)

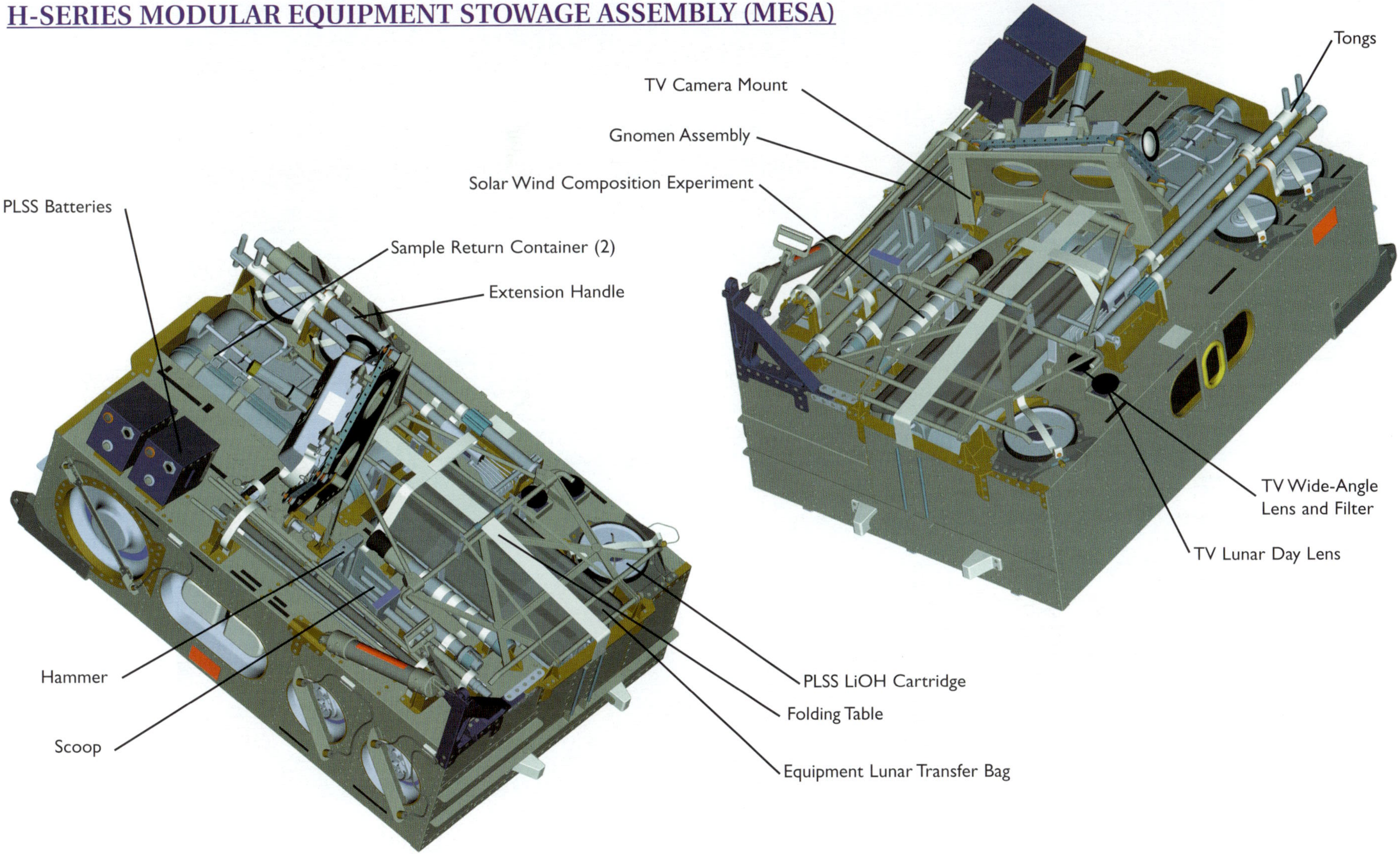

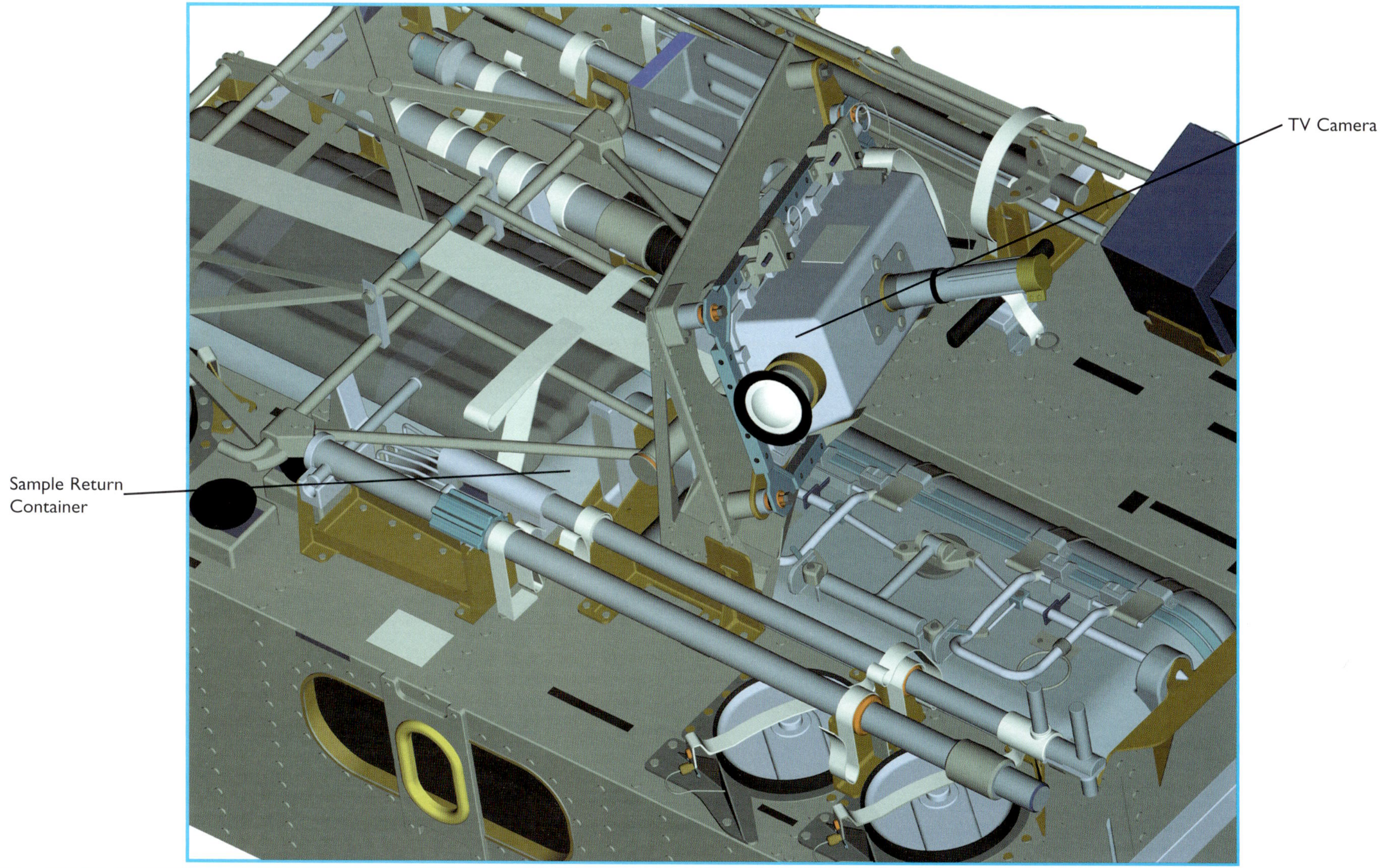
TV Camera
Sample Return Container

THERMAL COVERS OVER S-BAND ERECTABLE ANTENNA AND PLSS BATTERIES ARE REMOVED FOR ACCESS TO EQUIPMENT

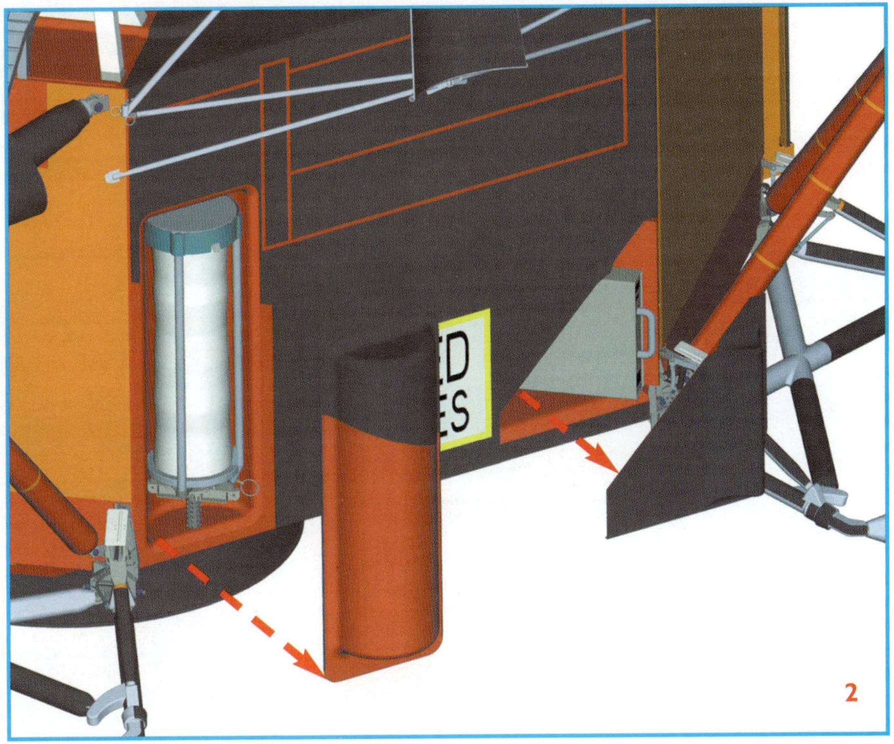

QUAD 2 APOLLO LUNAR SURFACE EXPERIMENTS PACKAGE (ALSEP) DEPLOYMENT

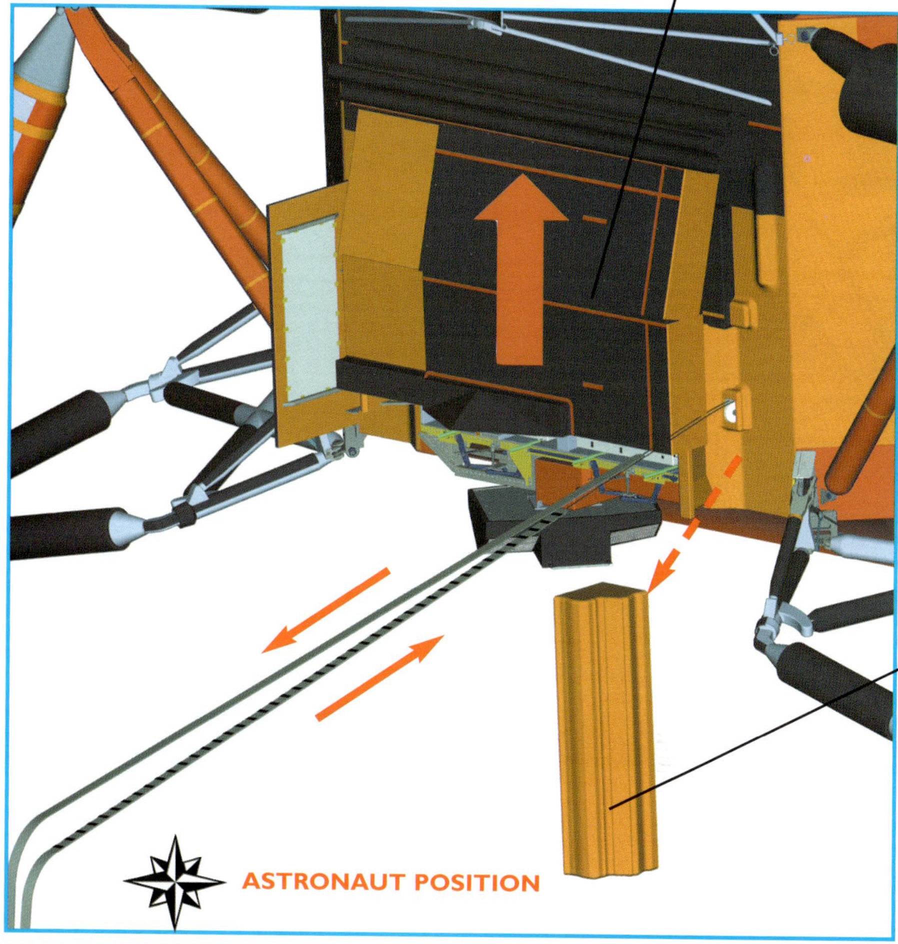

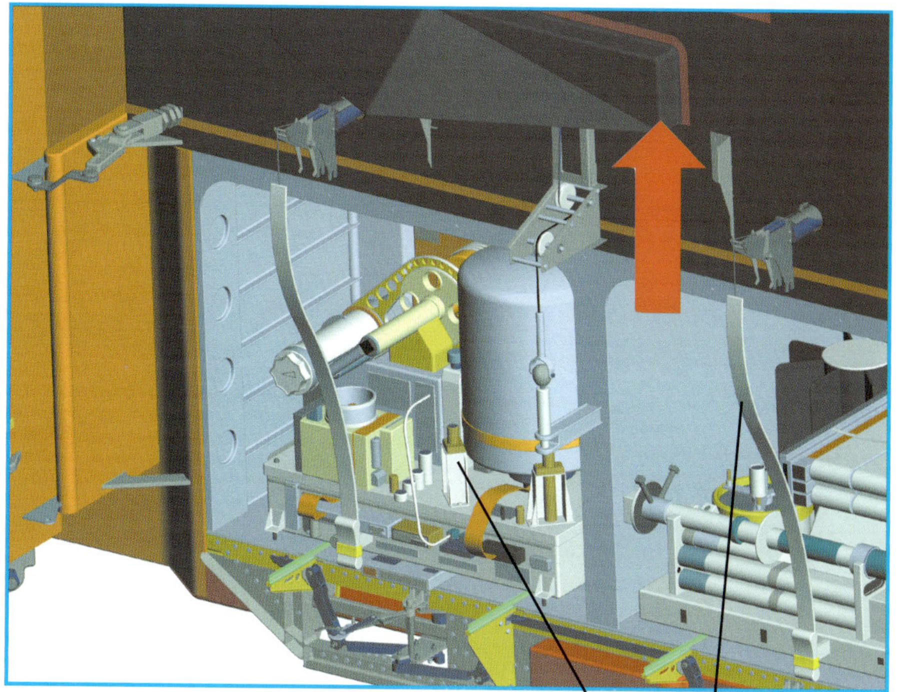

After removing the lanyard cover and disengaging the lower bay door locking claws, an astronaut would pull the ALSEP lanyards towards him first, causing the side door to open and then the bay door assembly to raise up and over the driver rollers on the top of the bay door actuating mechanism within Quad number 2.

BAY DOOR COMPLETELY RAISED

After the bay door was raised an astronaut would use the pallet lanyards to extract the Lunar Surface Scientific Experiment Packages from their stowage bay and lower them to the lunar surface.

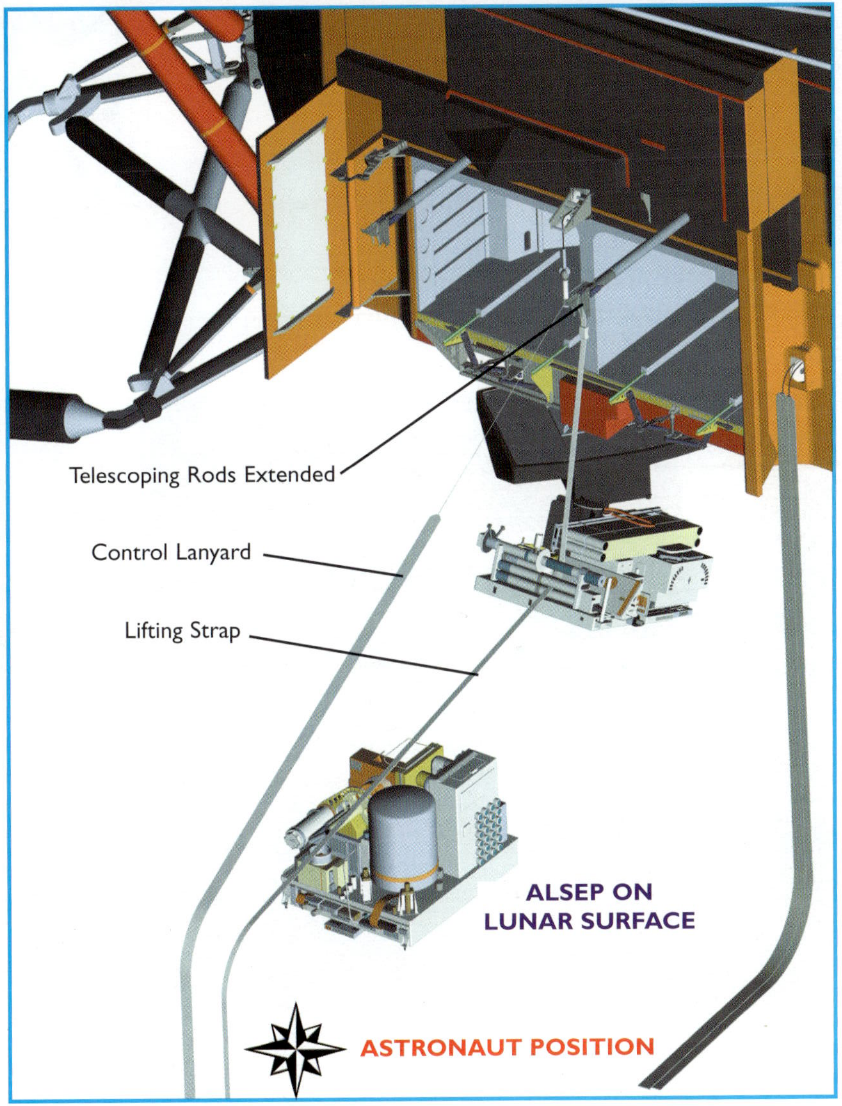

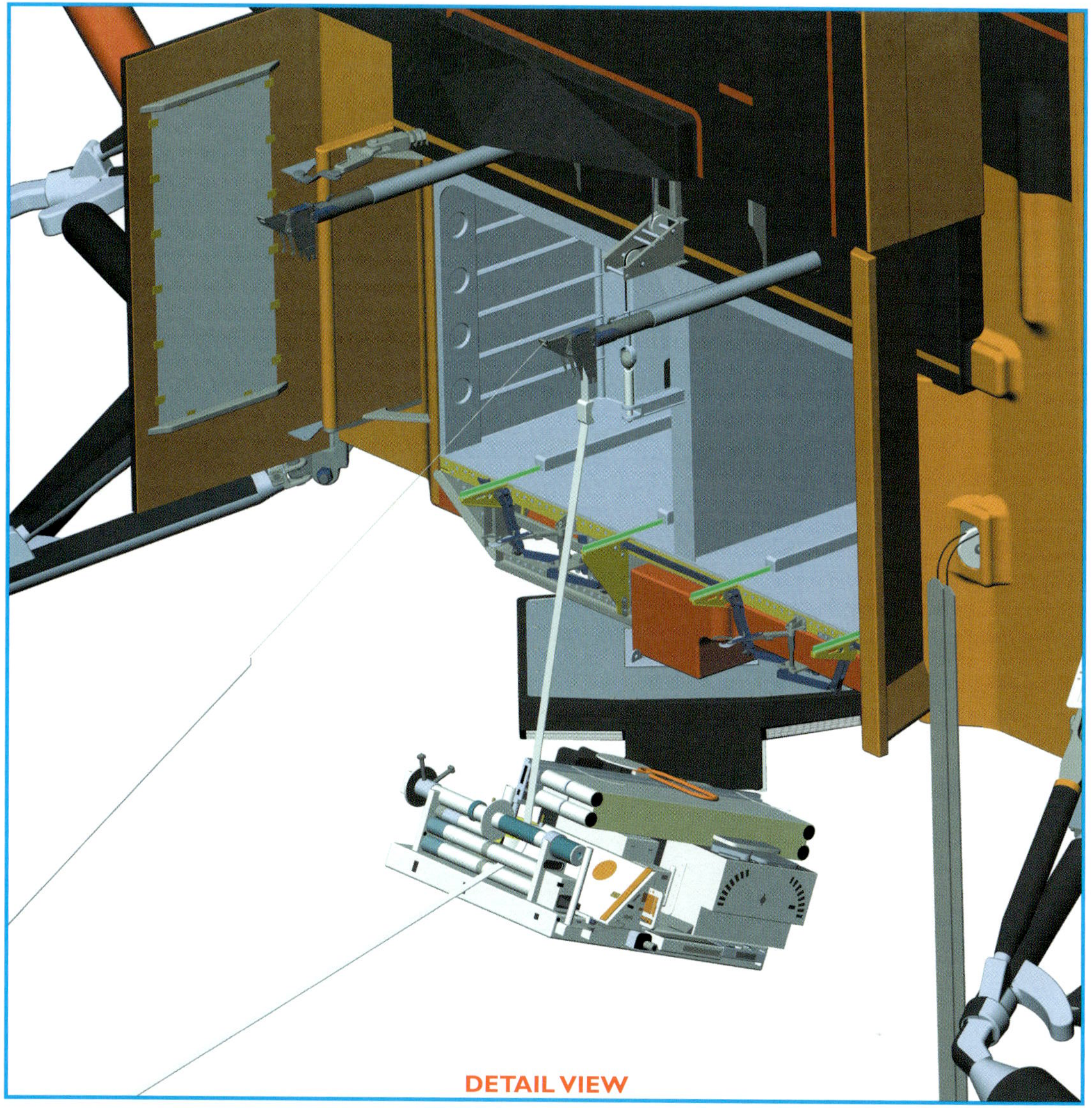

ASTRONAUT ASSEMBLING THE MAST CARRY BAR TO ALLOW HIM TO LIFT AND TRAVERSE WITH BOTH OF THE LUNAR SURFACE SCIENTIFIC EXPERIMENT PACKAGES

THE APOLLO 17 *ALSEP CONSISTED OF 5 EXPERIMENTS: A HEAT FLOW EXPERIMENT, AN EJECTA AND METEORITES EXPERIMENT, A SEISMIC PROFILING EXPERIMENT, AN ATMOSPHERIC COMPOSITION EXPERIMENT, AND A SURFACE GRAVIMETER.*

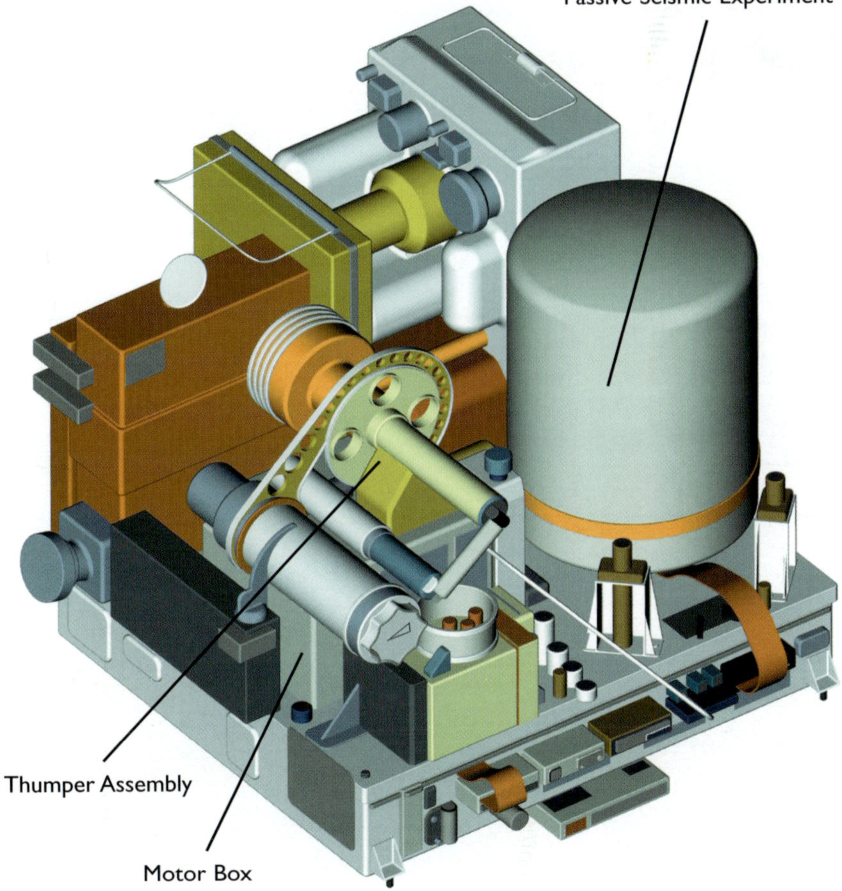

ALSEP SUBPACKAGE No. 1

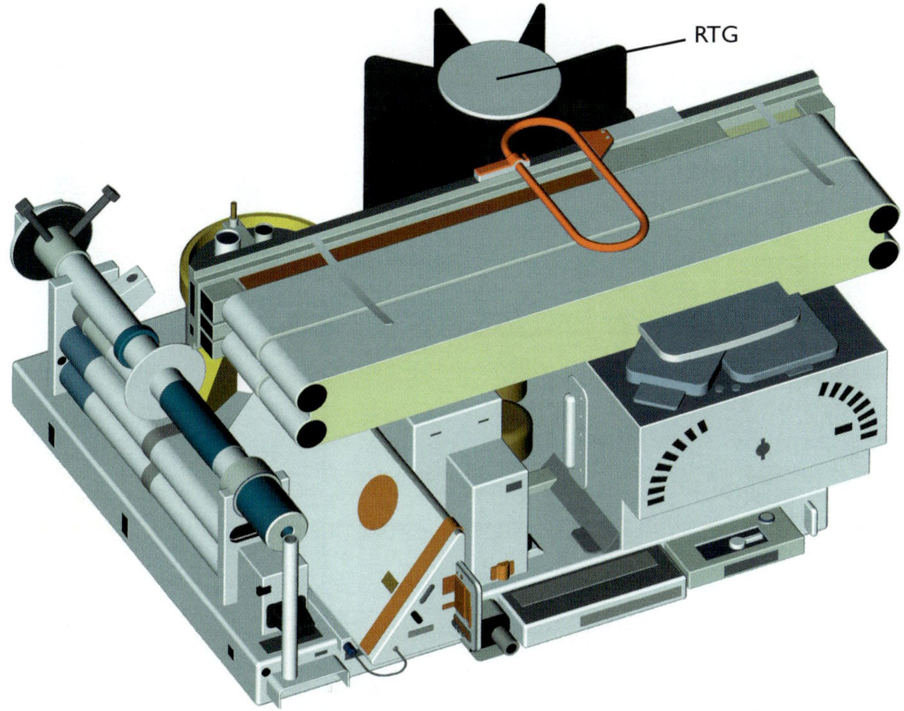

ALSEP SUBPACKAGE No. 2

QUAD 4 MESA DEPLOYMENT

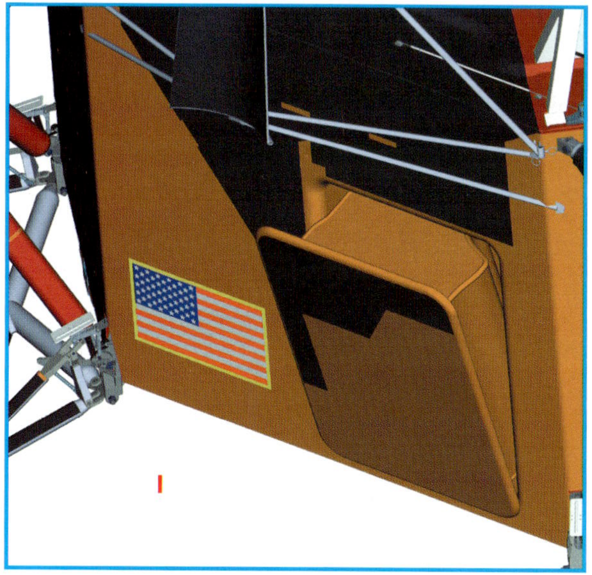
1

2

THERMAL COVERINGS REMOVED BY ASTRONAUTS

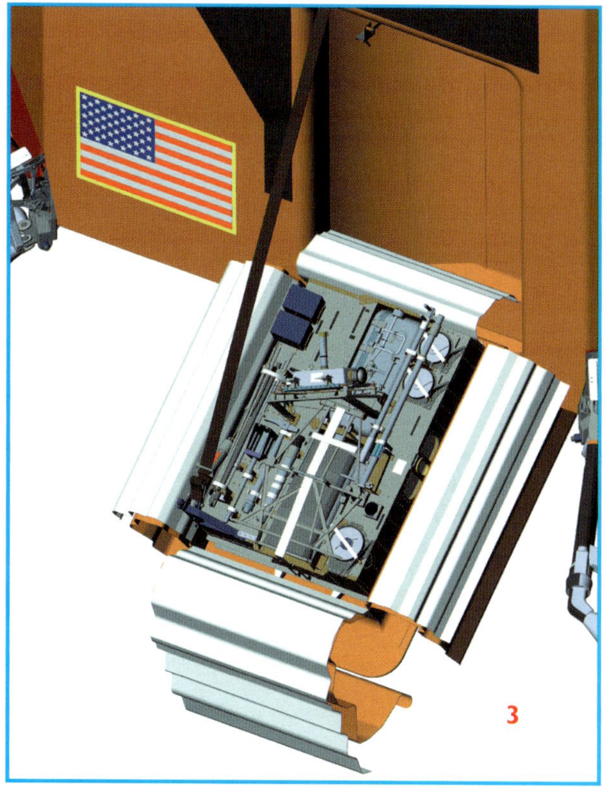
3

MAIN PROPULSION SYSTEM (MPS)

The main propulsion system (MPS) consisted of two separate, complete, and independent propulsion sections: the descent propulsion section and the ascent propulsion section. Each propulsion section performed a series of specific tasks during the lunar landing mission. The descent propulsion section provided propulsion for the LM from the time it separated from the Command and Service Modules until it landed on the lunar surface. The ascent propulsion section lifted the ascent stage off the lunar surface and boosted it into orbit. Both propulsion sections operated in conjunction with the reaction control system (RCS), which provided propulsion used mainly for precise attitude and translation maneuvers.

The fuel was a blend of hydrazine (N_2H_4) and unsymmetrical dimethyl-hydrazine (UDMH), commercially known as Aerozine 50. The proportions, by weight, were approximately 50% hydrazine and 50% dimethyl-hydrazine. The oxidizer was nitrogen textroxide (N_2O_4).

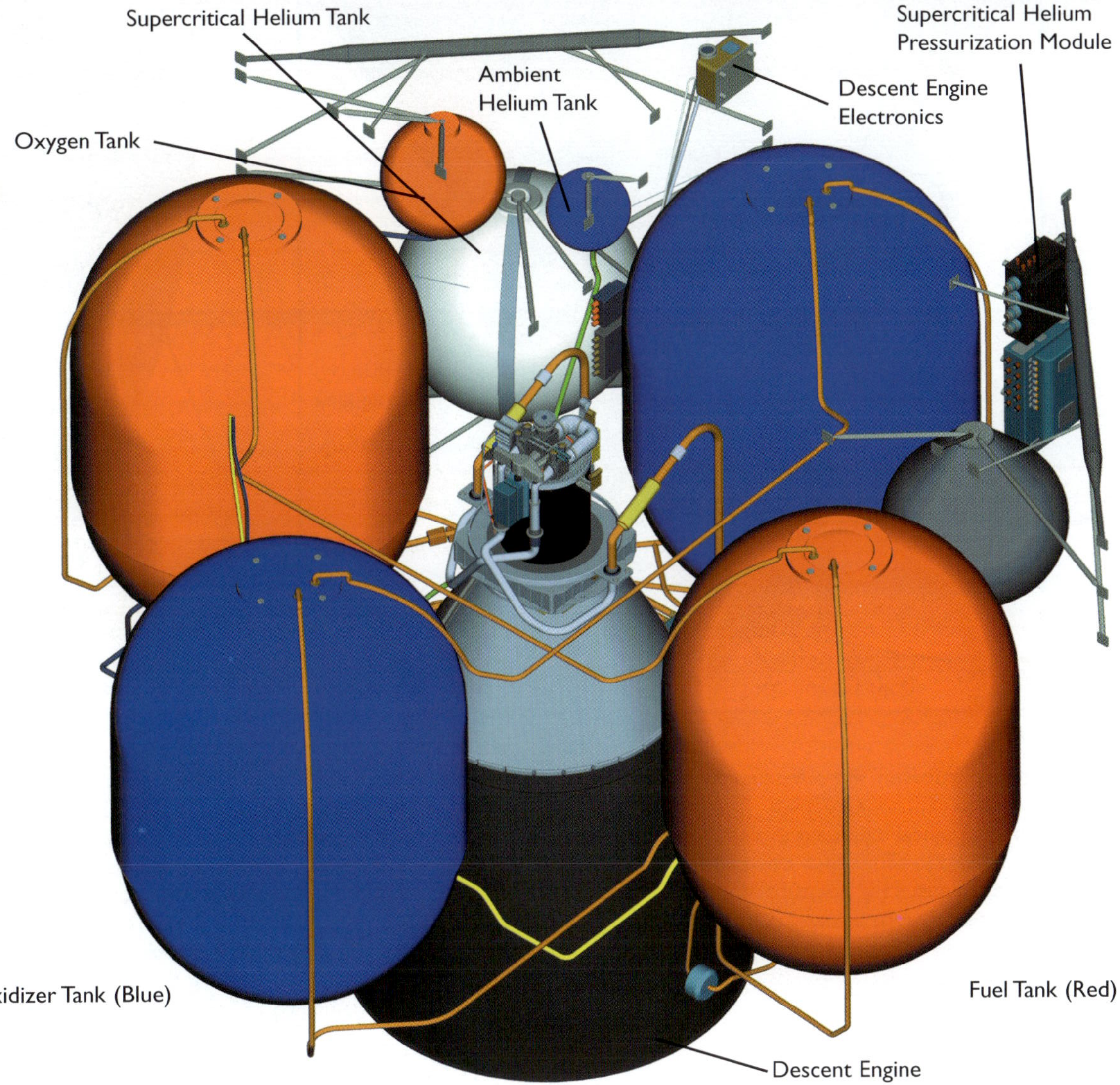

LUNAR MODULE DESCENT ENGINE

The Lunar Module's descent engine was probably the most technologically advanced piece of equipment in the entire Apollo / Saturn package. All other systems had been designed with proven technology. For a successful lunar landing, the descent engine had to be designed to easily restart, and be throttleable as well. The astronauts needed to have the ability to fully control their descent, to be able to hover over a landing site, and to change landing sites if need be. Hypergolic propellants were used for the descent propulsion system. Hypergolic means that the fuel and oxidizer would ignite and burn upon contact with one another, eliminating the need for an ignition system. To create the throttleable mechanism a helium injection system was incorporated. Inert gas was introduced into the propellant lines; this had the advantage of keeping the propellant flow constant. A constant propellant flow simplified the design of the injector, and the injected helium was available from the sub-system that pressurized the propellant tanks. The engine had a wide range of throttle, from 10% to 100%. The propellants had to move from the tanks to the engine if the engine was to fire. In the low-gravity space environment, the contents of the tanks would float around. To force the propellants to settle at the bottom of the tanks, where the outlet was located, the computer performed what was called a "ullage burn." Under the control of the computer, the reaction control system (RCS) performed a 7-second burn in the X (upward direction). This settled the propellants to the bottom of the tanks and the pressurized helium in the tanks then pushed the propellants into the fuel lines and to the engine.

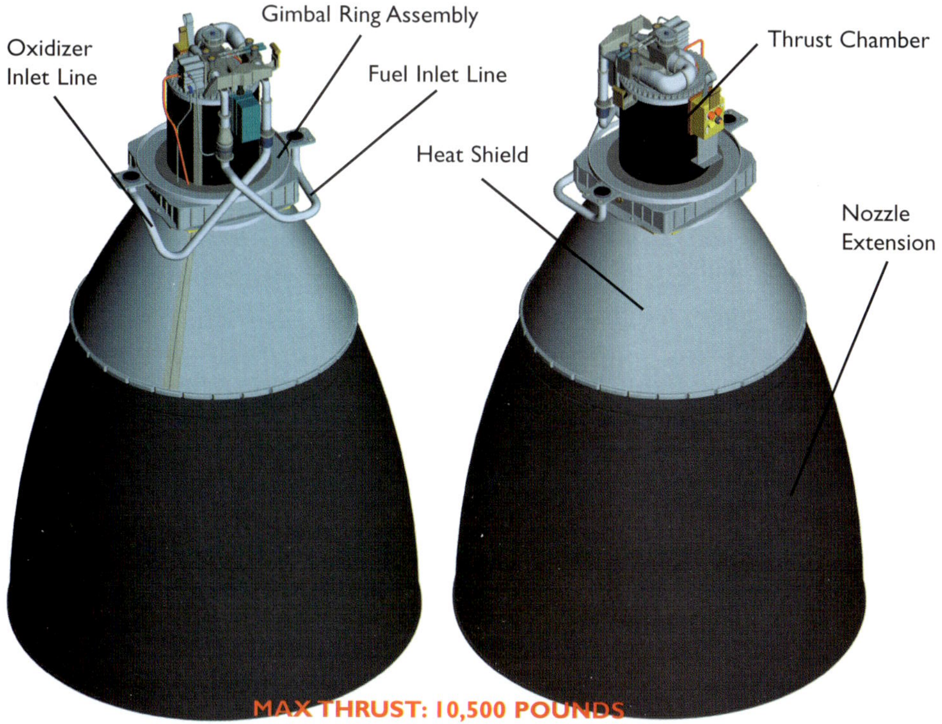

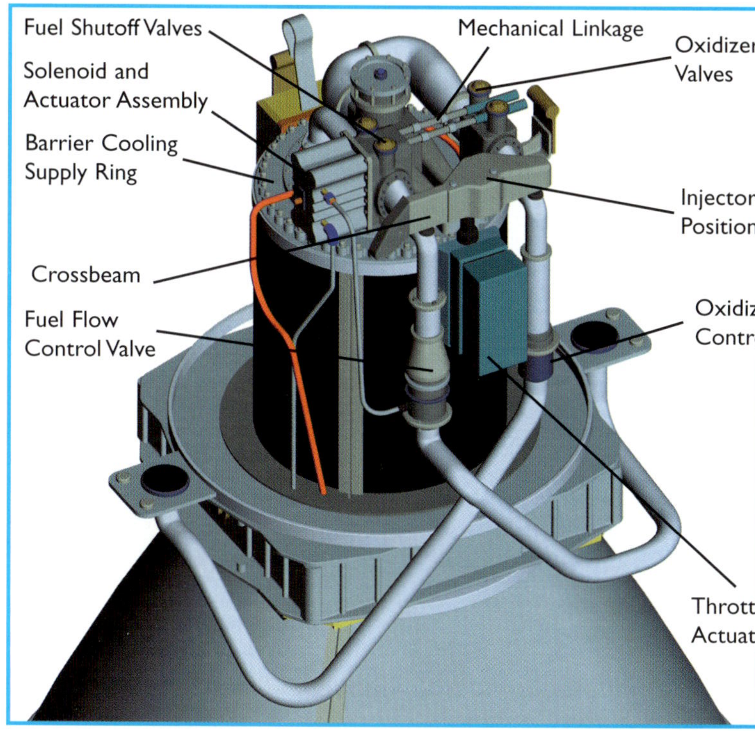

WHEN THE LUNAR MODULE RESTED ON THE LUNAR SURFACE, THE DISTANCE FROM THE BOTTOM OF THE NOZZLE EXTENSION TO THE LUNAR SURFACE WAS APPROXIMATELY 10 INCHES

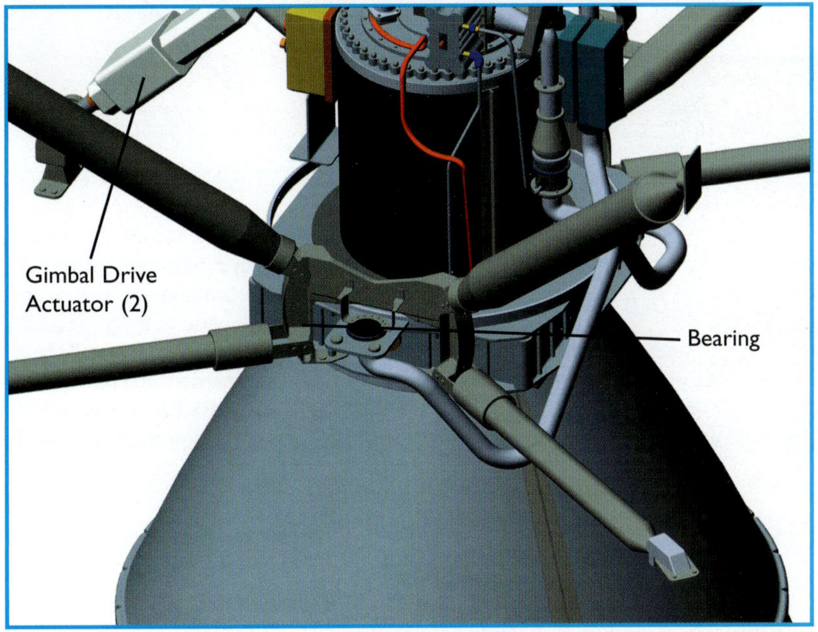

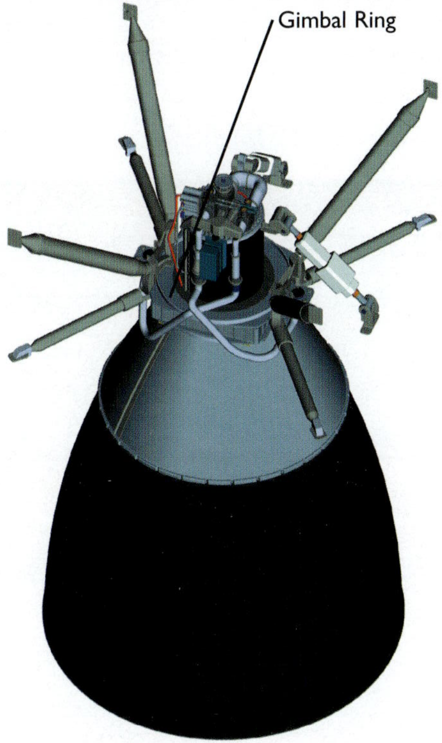

After initial pressurization of the descent engine section, the descent engine start required two separate and distinct operations: aiming and firing. The astronauts could perform engine arming or the LM guidance computer could automatically initiate it. When the astronauts set the switch to arm the descent engine, power was simultaneously routed to: open the actuator isolation valves in the descent engine; enable the instrumentation circuits in the descent propulsion section; and issue a command to the throttling controls to start the descent at the required 10% thrust level. The LM guidance computer and the abort guidance section received an engine-armed status signal. This signal enabled an automatic engine-on program in the Guidance, Navigation and Control Subsystem (GN&CS), resulting in a descent engine start. A manual start was accomplished when the commander pushed his engine start button (either astronaut can stop the engine; engine-stop buttons were included at both flight stations.) The descent propulsion section consisted of an ambient and supercritical helium tank with associated helium pressurization components; two fuel and two oxidizer tanks with associated feed components; and a pressure-fed ablative, throttleable rocket engine. The engine had a nominal thrust of 9,900 pounds and could also be operated in a range of 1,280 to 6,400 pounds-thrust. The descent engine was mounted in the center of the descent stage cruciform. Fuel and oxidizer entering the engine assembly where routed through flow control valves to the propellant shutoff valves. A total of eight propellant shutoff valves were used. They were arranged in series-parallel redundancy – four in the fuel line and four in the oxidizer line. To prevent rough starts, the engine was designed to allow the oxidizer to reach the injector first. The propellants were then injected into the combustion chamber, where hypergolic ignition occurred.

The combustion chamber of the descent engine consisted of an ablative-cooled chamber section, nozzle throat, and nozzle divergent section. The ablative sections were enclosed in a continuous titanium shell and jacketed in a thermal blanket composed of aluminized nickel foil and glass wool. A seal prevented leakage between the combustion chamber and nozzle extension. The nozzle extension was a radiation-cooled, crushable skirt; it could collapse a distance of 28 inches on lunar impact so as not to affect the stability of the Lunar Module. The extension was made of Columbian coated with aluminide. It was thatched to the combustion chamber case at a nozzle ratio of 16 to 1 and extended to an exit ratio of 54 to 1.

The gimbal ring was located at the plane of the combustion chamber throat. It consisted of a rectangular beam frame and four trunnion subassemblies. The gimbal drive actuators, under control of the descent engine control assembly, tilted the descent engine in the gimbal ring along the pitch and roll axes so that the engine thrust vector went through the LM center of gravity. One actuator controlled the pitch gimbal and the other the roll gimbal. Each gimbal drive actuator consisted of a single-phase motor, a feedback potentiometer, and associated mechanical devices. They could extend or retract two inches from the mid-position to tilt the descent engine a maximum of six degrees along the X and Y axes.

ENGINE WELL DETAIL VIEW

MAJOR EXPLOSIVE DEVICES EQUIPMENT LOCATIONS

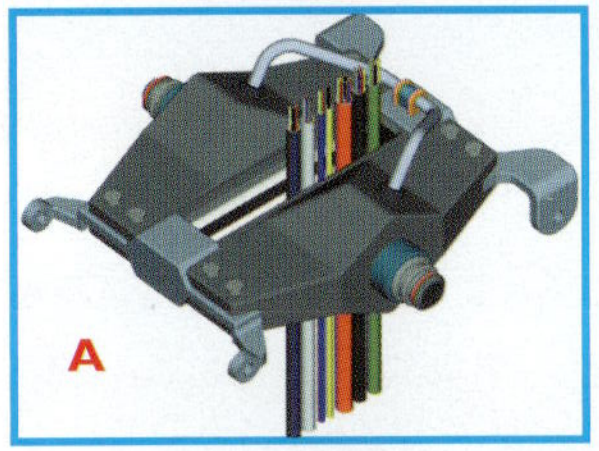

A

Interstage Umbilical Cutter (Guillotine)

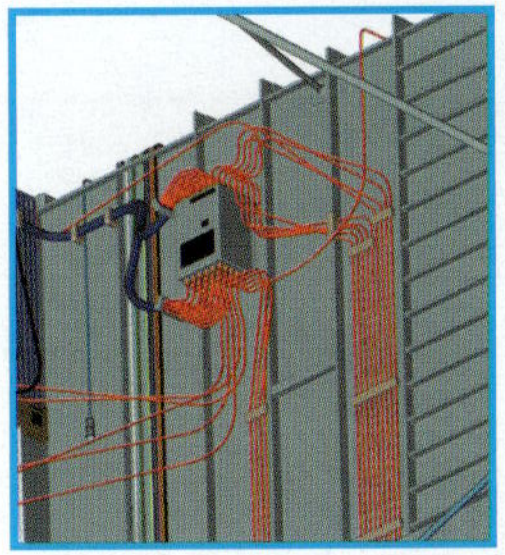

B

Explosive Devices Battery (Descent and Ascent Stage)

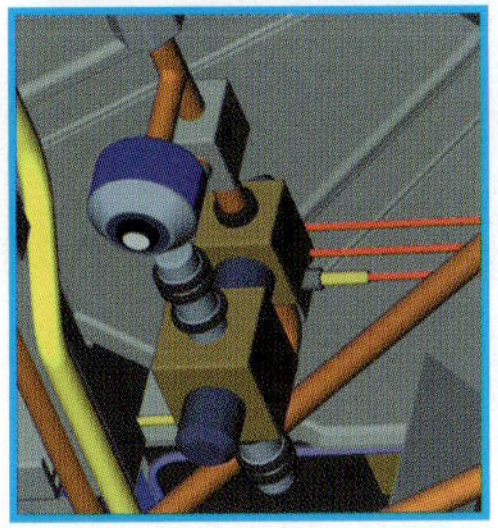

C

Ascent Propulsion Compatibility Valves

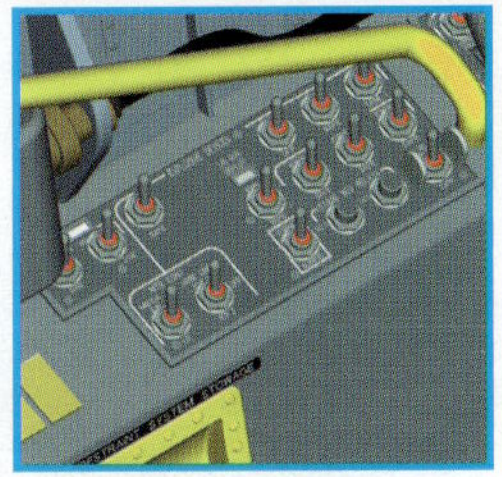

D

Explosive Devices Control Panel

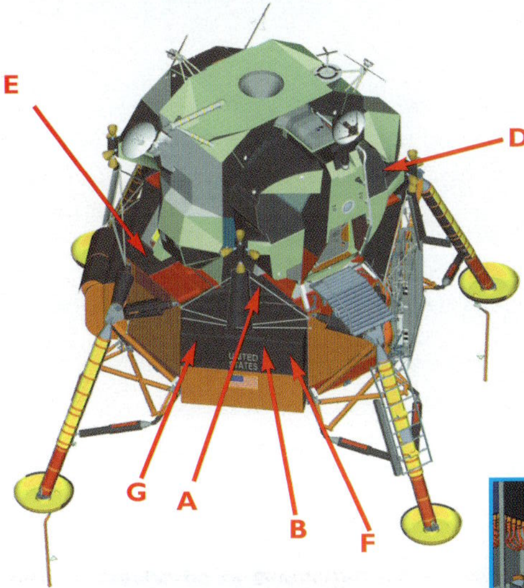

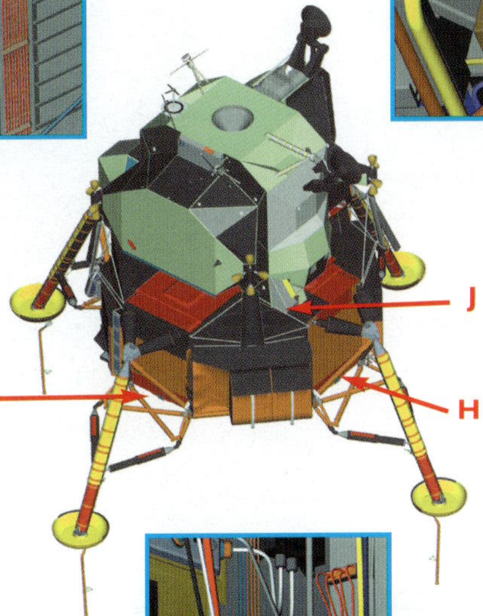

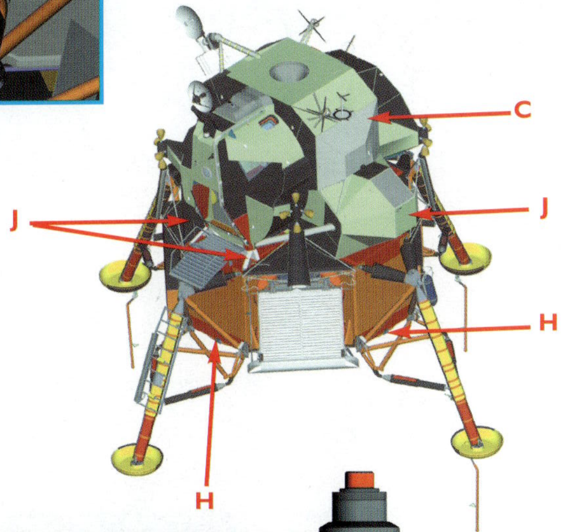

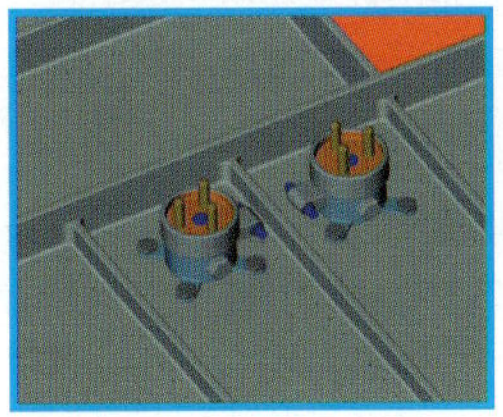

E

Circuit Interrupters

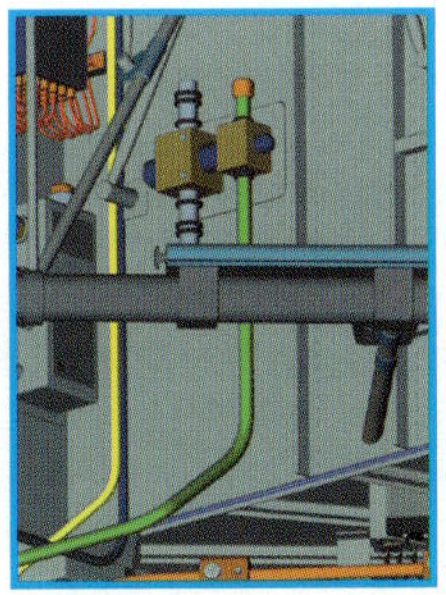

F

Descent Propulsion Helium Isolation Valve

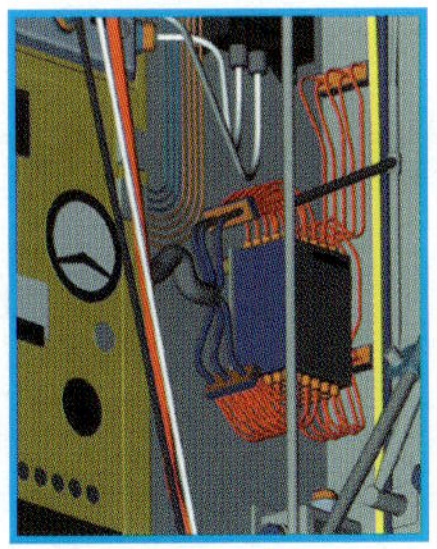

G

Explosive Devices Relay Box(Descent And Ascent Stage)

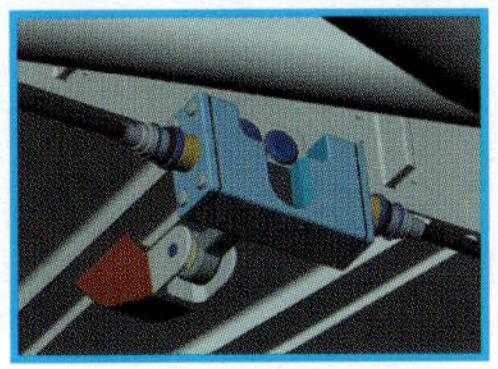

H

Landing Gear Uplock (4 Places)

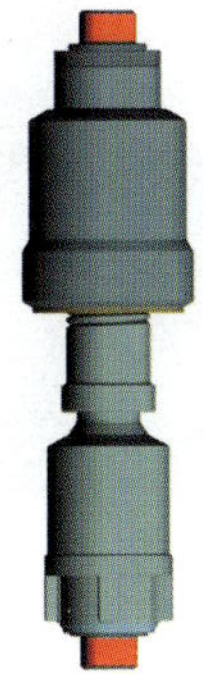

J

Interstage Structural Connection-Nut and Bolt Assembly (4 Places)

– Scott P. Sullivan –

INTERSTAGE UMBILICAL CUTTER (GUILLOTINE)

At the end of the lunar mission, the ascent stage separated from the descent stage to rendezvous with the Command and Service Modules in orbit around the Moon. To separate the two stages the connection between the four hard-point connections, circuit interrupters, and the interstage umbilical needed to be broken. This was accomplished with explosive devices. The commander would first set the control switches to initiate a controlled sequence of stage separation. All signal and electrical power was terminated between the two stages. After the circuit interrupters and explosive nuts and bolts had been separated, an explosive guillotine automatically severed the wires, cables, and water lines between the two stages.

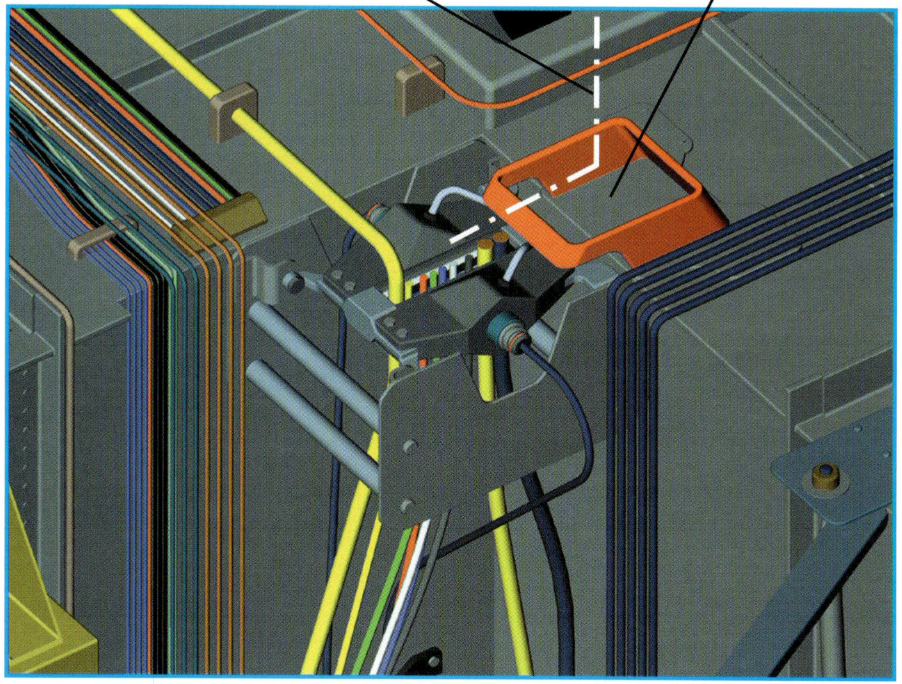

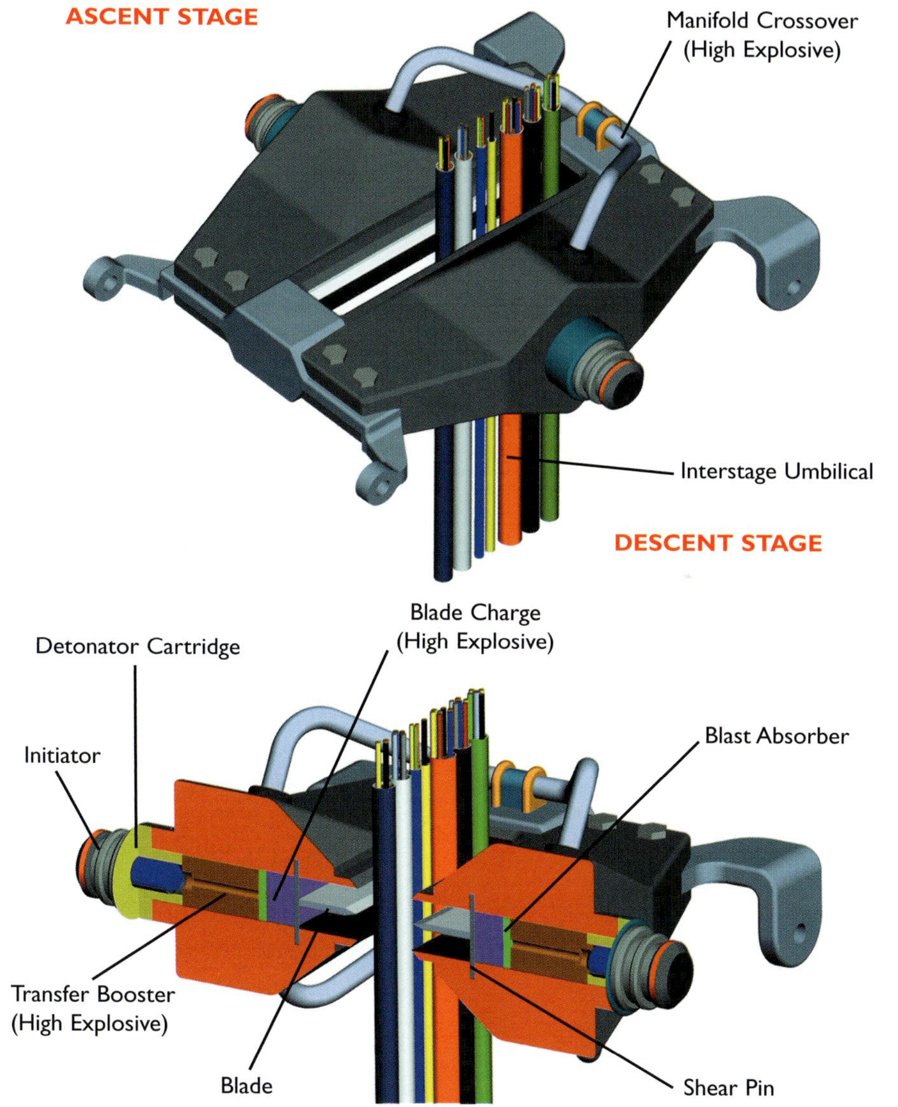

INTERSTAGE STRUCTURAL CONNECTION NUT AND BOLT ASSEMBLY

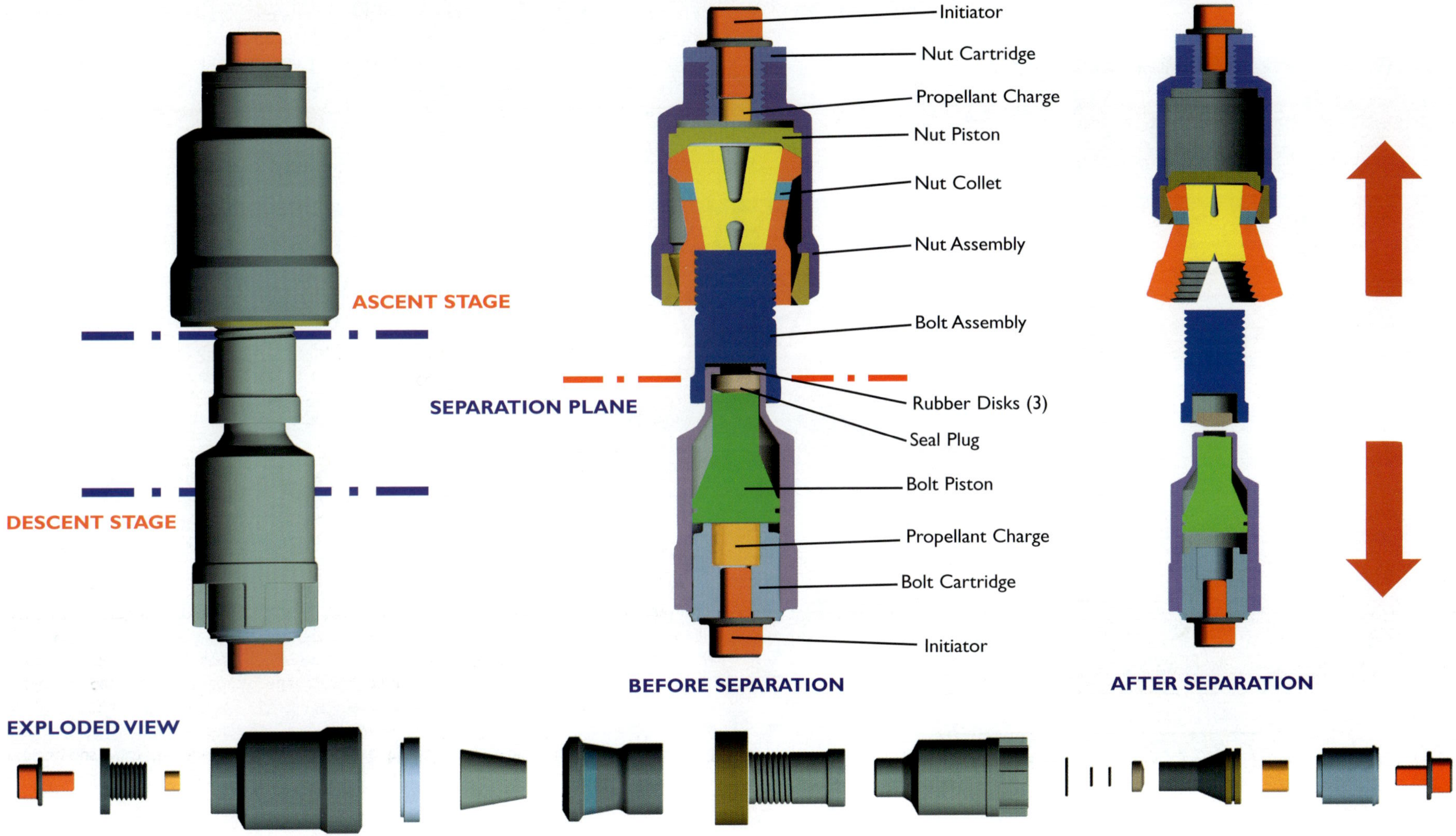

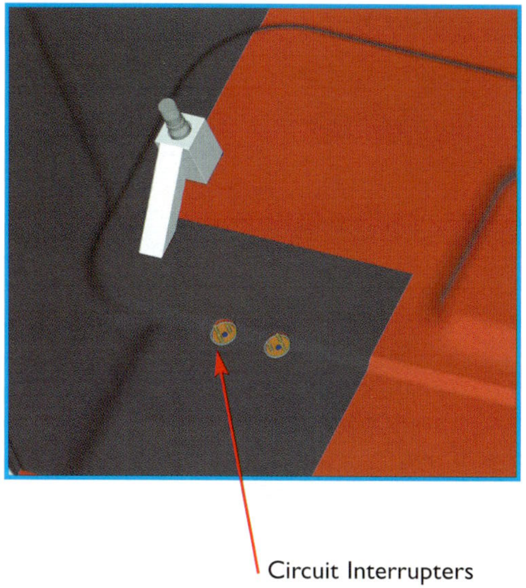

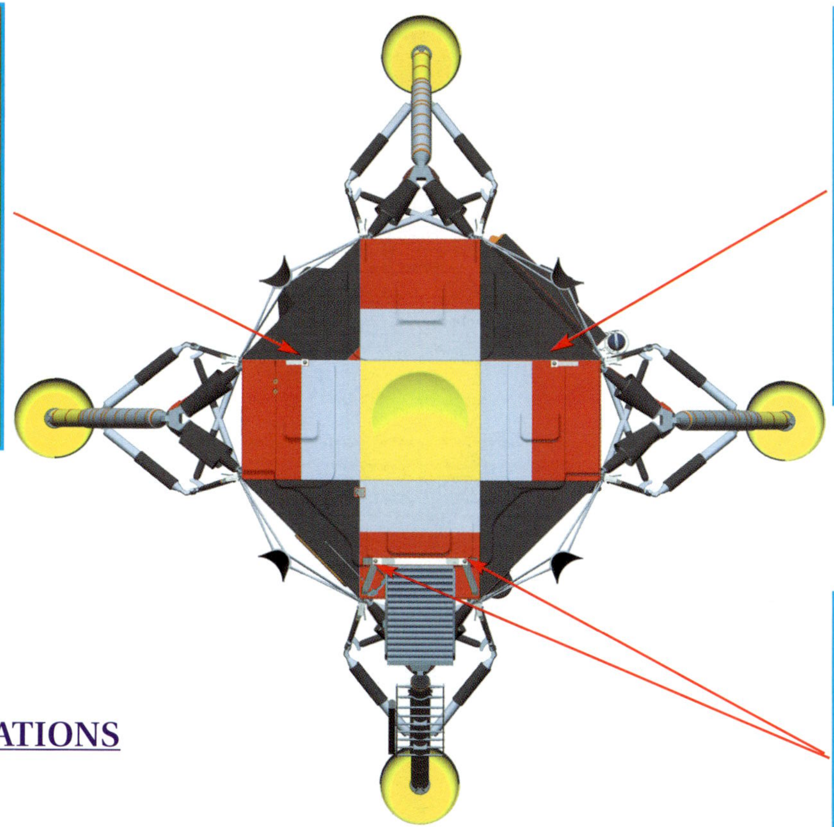

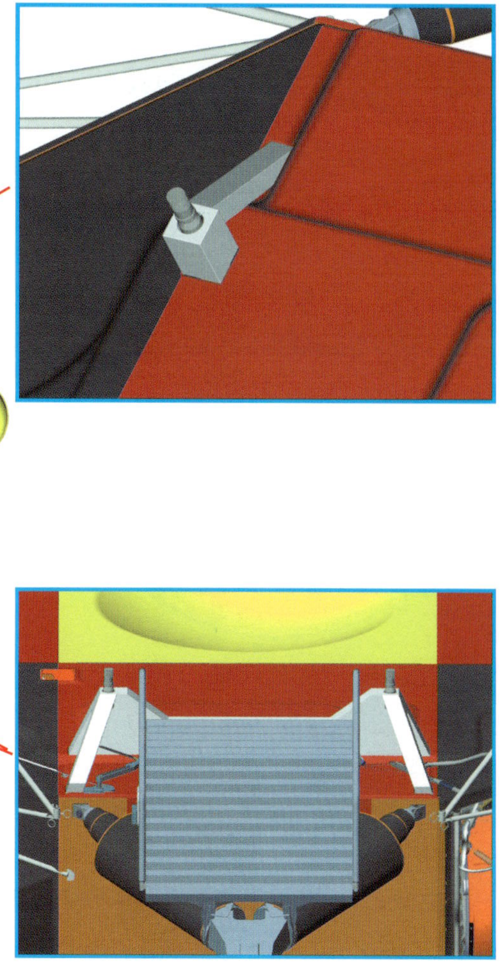

INTERSTAGE FITTINGS LOCATIONS

Two of the four interstage fittings for attachment of the ascent stage to the descent stage were mounted on the forward compartment beams, behind the egress platform. The other two fittings were on the aft beam of the side compartments. These "hard point" connections were armed with explosive devices which where used to separate the ascent stage from the descent stage at the end of the lunar mission.

LANDING GEAR UPLOCK

One uplock assembly was attached to each of the four landing gear assemblies. It consisted of a fixed link (strap) and two end detonator cartridges in a single case. The fixed link, attached to the primary strut and the descent stage structure, held the landing gear in its retracted (stowed) position. When the commander operated the landing gear deployment switch, it activated an electrical circuit which explosively severed the fixed link to permit the deployment mechanism to extend the landing gear. When detonated, either end cartridge had sufficient energy to sever the fixed link.

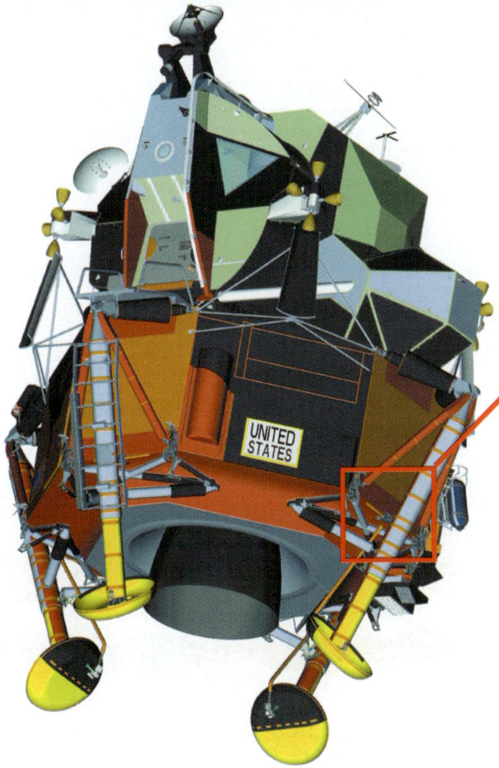

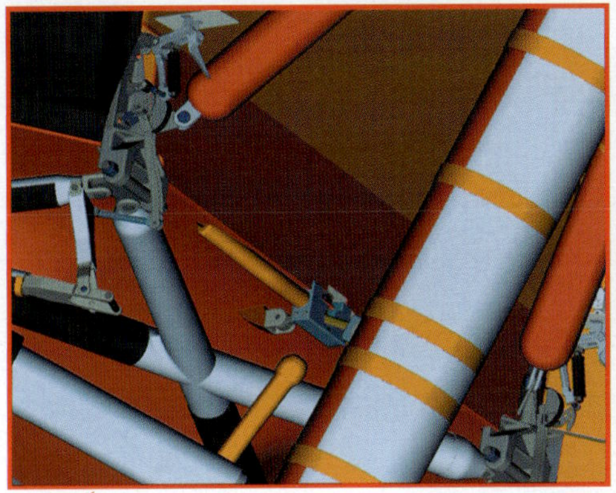

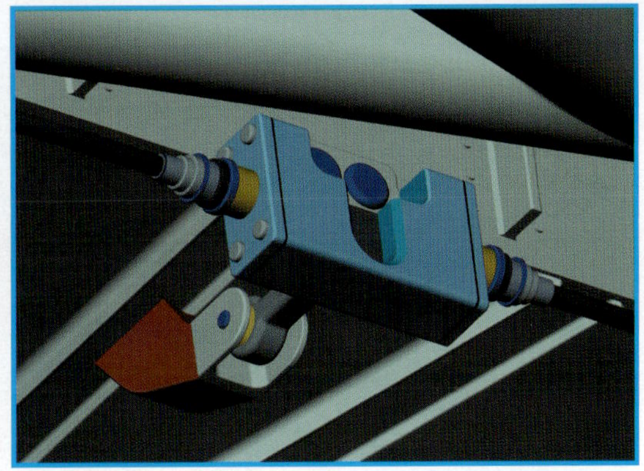

THERMAL COVERING NOT SHOWN

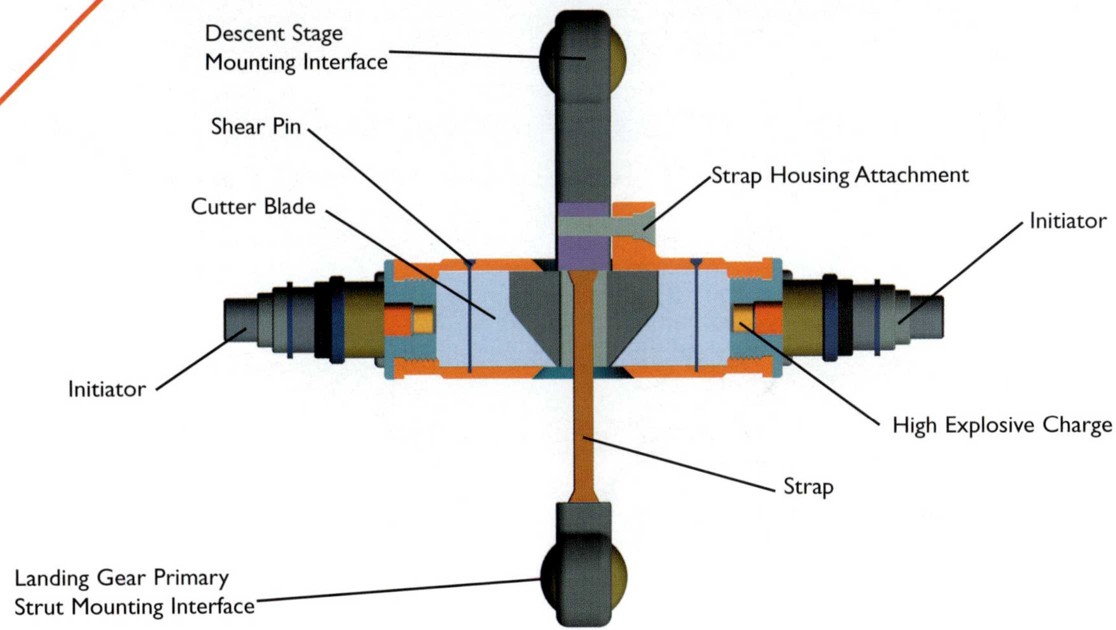

DESCENT STAGE PLUMBING

Lunar Module Chronology

November 15, 1961
John C. Houbolt sends a letter to NASA outlining Lunar Orbit Rendezvous and stating deficiencies in other rendezvous programs.

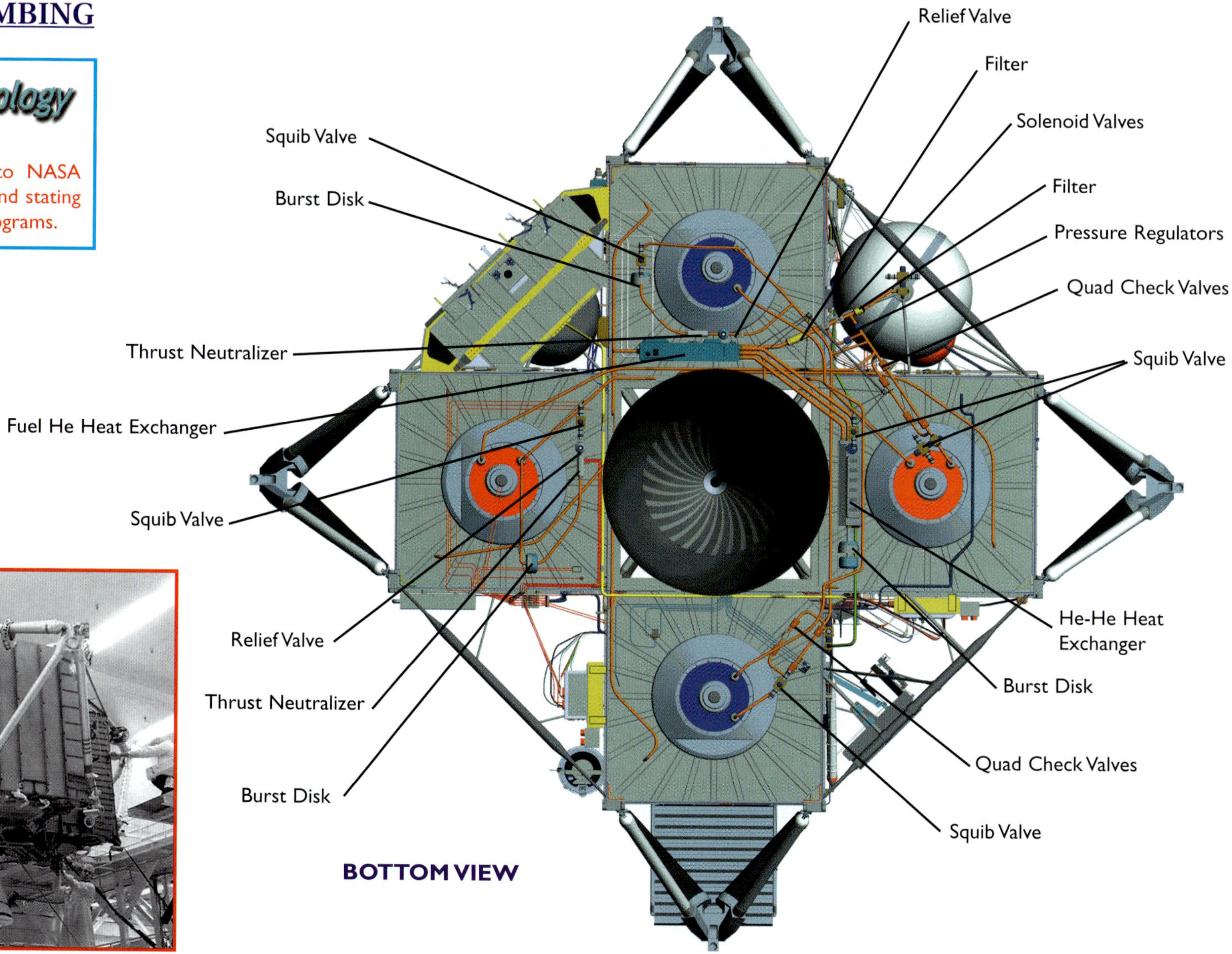

BOTTOM VIEW

DESCENT STAGE ELECTRICAL

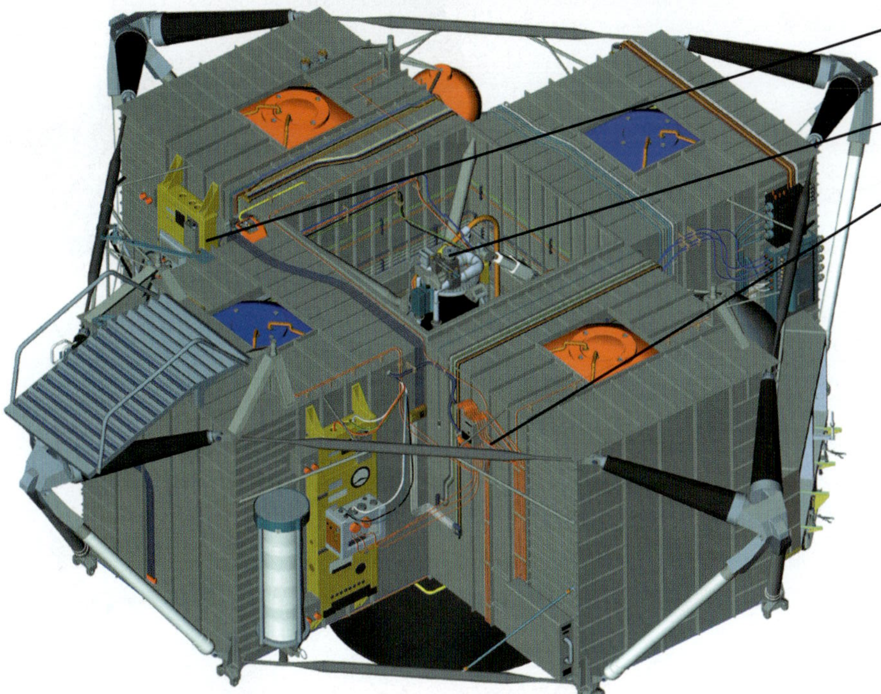

Lunar Module Chronology

January 01, 1962
Grumman conducts its own study on Lunar Orbit Rendezvous.

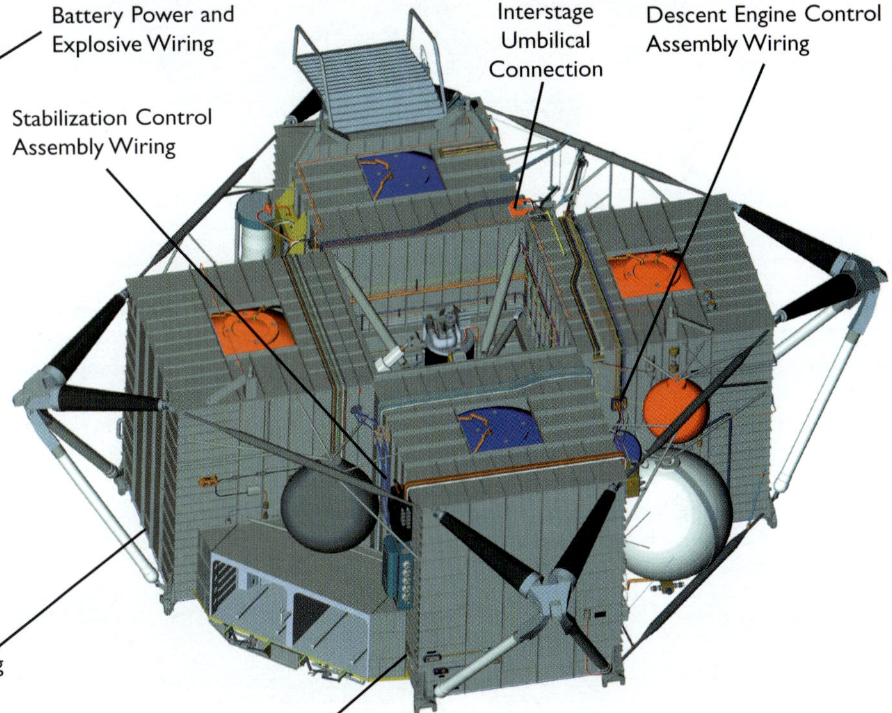

LANDING RADAR ANTENNA ASSEMBLY

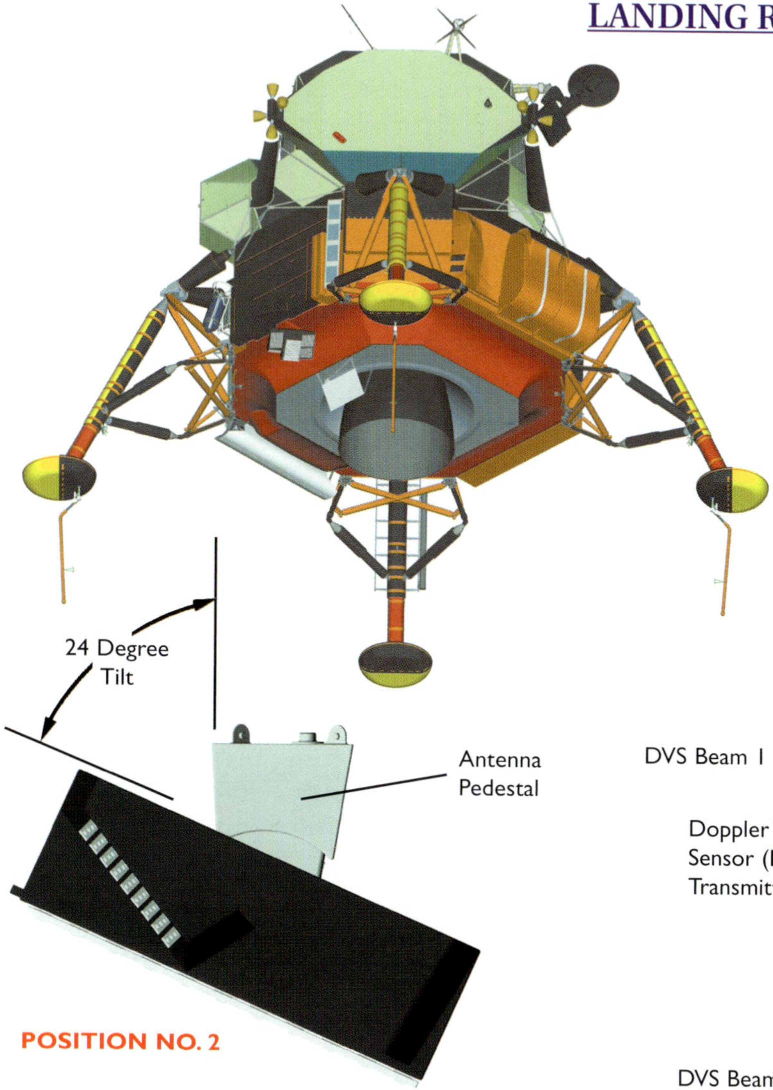

POSITION NO. 2

POSITION NO. 1

The landing radar, located in the descent stage, provided altitude and velocity data during descent. The primary guidance and navigation section calculated control signals for lunar descent rate, hovering, and soft landing. Slant range data began at approximately 40,000 feet above the lunar surface, and velocity data at approximately 35,000 feet. The landing radar used microwave beams: three to measure velocity by Doppler shift continuous wave, and one to measure altitude by continuous-wave frequency modulation. The antenna pedestal has a tilt mechanism for tilting the antenna to either of 2 positions.

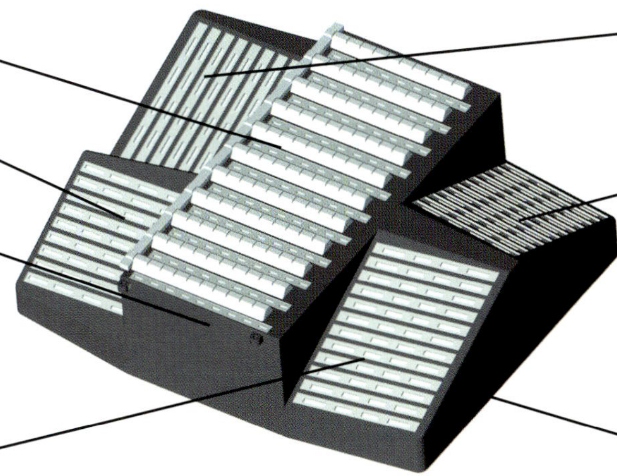

LANDING GEAR ASSEMBLY

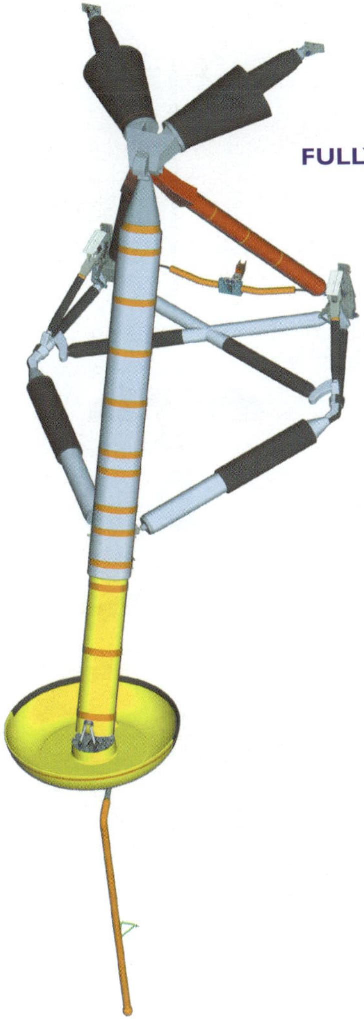

FULLY DEPLOYED (LOCKED)

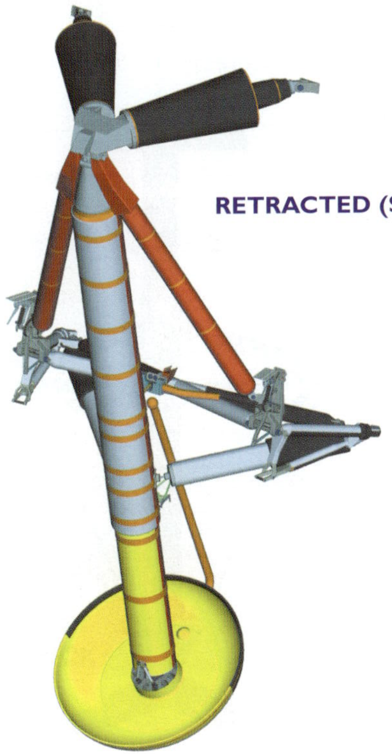

RETRACTED (STOWED)

OUTBOARD VIEW OF DOWNLOCK ASSEMBLY

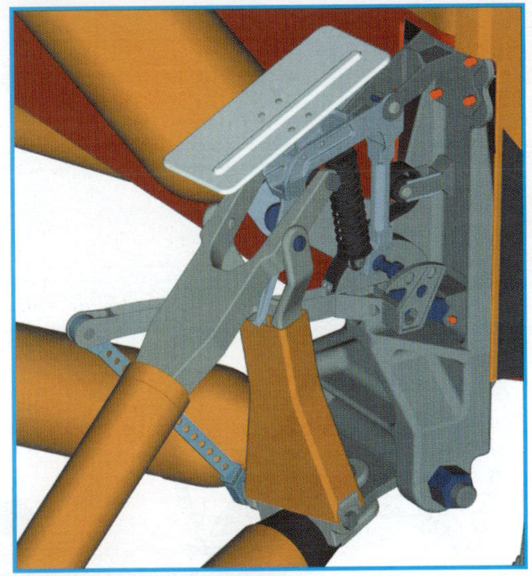

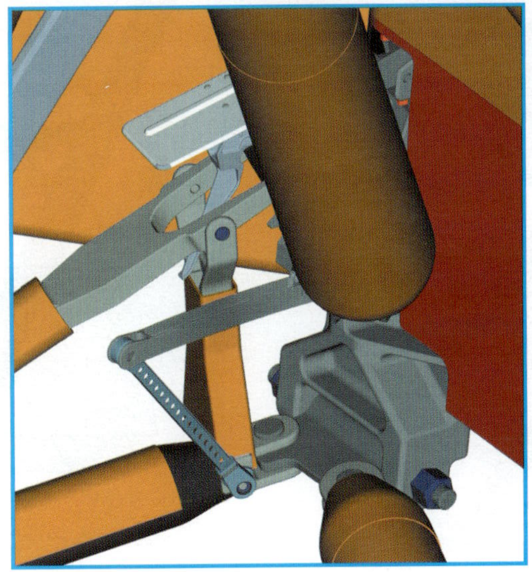

INBOARD VIEW OF DOWNLOCK ASSEMBLY

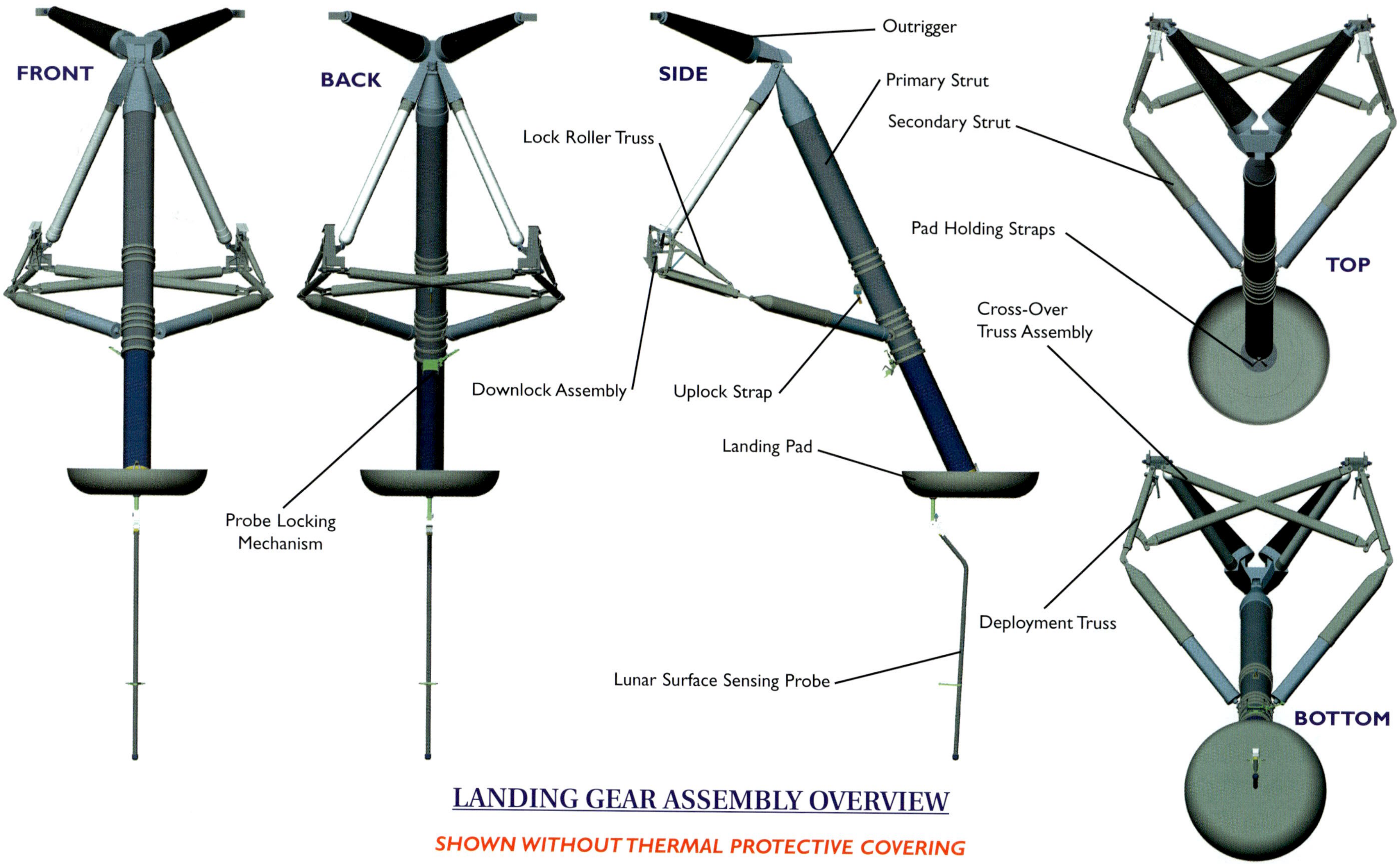

LANDING GEAR ASSEMBLY OVERVIEW

SHOWN WITHOUT THERMAL PROTECTIVE COVERING

Mounting Interface with LM Structure

The landing gear provided required impact attenuation when the Lunar Module landed on the lunar surface; it also prevented tip-over of the LM on the lunar surface with a 6 degree slope having 24 inch depressions (protuberance);, and it supported the LM during the lunar stay and at ascent stage launch. When the LM reached a height of 5 feet above the lunar surface, the surface sensing probes at the base of the landing gear assembly made contact with the lunar surface causing the contact light to activate. This signaled the commander to shut off the descent engine, and the Lunar Module fell to the lunar surface crushing the surface sensing probes beneath it.

At Earth launch the landing gear was retracted into its stowed position to allow it to fit into the confines of the Saturn Lunar Module adapter, thus reducing its overall size. The landing gear remained retracted until the docked LM and CSM attained lunar orbit and the astronauts transferred to the LM. Before the LM was separated from the CSM, the commander in the LM operated the landing gear deployment switch to extend the gear. At that time the landing gear uplocks were explosively released, allowing powerful springs in the deployment mechanisms to extend the gear. Once extended, the gear was locked into position by the downlock mechanism.

The cantilever landing gear consisted of four assemblies, each connected to an outrigger that extended from the ends of the LM's structural parallel beams. The landing gear assemblies were located at the front, rear, left, and right sides of the descent stage. Each assembly consisted of struts; a landing pad; lock and deployment mechanisms; and on all but one of the landing gear, a lunar surface sensing probe. There was no probe on the forward landing gear beneath the forward hatch, (the gear with the ladder); this was to avoid having the crushed probe interfere with the astronaut's EVA (extra-vehicular activity) around the gear landing pad. A ladder was affixed to the forward landing gear to allow the astronauts to exit and enter the LM. The landing gear could withstand a 10-foot per second vertical velocity (or drop) of the LM when the horizontal velocity was zero feet per second (or static); a 7-foot per second vertical velocity with a horizontal velocity not exceeding 4 feet per second; and a vehicle attitude within 6 degrees of the local horizontal when the rate of attitude change was 2 degrees per second or less.

The landing gear primary strut had a footpad at the bottom that was 37 inches in diameter and was attached to the strut with a ball-socket fitting, allowing radial movement. The landing pad was an aluminum honeycomb construction and its large diameter ensured minimal penetration of the LM on lunar impact. During launch, four straps held the pads in a fixed position on the strut. The straps would shear or bend on impact with the lunar surface, permitting the pad to conform to the irregularities of the lunar surface.

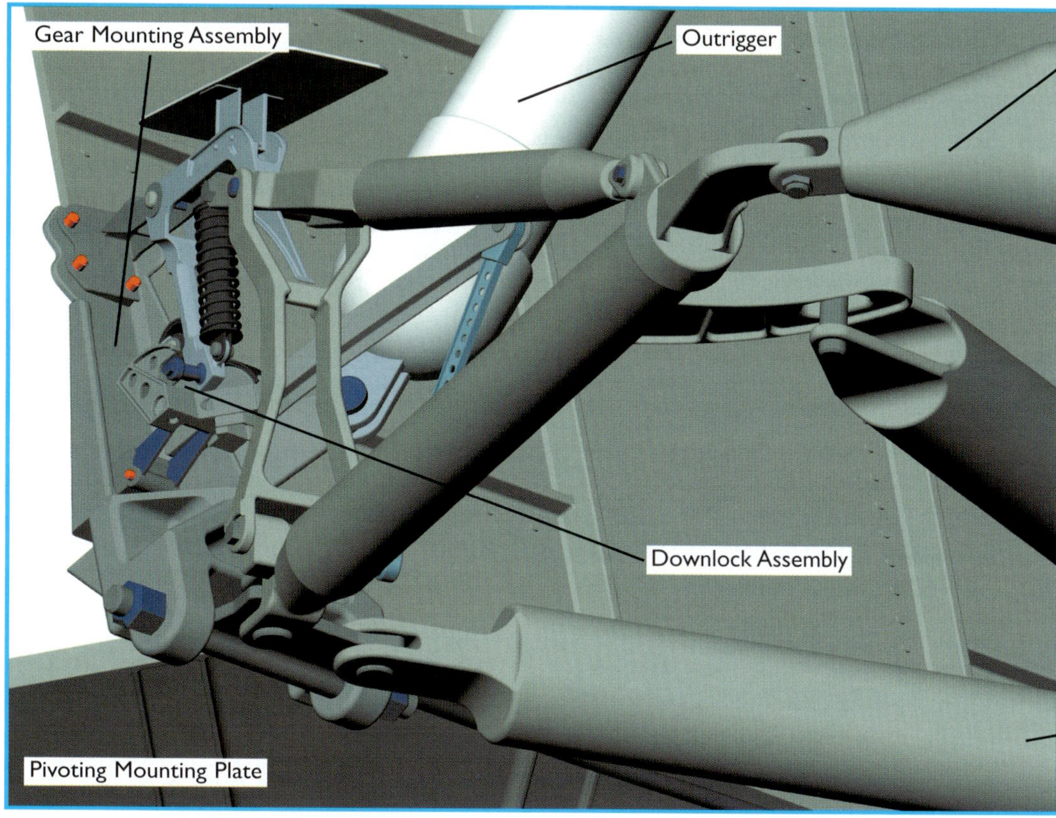

VIEW FROM UNDER THE LANDING GEAR ASSEMBLY, LOOKING UP

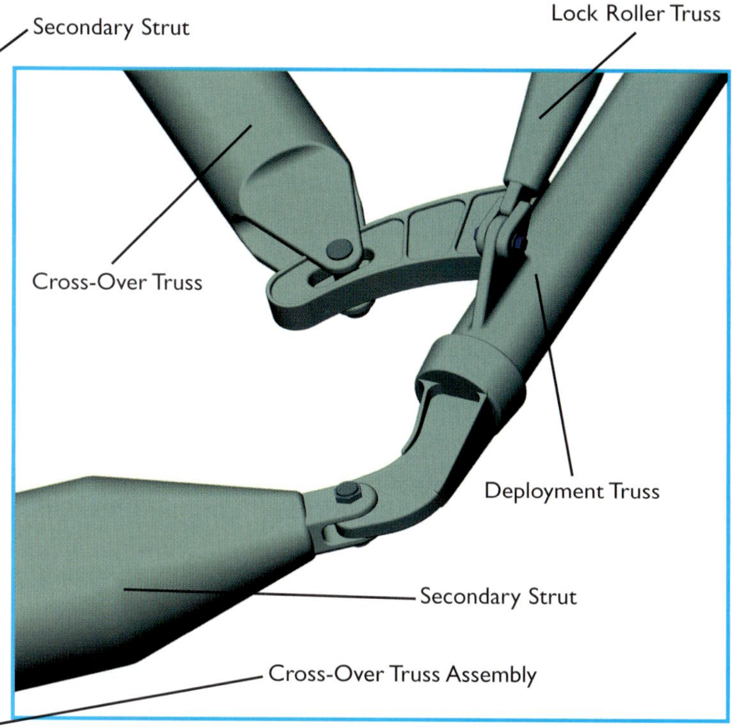

DETAILS OF THE INTERFACE BETWEEN THE SECONDARY STRUT, THE CROSS-OVER TRUSS ASSEMBLY, AND THE DEPLOYMENT TRUSS, AS VIEWED FROM THE TOP OF ASSEMBLY

Each landing gear assembly had two secondary struts. The outboard end of each strut was attached to the primary strut with a ball socket-type assembly; the inboard ends were attached to a deployment truss assembly. Each strut was a piston-cylinder-type device that contained compressible aluminum honeycomb cells capable of absorbing compressive and tensile loads. The design and location of the secondary struts in relation to the primary strut enabled the LM to land on a non-symmetrical surface, or to land when it was moving laterally over the lunar surface.

The deployment portion of the deployment and downlock mechanism consisted of a truss assembly, one clock-like deployment spring, one extension-type spring, and a connecting linkage. The truss, connecting the secondary struts indirectly to the descent stage structure, comprised two side frames assemblies, separated by a cross member. The deployment springs were indirectly attached to the side frame assemblies through connecting linkages. The downlock portion of the mechanism consisted of a spring-loaded lock and a cam follower.

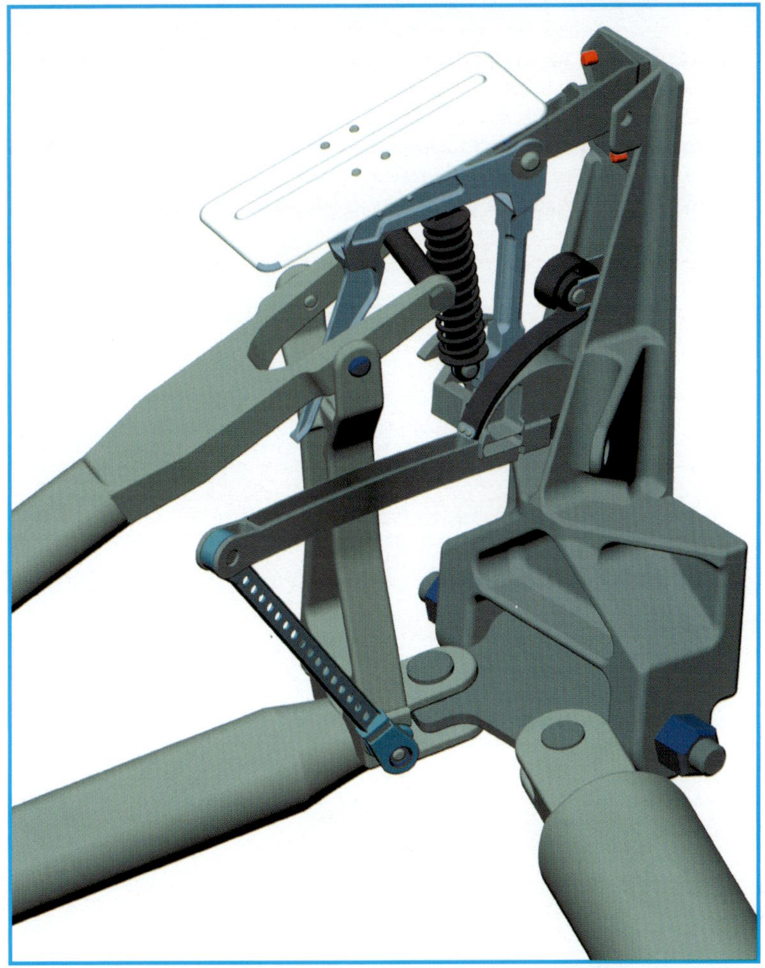

INBOARD VIEW (OUTRIGGER NOT SHOWN)

DETAIL VIEWS OF DOWNLOCK ASSEMBLY
IN LOCKED POSITION

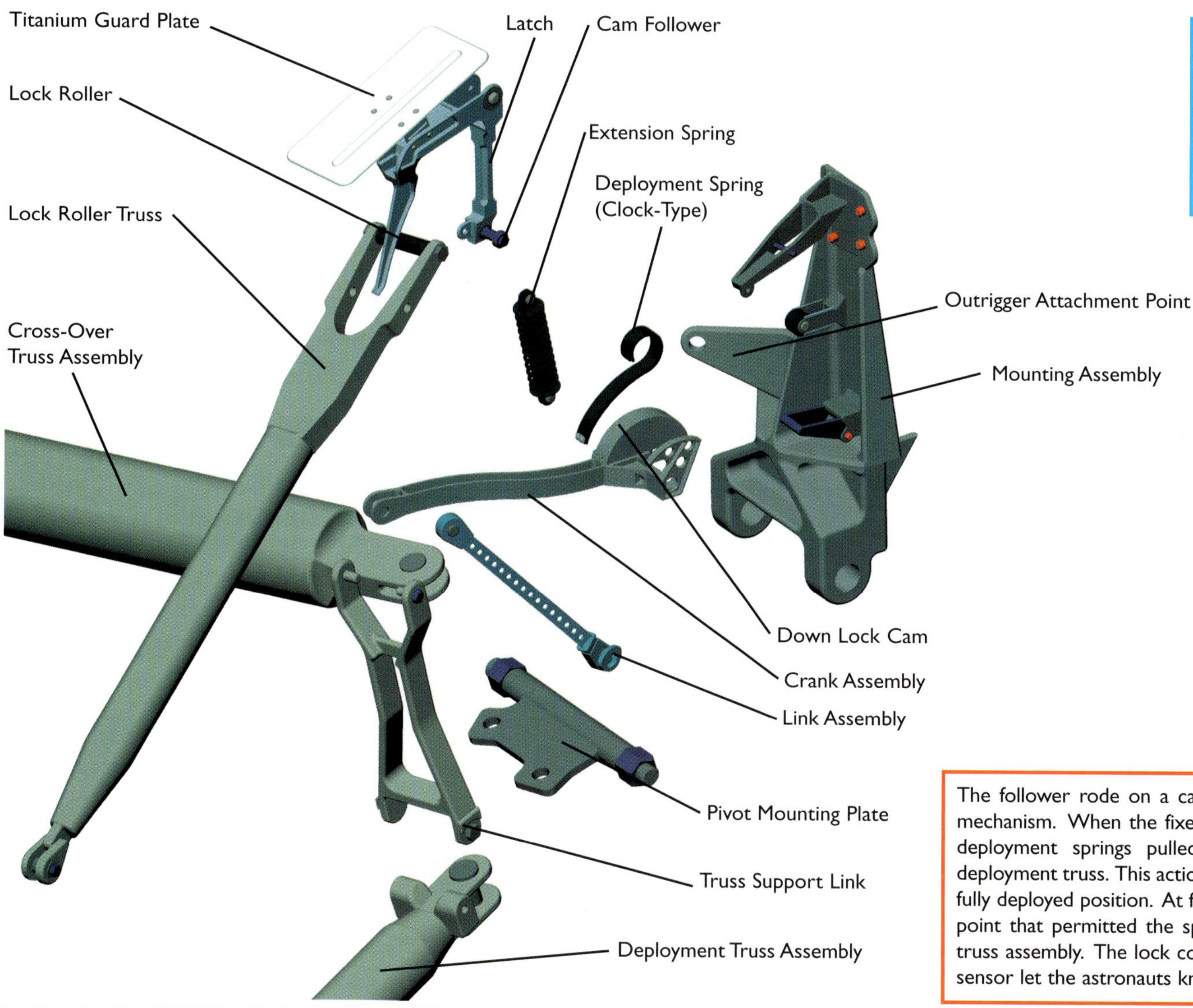

Lunar Module Chronology

February 6, 1962
John C. Houbolt presents his plan for Lunar Orbit Rendezvous to the Manned Space Center Flight Management Council.

The follower rode on a cam attached to the deployment portion of the mechanism. When the fixed link of the uplock assembly was severed, the deployment springs pulled the connecting linkage and, indirectly, the deployment truss. This action drove the landing gear from the stowed to the fully deployed position. At full gear deployment the cam follower reached a point that permitted the spring-loaded lock to snap over a roller on the truss assembly. The lock could not be opened. A landing gear deployment sensor let the astronauts know that the gear was fully deployed.

PRIMARY STRUT IMPACT ATTENUATION

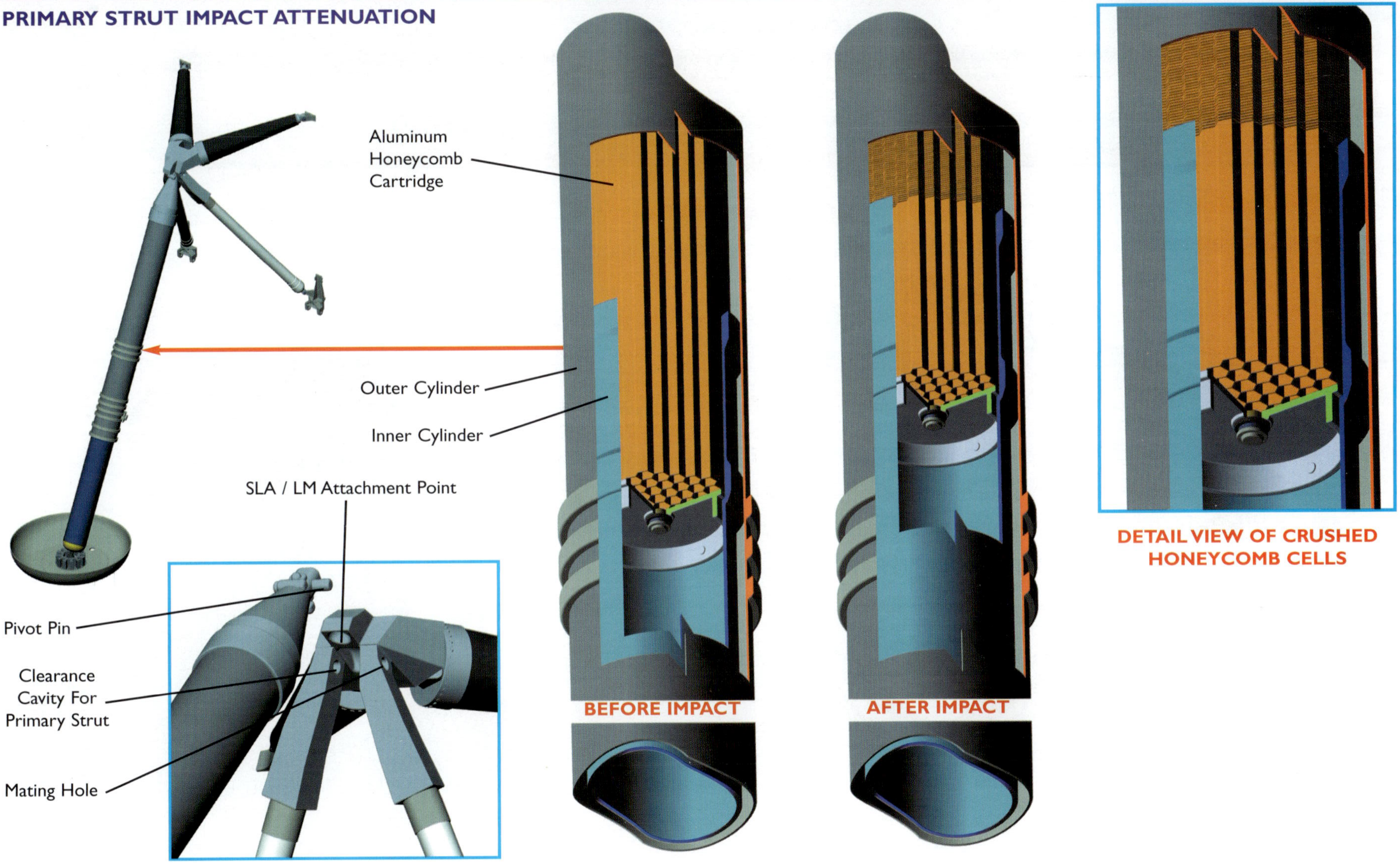

DETAIL VIEW OF CRUSHED HONEYCOMB CELLS

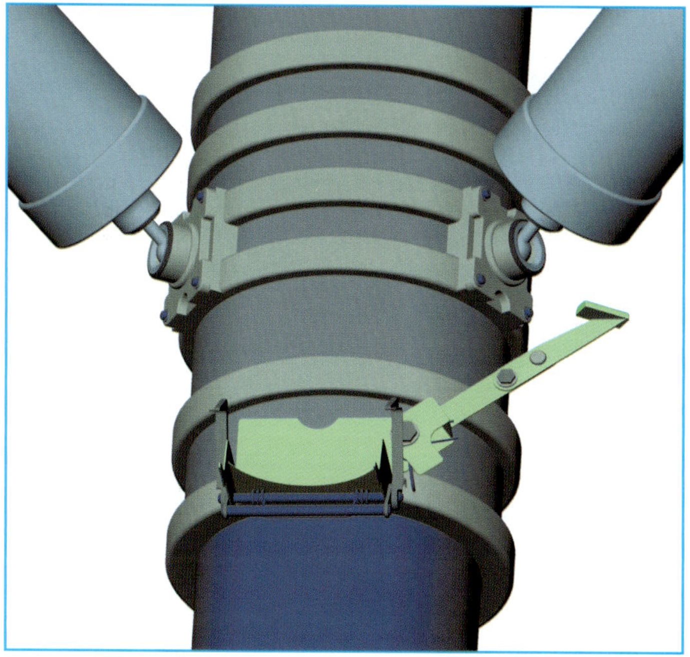

SECONDARY STRUT BALL SOCKET INTERFACE WITH PRIMARY STRUT

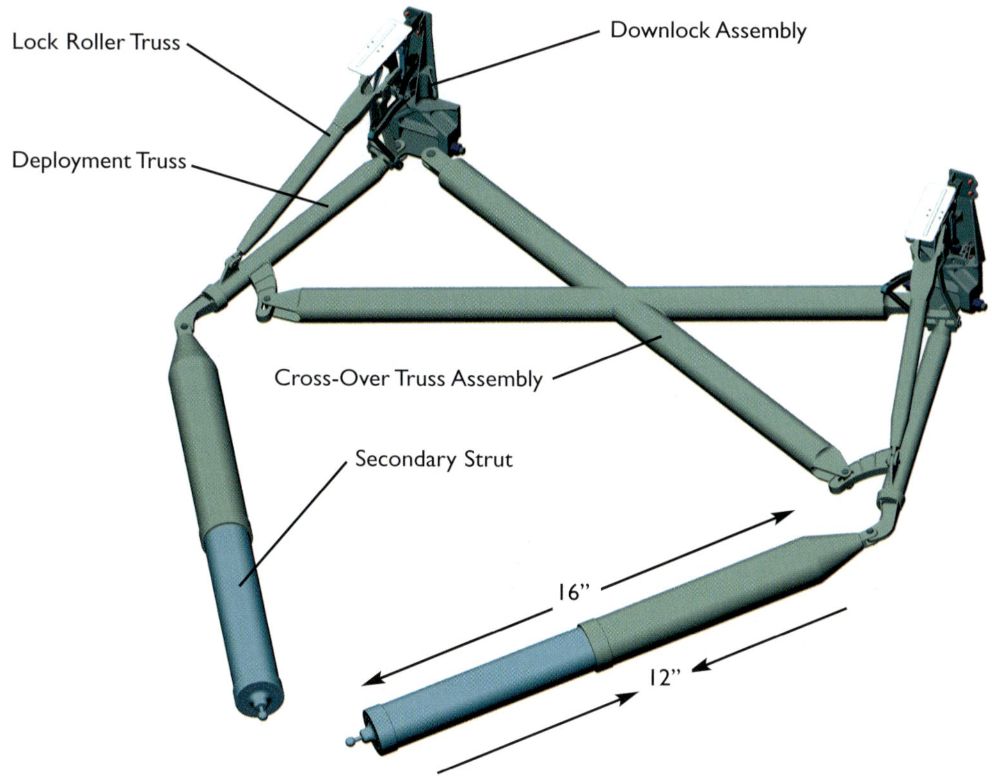

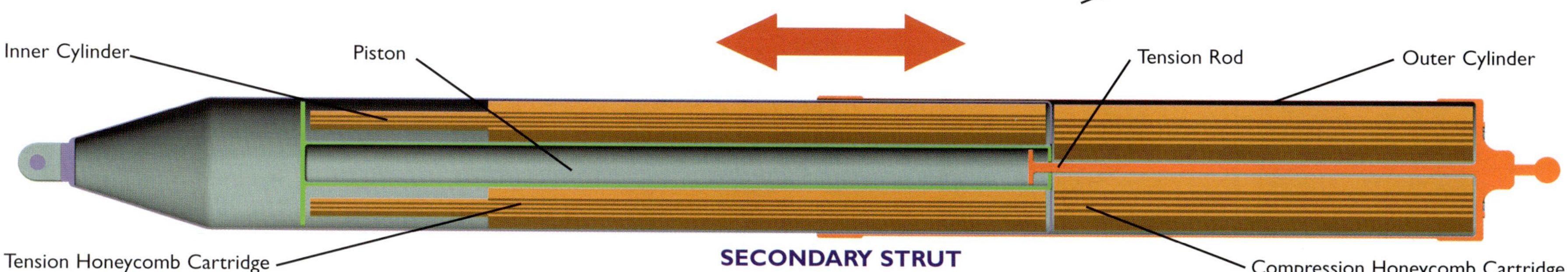

SECONDARY STRUT

LUNAR SURFACE SENSING PROBE CRUSHED BENEATH THE LANDING GEAR FOOT PAD

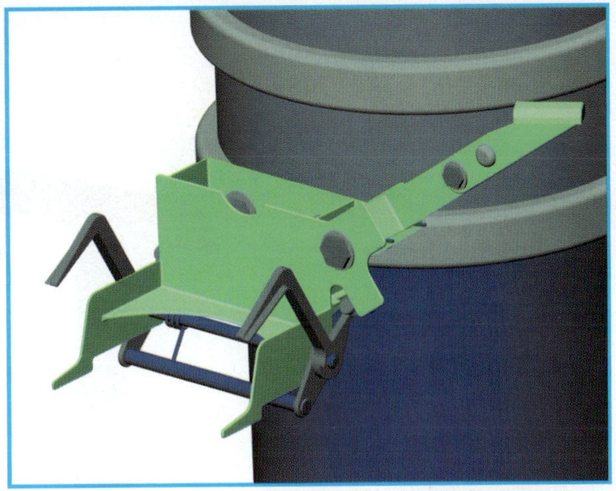

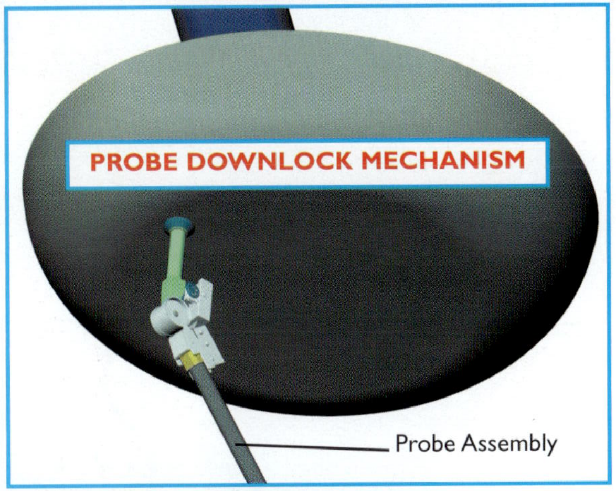

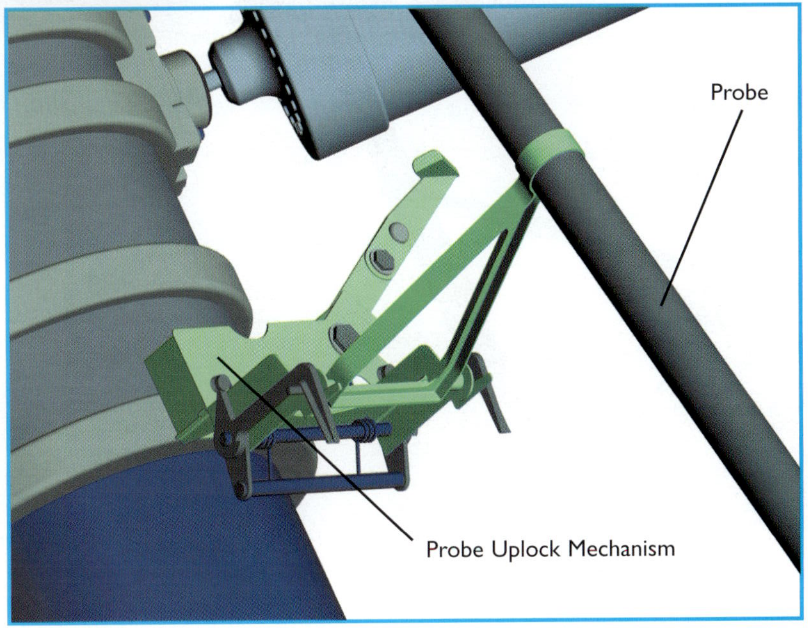

An electromechanical lunar surface-sensing probe was attached to each of the landing gear footpads, with the exception of the forward gear assembly. The probes were retained in the stowed position, against the primary strut, until landing gear deployment. During deployment, mechanical interlocks were released permitting spring energy to extend the probes so that the probe heads were approximately 5 feet below the surface of the bottom of the footpad. When the probe touched the lunar surface, pressure on the probe head completed the electrical circuit, advising the astronauts to shut off the descent engine. This shutdown point, which determined the LM velocity at impact, was a tradeoff between gear design weight and the thermal and thrust reactions caused by the descent engine operating near the lunar surface. Each probe had an indicator plate attached to it which, when aligned, indicated that the probe was fully extended.

STOWED CONFIGURATION

LANDING GEAR HALF DEPLOYED

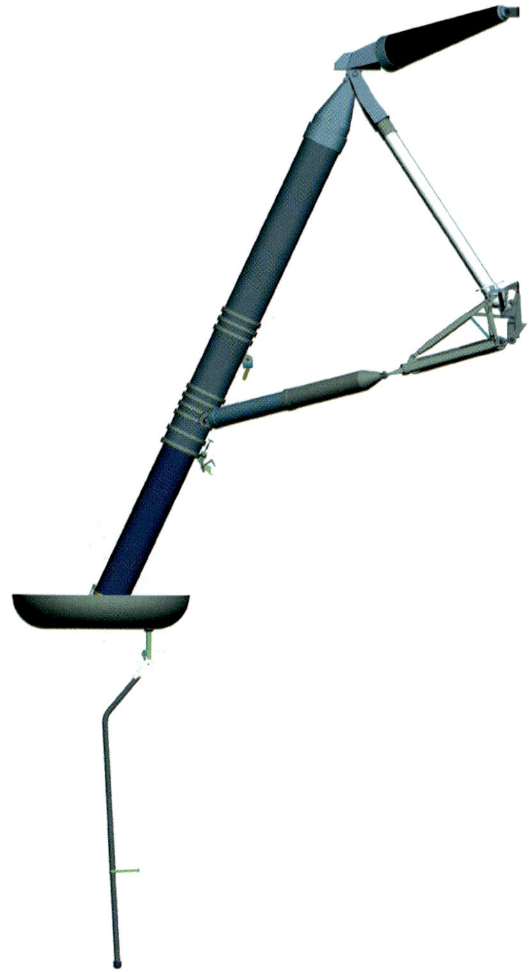

LANDING GEAR FULLY DEPLOYED (LOCKED)

LANDING GEAR DEPLOYMENT SEQUENCE

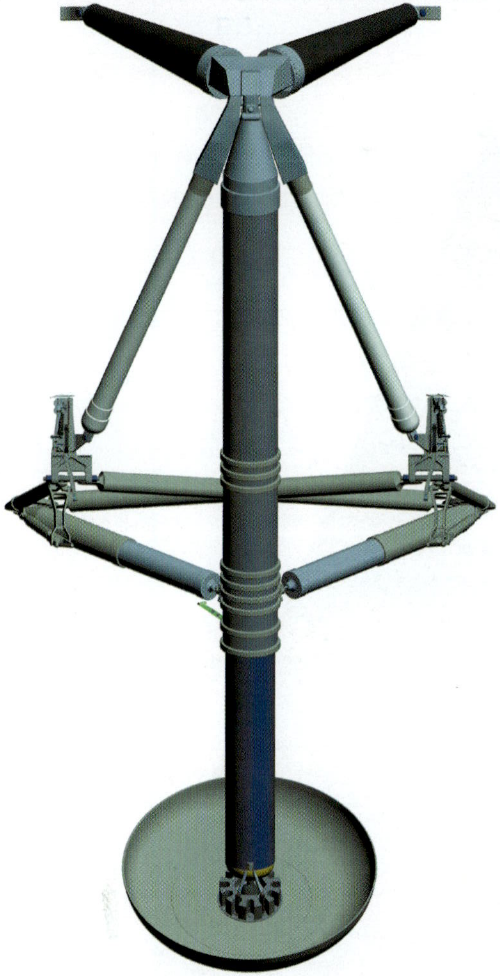

FRONT

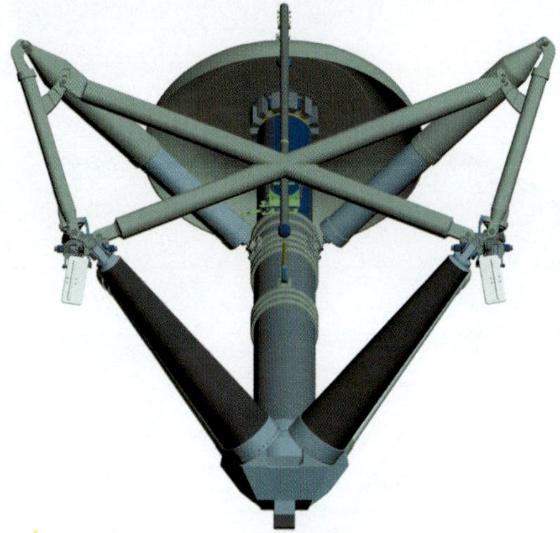

TOP

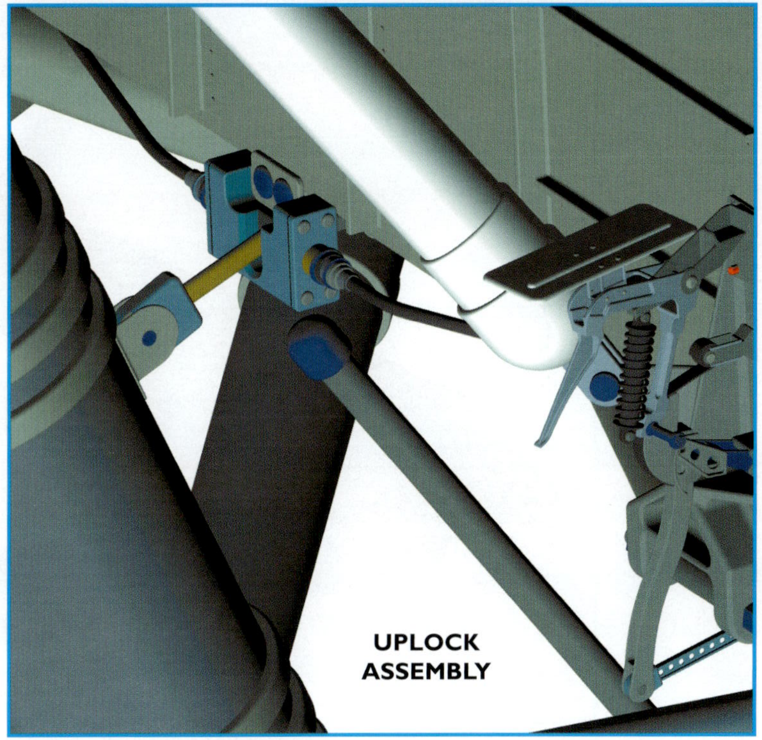

LANDING GEAR IN STOWED (RETRACTED) POSITION

Lunar Module Chronology

March 29, 1962
Chance Vought Corporation briefed on Lunar Orbit Rendezvous in connection with the study on spacecraft rendezvous.

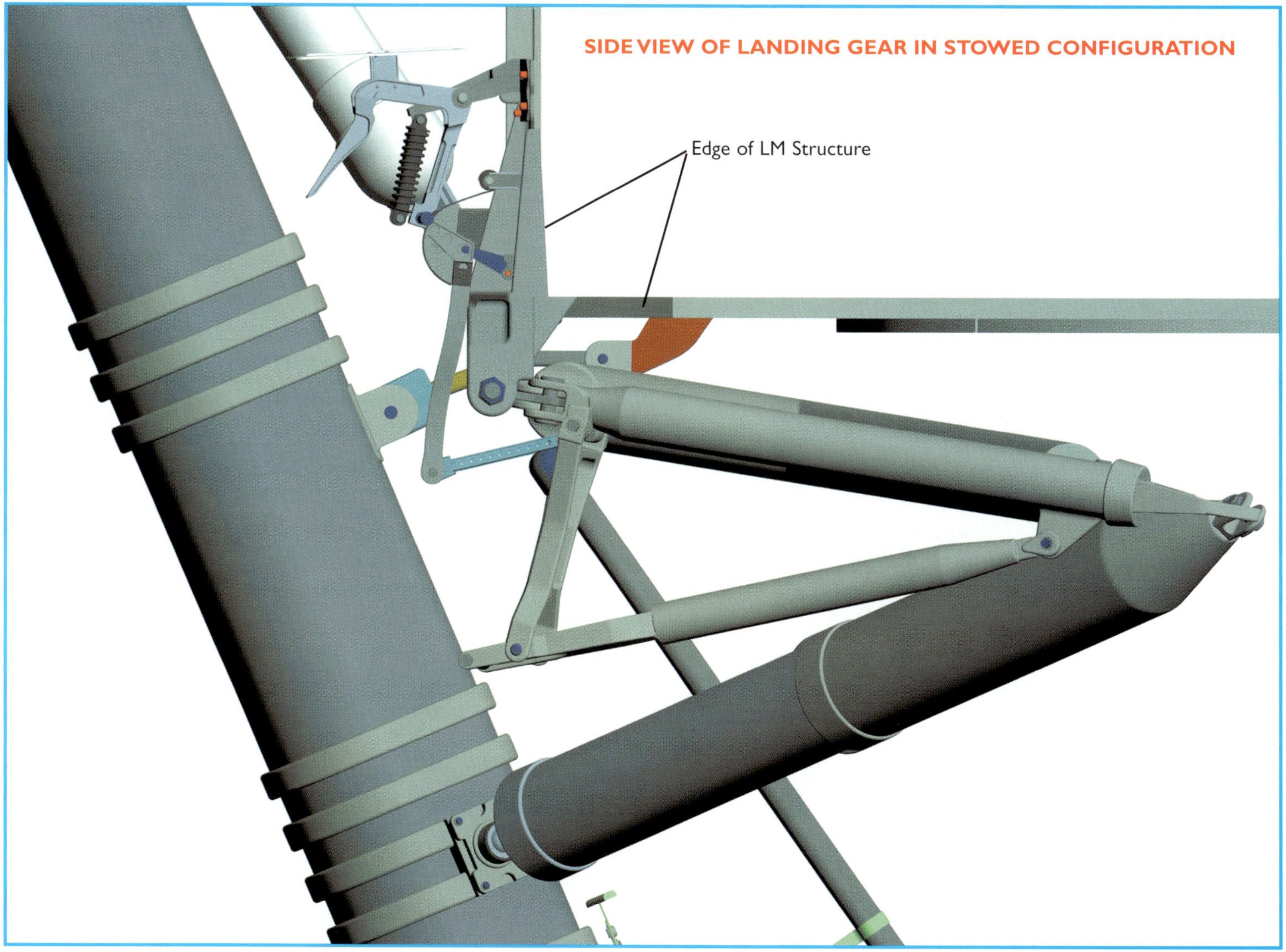
SIDE VIEW OF LANDING GEAR IN STOWED CONFIGURATION
Edge of LM Structure

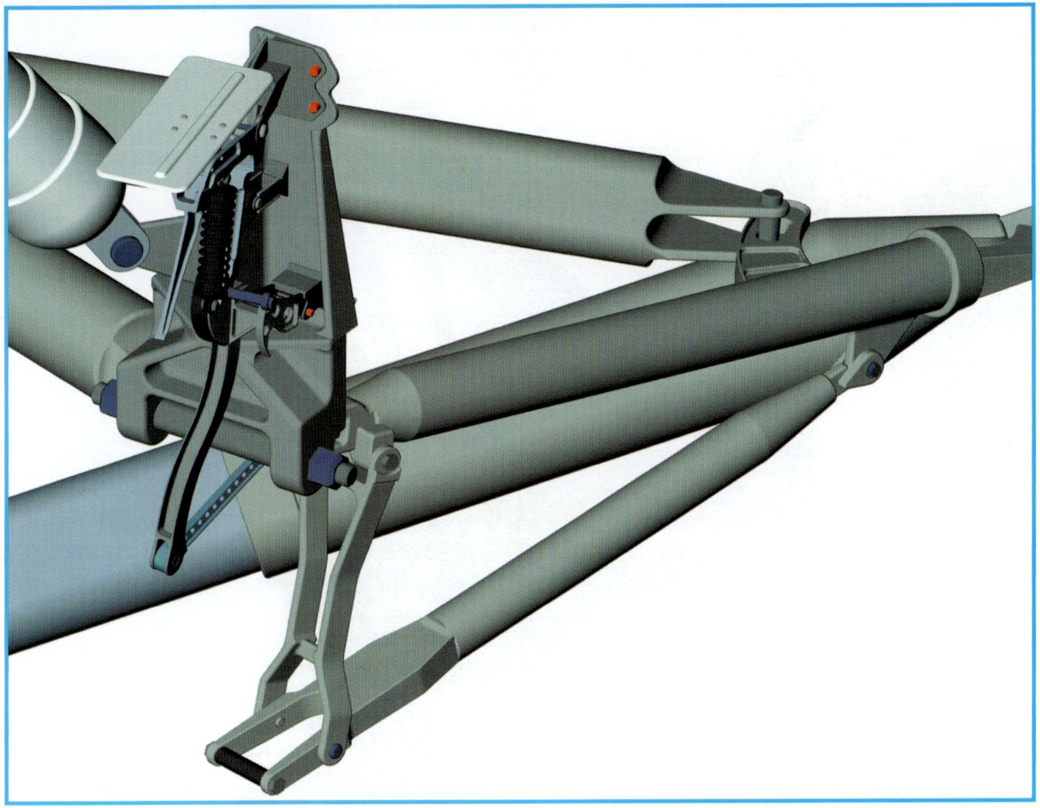

Lunar Module Chronology

May 6, 1962
Preliminary Statement of Work for the Apollo Lunar Excursion Module completed, although the mission mode had not yet been selected.

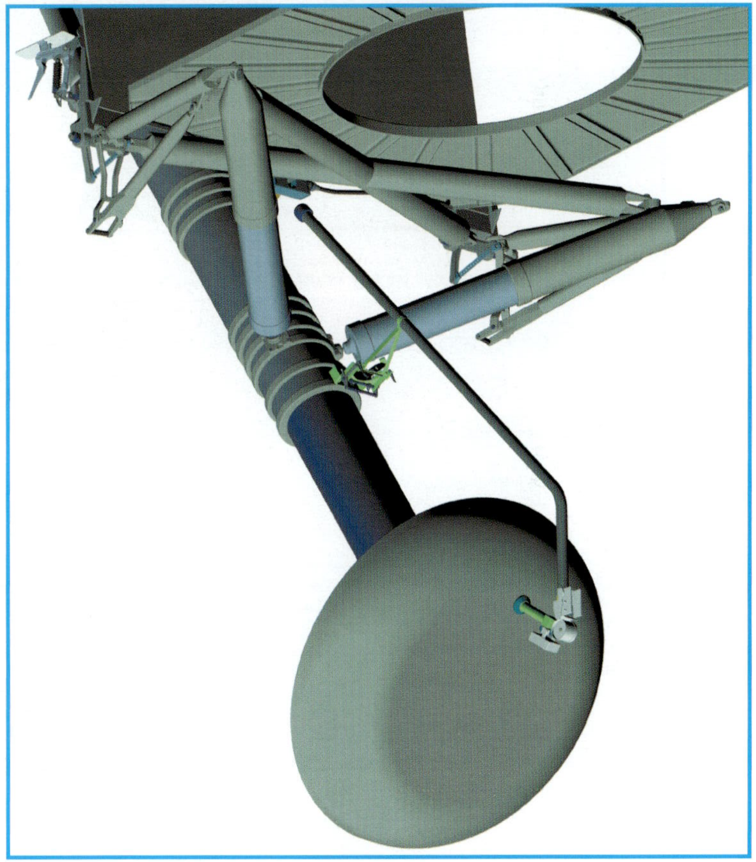

Lunar Module Chronology

April 24, 1962
Indecisions on the lunar mission mode cause delays in Apollo program.

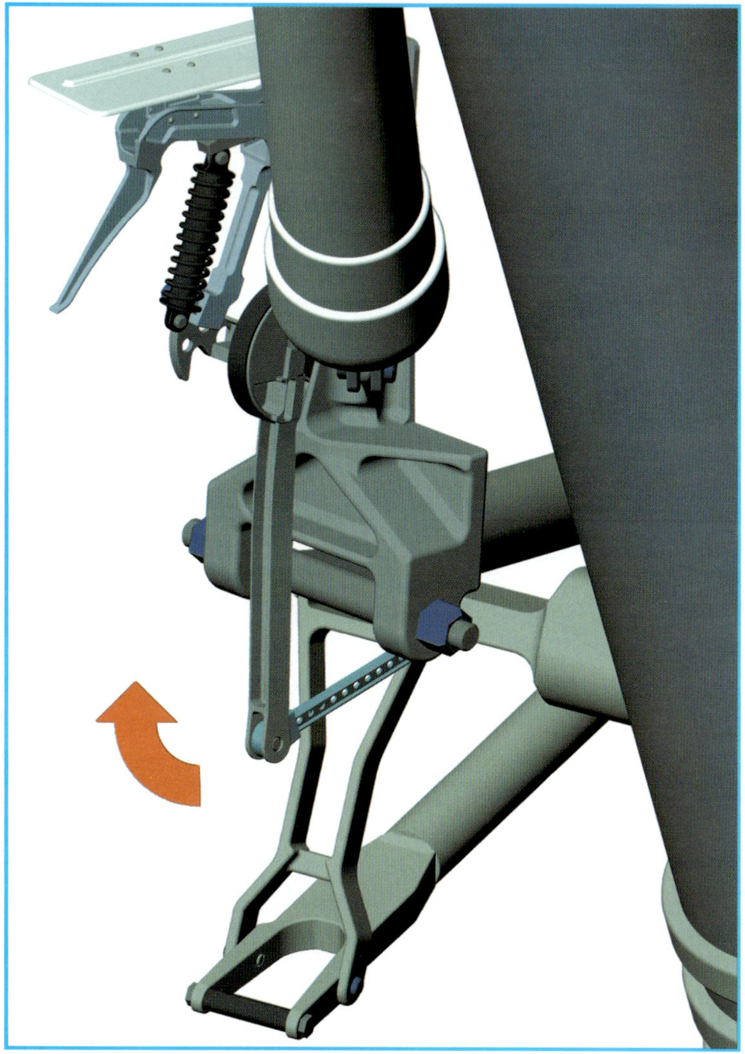

INBOARD VIEW

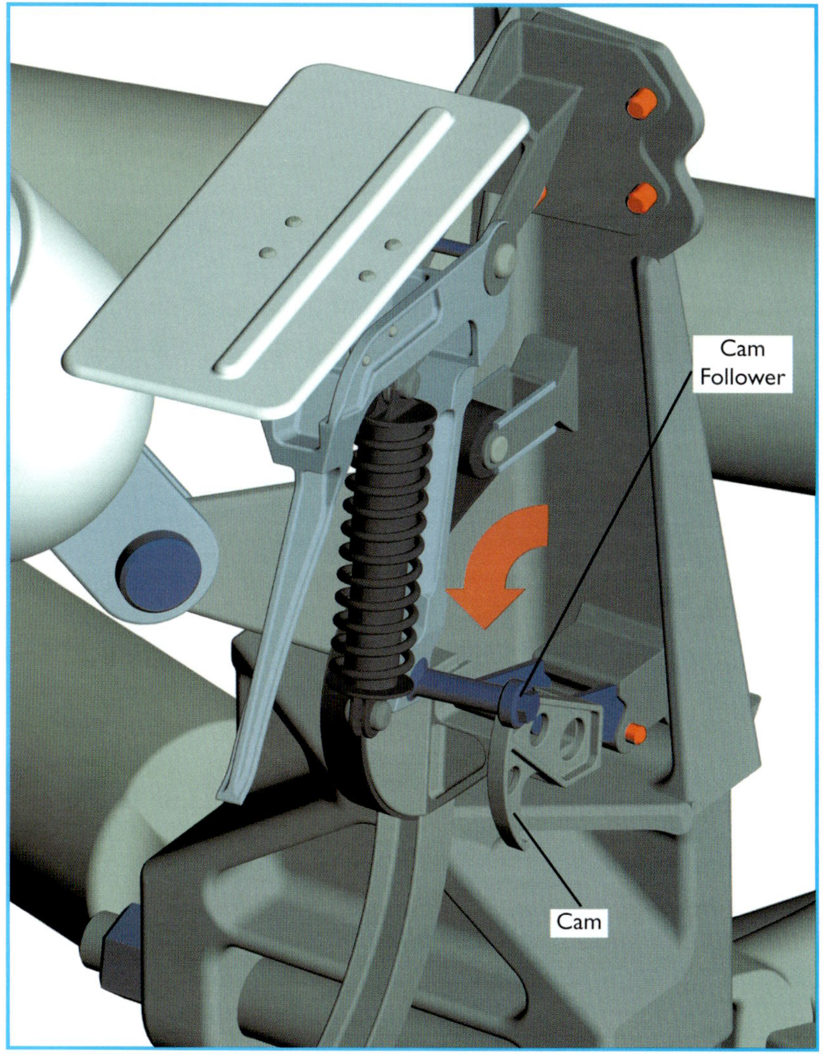

INBOARD VIEW

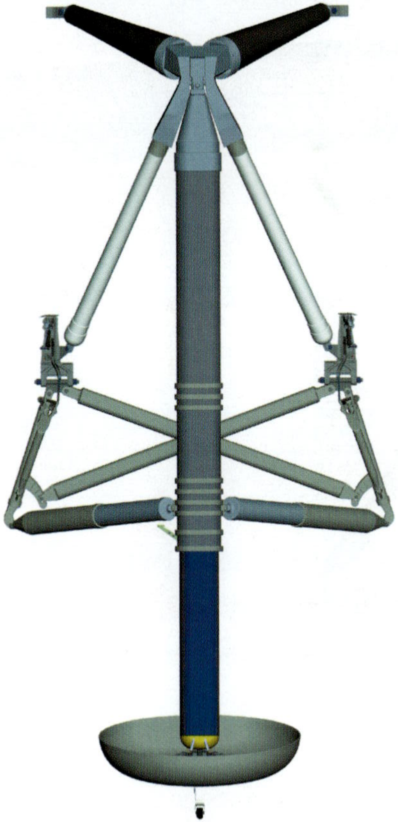

FRONT

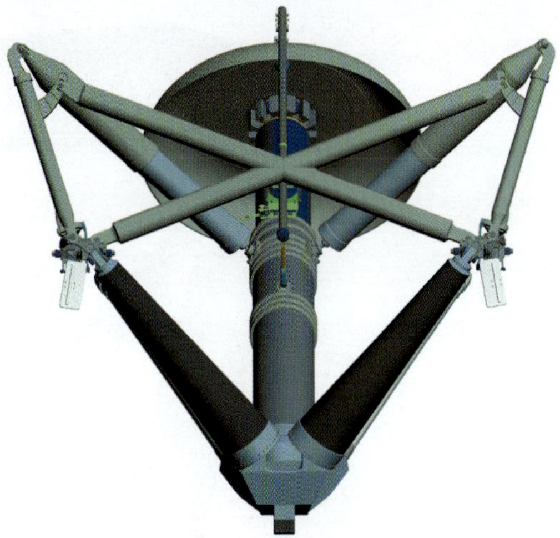

TOP

LANDING GEAR HALF DEPLOYED

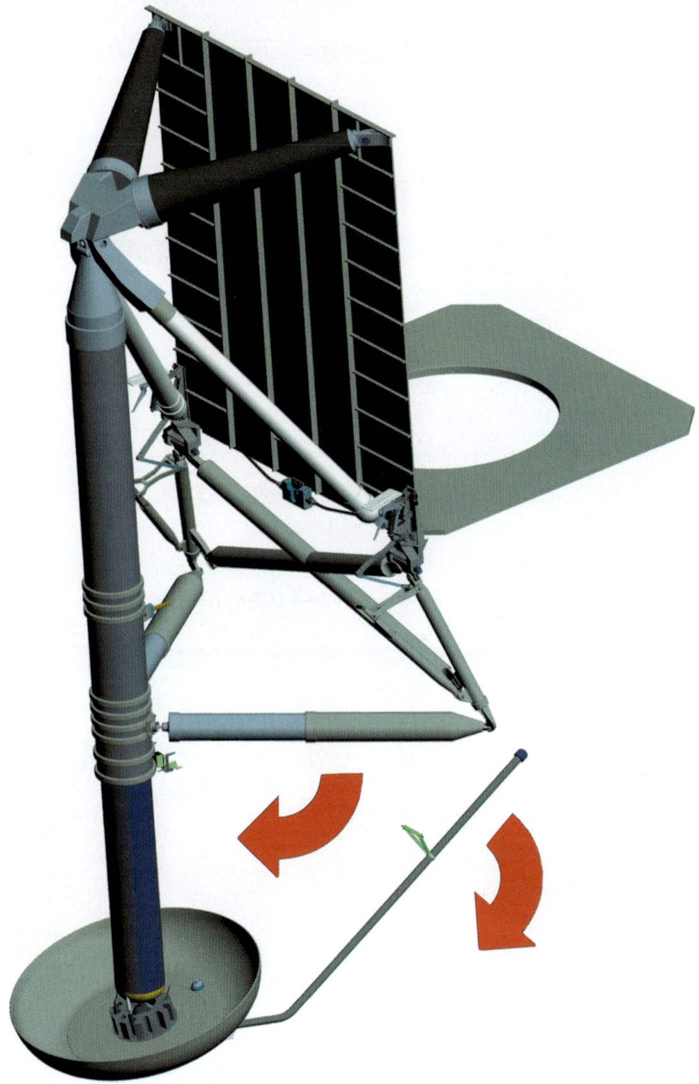

Lunar Module Chronology

July, 25 1962
Invitations to bid on the Apollo Lunar Module contract sent out by NASA.

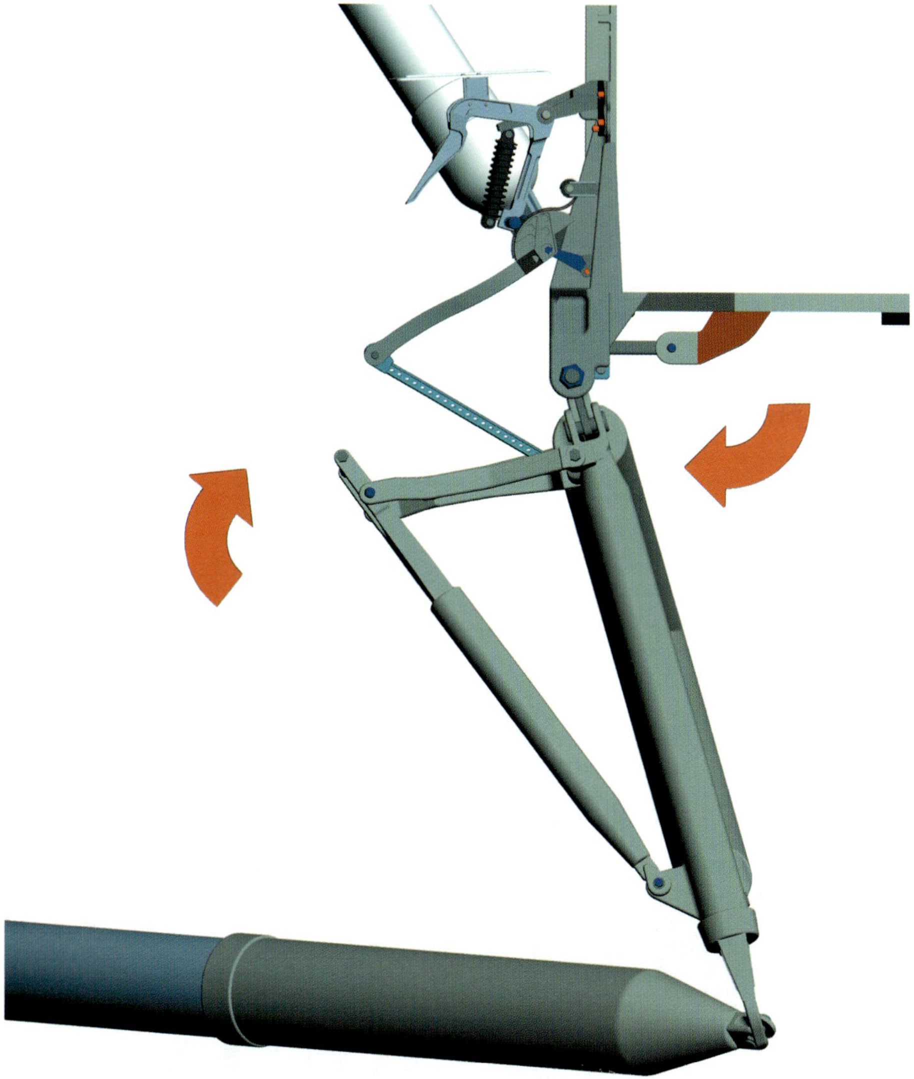

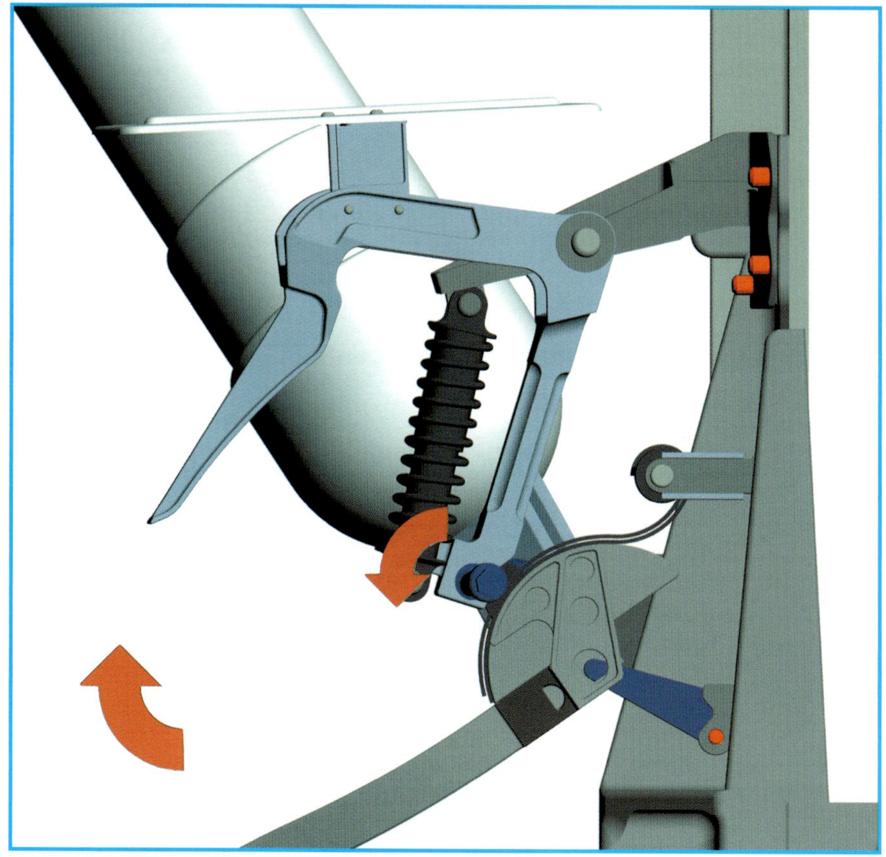

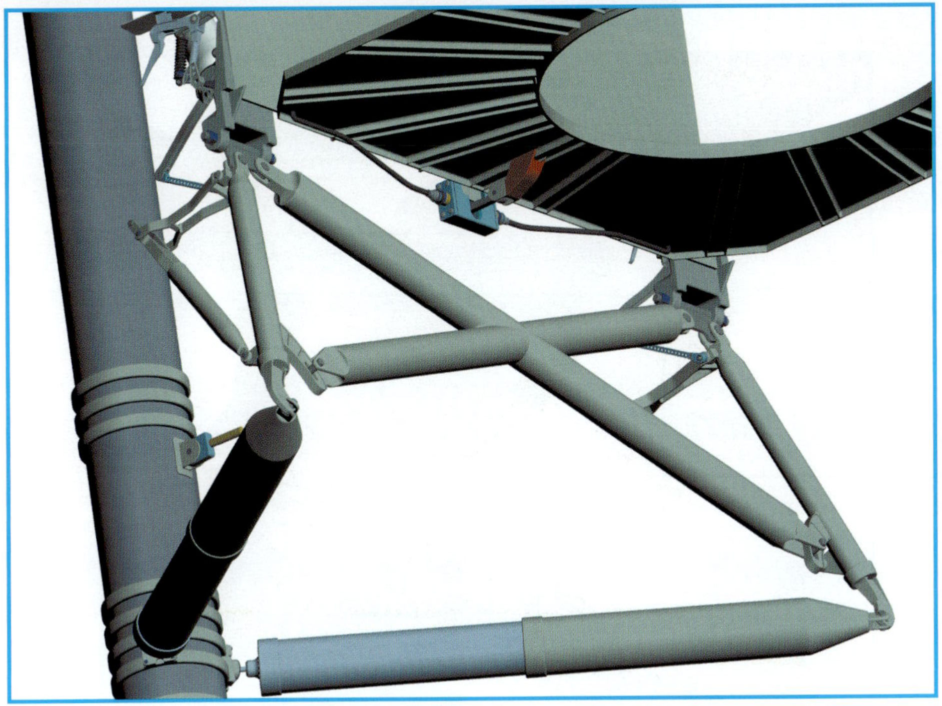

Lunar Module Chronology

August 14, 1962
Lunar Module added to Apollo CSM Statement of Work.

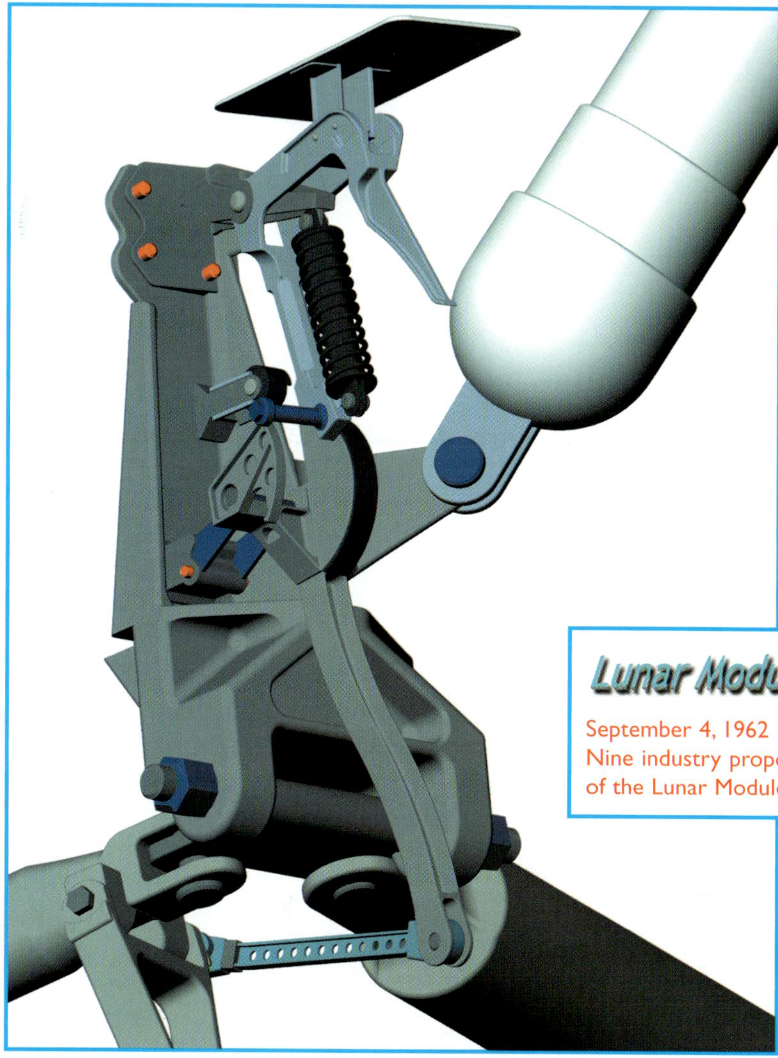

OUTBOARD VIEW

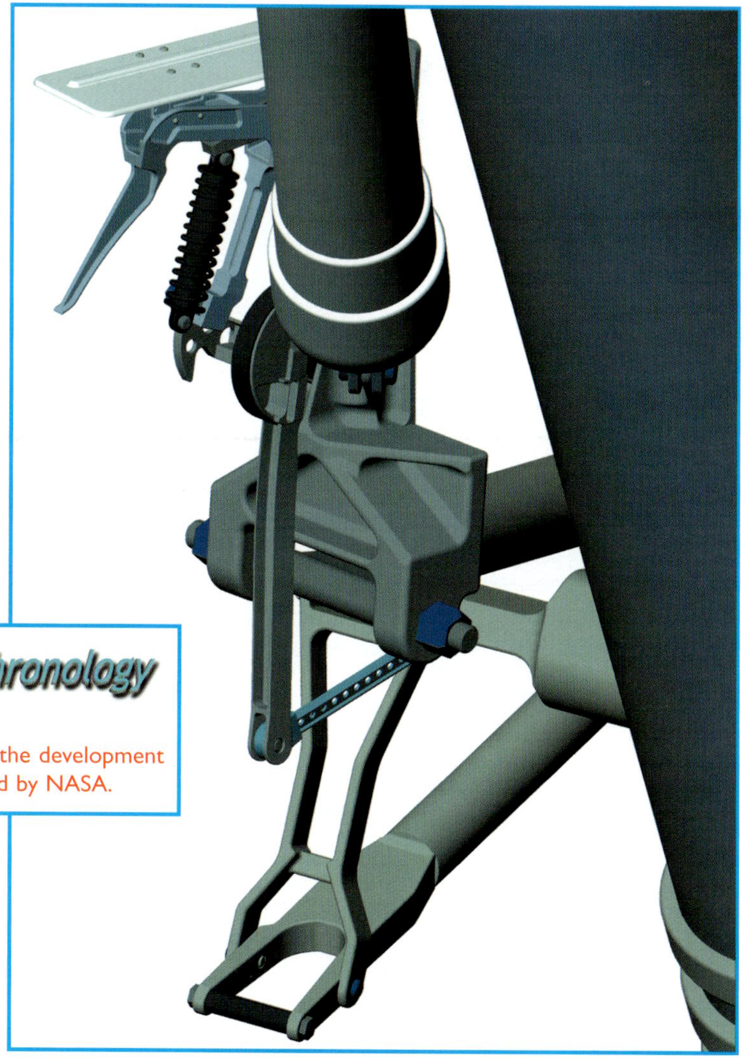

INBOARD VIEW

Lunar Module Chronology

September 4, 1962
Nine industry proposals for the development of the Lunar Module received by NASA.

RADIOISOTOPE THERMOELECTRIC GENERATOR (RTG) CASK

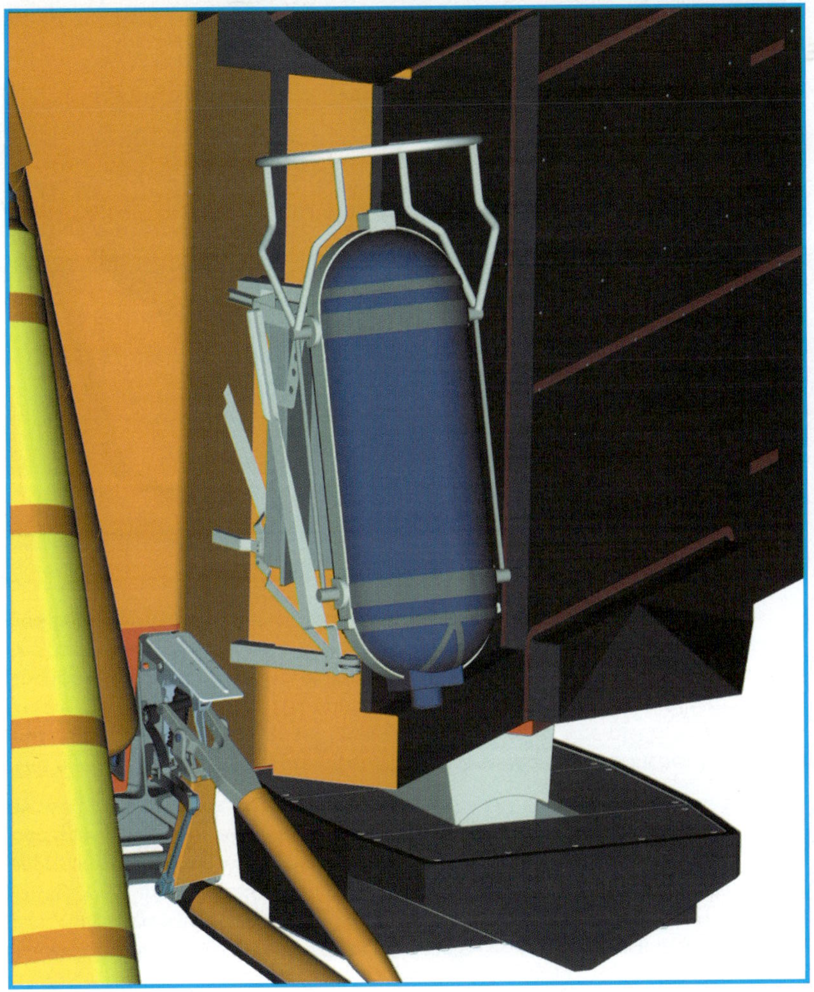

RADIOISOTOPIC THERMOELECTRIC GENERATOR CASK ASSEMBLY

The Radioisotope Thermoelectric Generator (RTG) was used to power the ALSEP. Electrical energy was generated through thermoelectric action. The RTG provided a minimum of 16 volts at 56.2 watts to a power-conditioning unit. The radioisotopic fuel capsule emitted nuclear radiation and approximately 1,500 thermal watts continuously. The surface temperature of the fuel capsule was approximately 1,400 degrees F. The capsule was stored in a graphite cask, which was externally mounted to the descent stage in Quadrant number 2.

ASTRONAUT TRAINING FOR REMOVAL OF FUEL ELEMENT FROM RTG CASK

SPACECRAFT EGRESS

The ladder, attached to the primary strut of the forward landing gear assembly, had nine rungs between two railings. The rungs were spaced nine inches apart. The railings were approximately 20 inches between centers. The top of the ladder was approximately 18 inches below the forward end of the egress platform, on the outrigger. The ladder extended down to within 30 inches of the landing pad. This allowed the primary strut to telescope when the LM impacted in lunar surface.

The egress platform was approximately 32 inches wide and 45 inches long and was mounted over the forward landing gear outrigger. The platform was just below the forward hatch. The upper surface was corrugated to facilitate hand and foot holds. The platform, in conjunction with the ladder below it, provided the astronauts with a means of access between the LM and the lunar surface and between the LM interior and space for any EVA.

COMMEMORATIVE PLAQUE THAT WAS UNIQUE FOR EACH MISSION

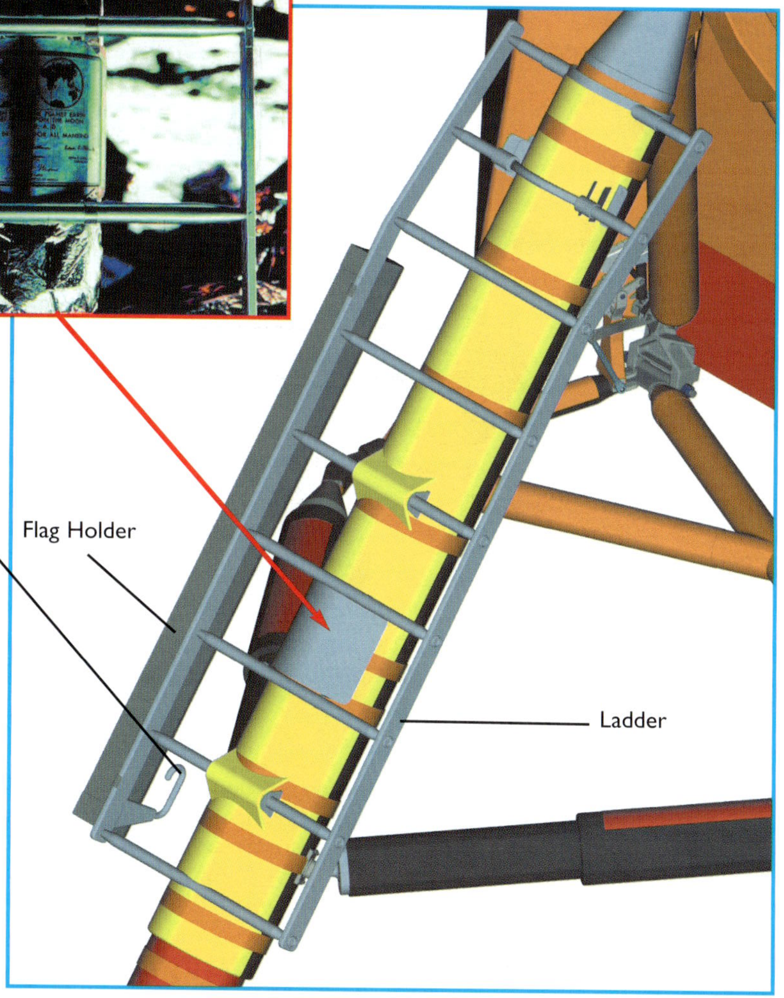

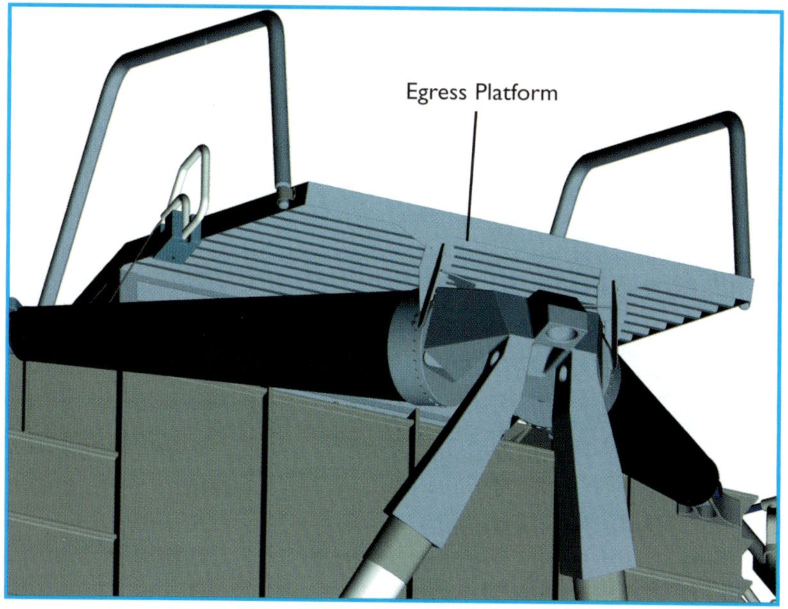

H-SERIES OUTRIGGER VENTING

The H-Series LM incorporated a perforated venting fixture inside the thermal covering over the landing gear outrigger. The fixture allowed heat to escape from the cone-like sleeve around the outrigger.

The outrigger in the +Y axis had an umbilical connection that was attached to the SLA until the Command and Service Modules docked and extracted it.

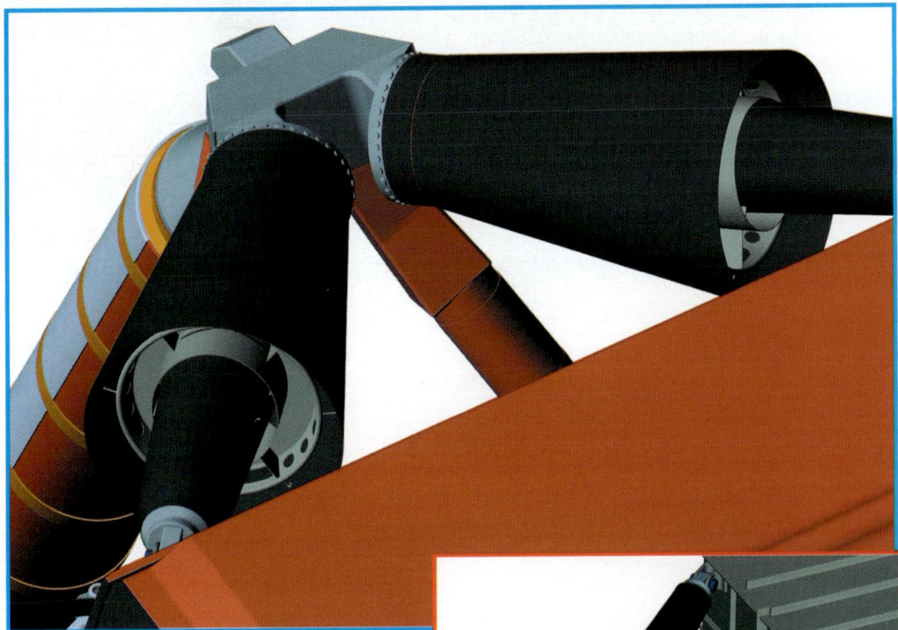

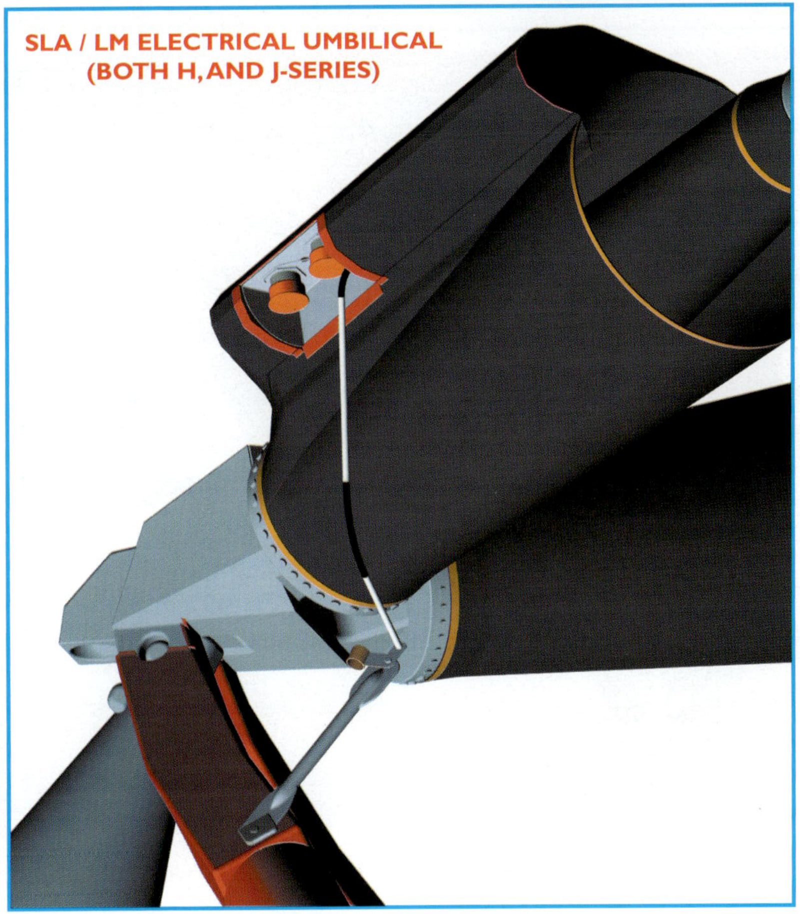

PLUME DEFLECTOR

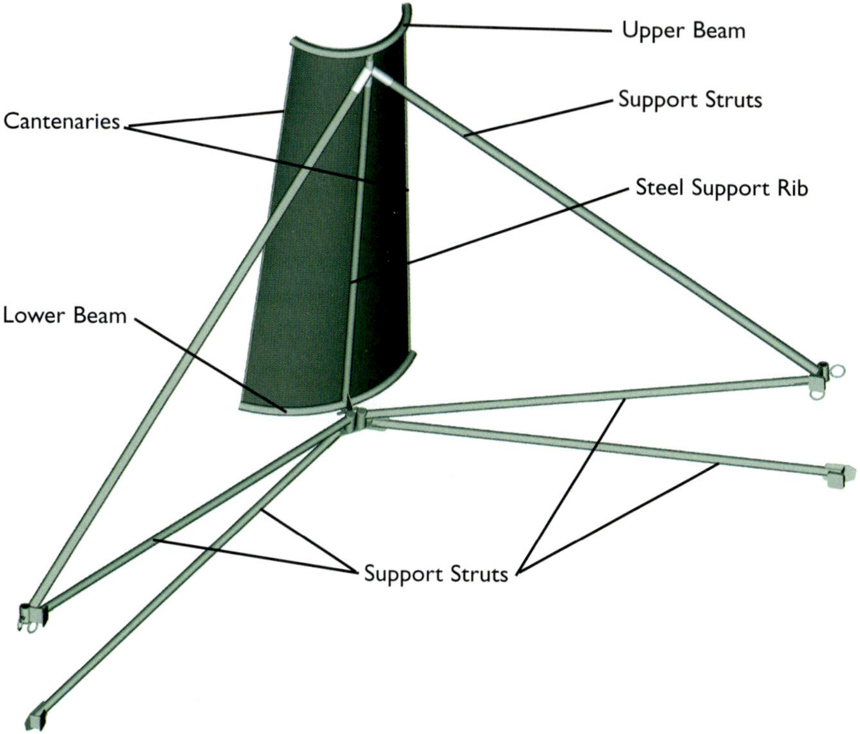

LM 3 (*APOLLO 9*) WITHOUT PLUME DEFLECTORS

Plume deflectors where not present on Lunar Modules 1 through 4. It was realized, at the last minute, that a device to eliminate damage to the descent stage created by the ascent stage's RCS engine thrust plume would be required. The plume deflectors where designed and fabricated and then installed on LM-5 (*Apollo 11*), while the spacecraft was on the launch pad, by technicians who had to crawl around in the cramped quarters of the Saturn V's Lunar Module adapter.

Lunar Module Chronology

October 24, 1962
Final manned lunar landing report by McDonnell Corporation and STL received by NASA.

THE J-SERIES DESCENT STAGE

The last three Apollo missions incorporated the J-Series Lunar Module. The J-Series version had been designed with extended lunar stay and equipment as the overall design intent. The incorporation of the lunar roving vehicle was also only seen on the J-Series LM. The basic dimensions remained the same, although the weight increased by 3,000 pounds.

Lunar Module Chronology

November 7, 1962
NASA selects Grumman to build the Lunar Excursion Module.

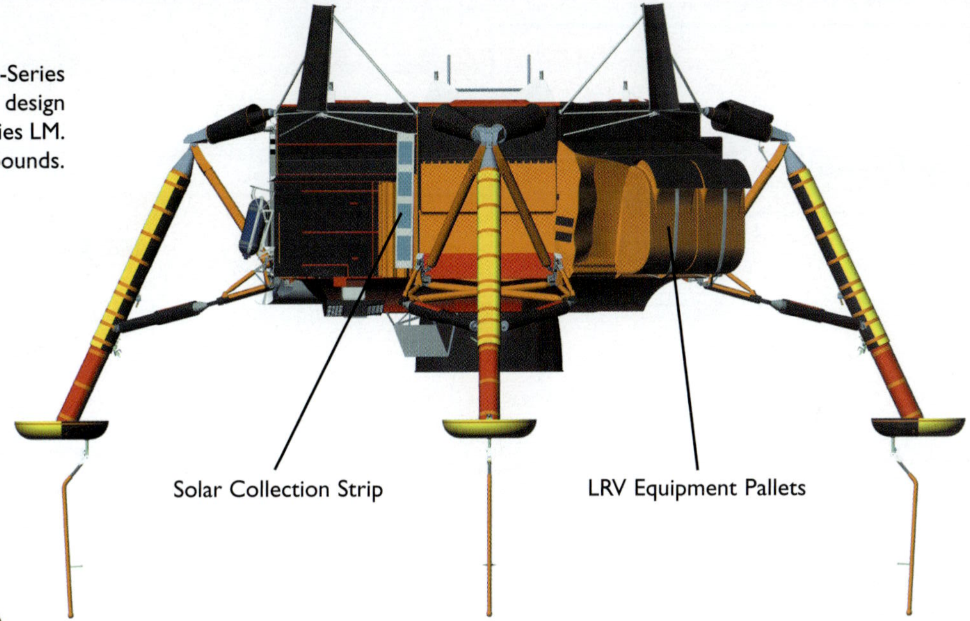

BACK

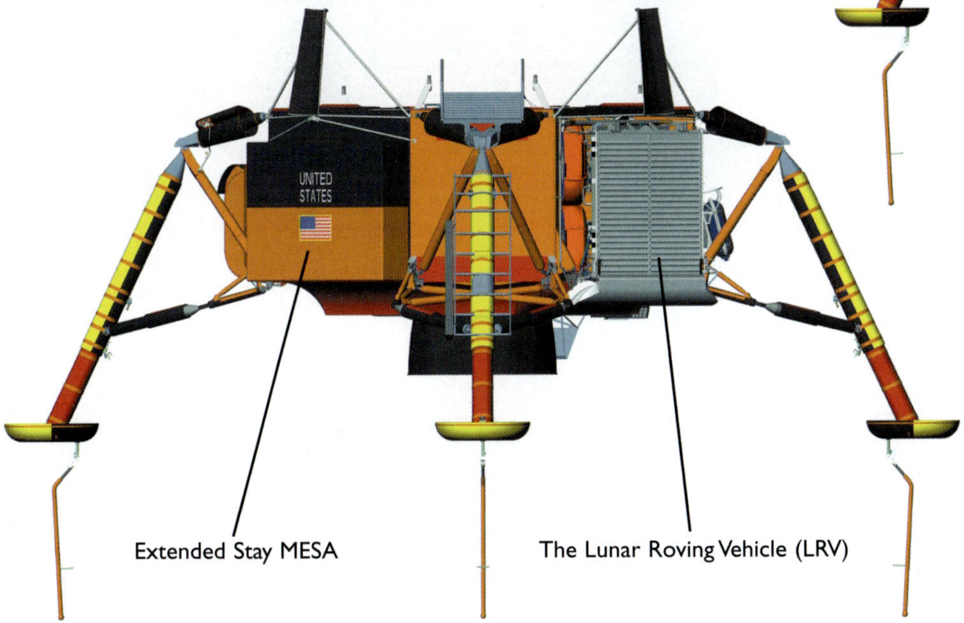

FRONT

MAJOR SUBSYSTEMS:
- (A) Provide Velocity Change for Lunar Deorbit and Lunar Landing (Throttleable)
- (B) Protect Ascent Stage from Landing Damage
- (C) Provide Ascent / Descent Stage Staging
- (D) Provide Ascent Stage Launch Platform
- (E) Stow Lunar Scientific Equipment
- (F) Provide Attachment for Landing Gear
- (G) Support LM in the SLA

Lunar Module Chronology

January 30, 1963
Selection of four companies as major subcontractors for Apollo Lunar Module.

WEIGHT FULLY FUELED WITH CREW: 36,244 POUNDS (*APOLLO 17*)

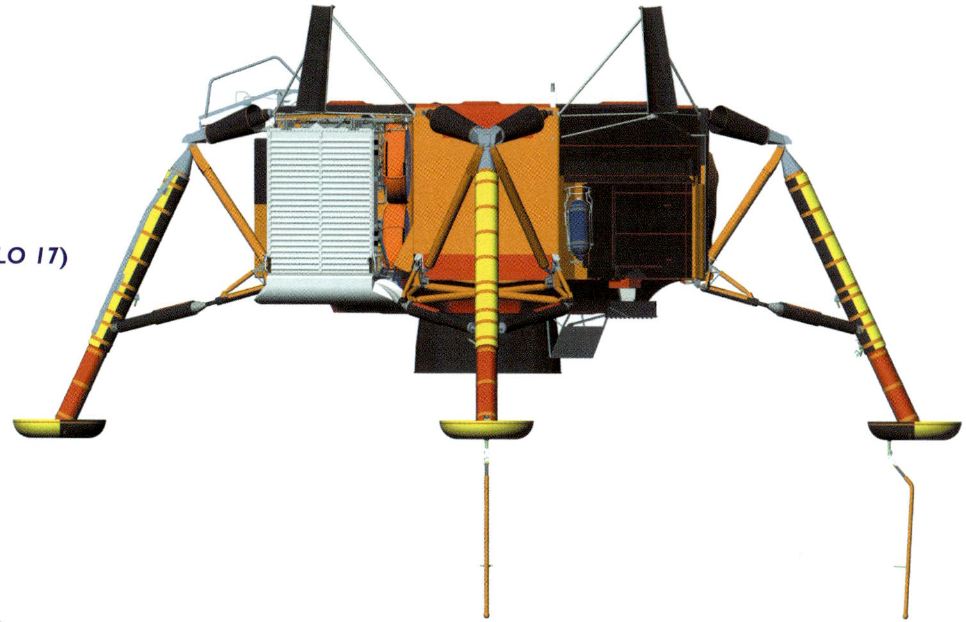

RIGHT

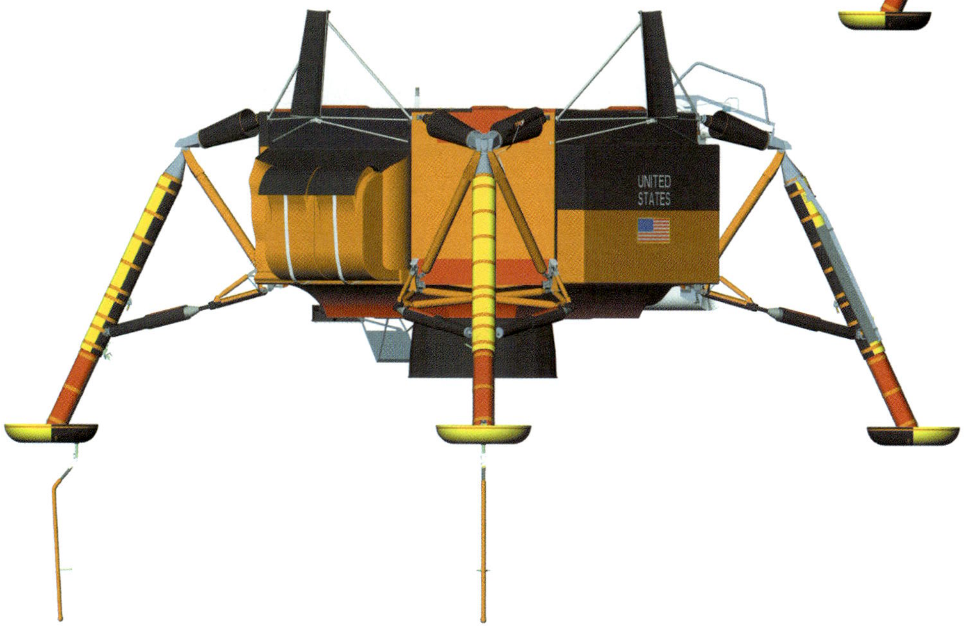

LEFT

Lunar Module Chronology

March 11, 1963
First Apollo Lunar Module "fire-in-the-hole" model test.

Lunar Module Chronology

March 17, 1963
Preliminary design freeze on specific parameters of the Lunar Module.

BOTTOM

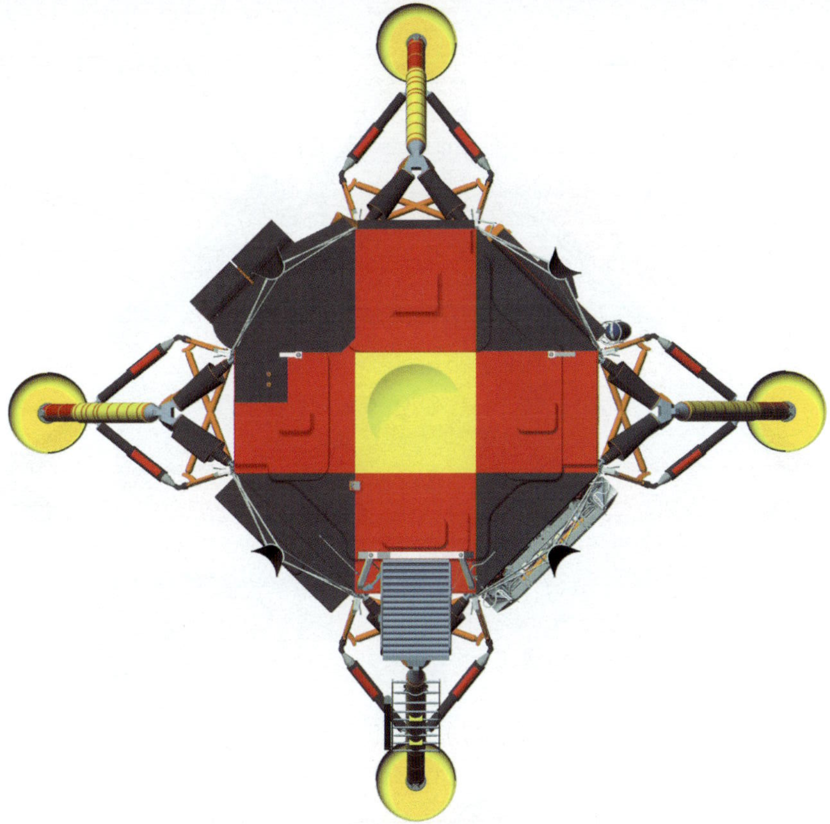

TOP

Lunar Module Chronology

May 1, 1963
Rocketdyne given the go-ahead for Apollo LM descent engine.

BASIC DIFFERENCES BETWEEN H-SERIES AND J-SERIES DESCENT STAGES

The major differences between the H-Series and J-Series Lunar Modules was the addition of the lunar roving vehicle in Quad 1, the relocation of the LM batteries to the direct rear of the LM structure to maintain the Lunar Module's center of gravity, the addition of two LRV equipment pallets in Quad 3, and the addition of an extra water tank, oxygen tank, waste containment container, as well as an extended stay MESA in Quad 4.

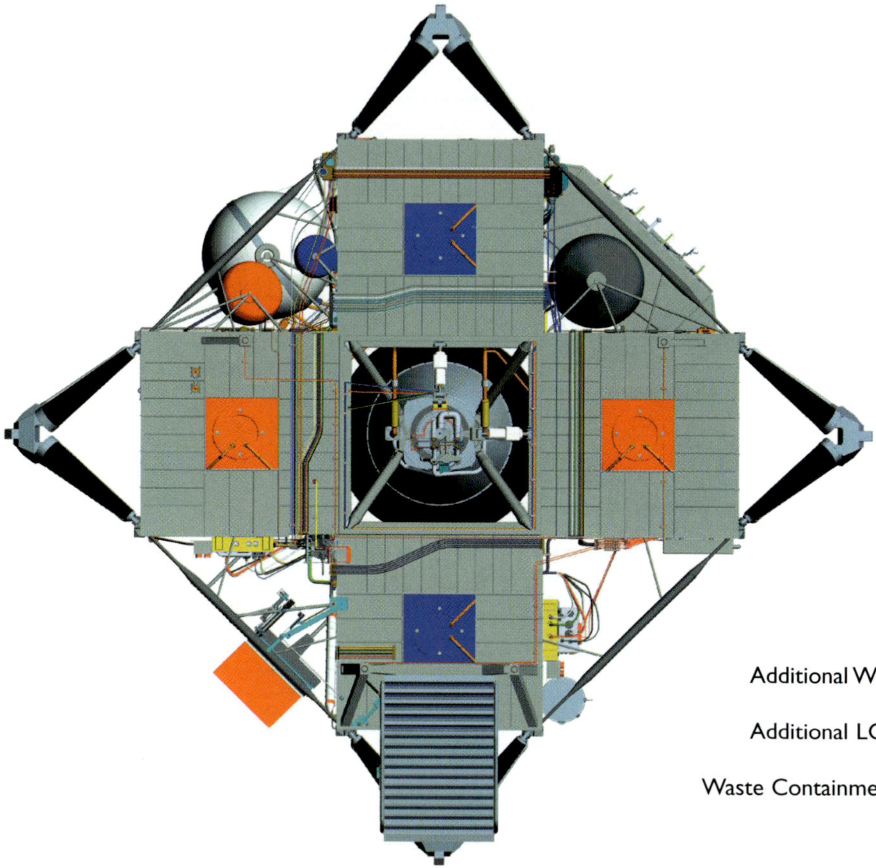

H-SERIES

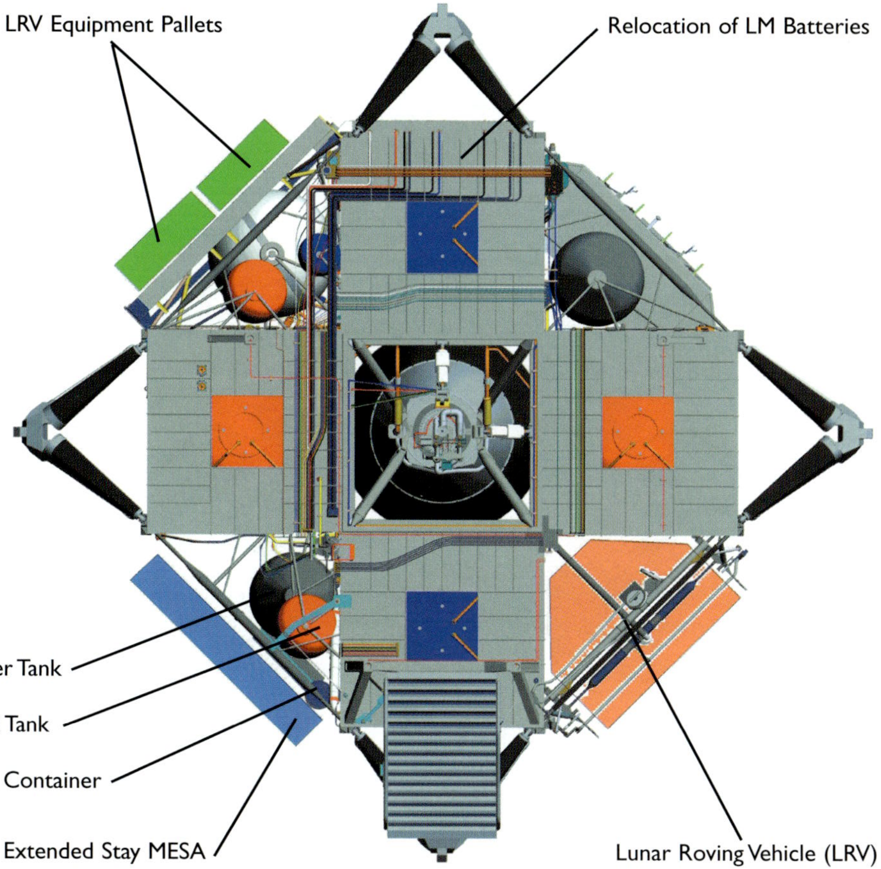

J-SERIES

QUAD I – ADDITION OF LRV

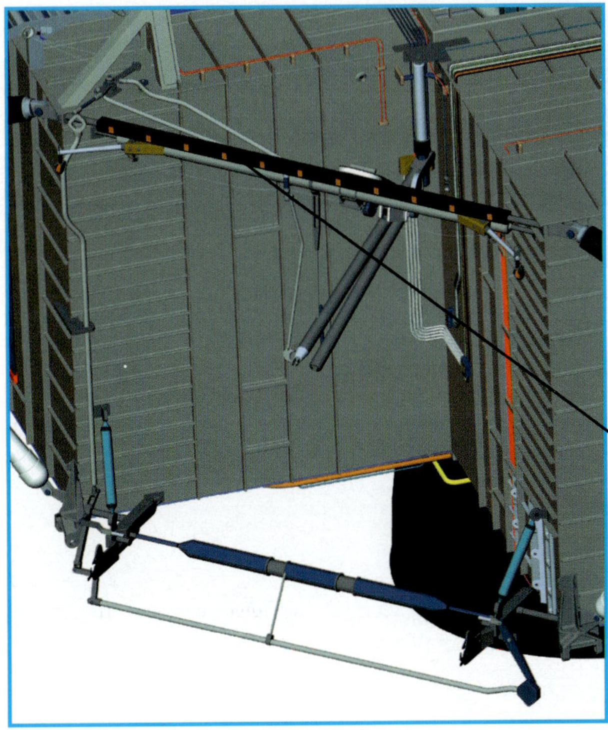

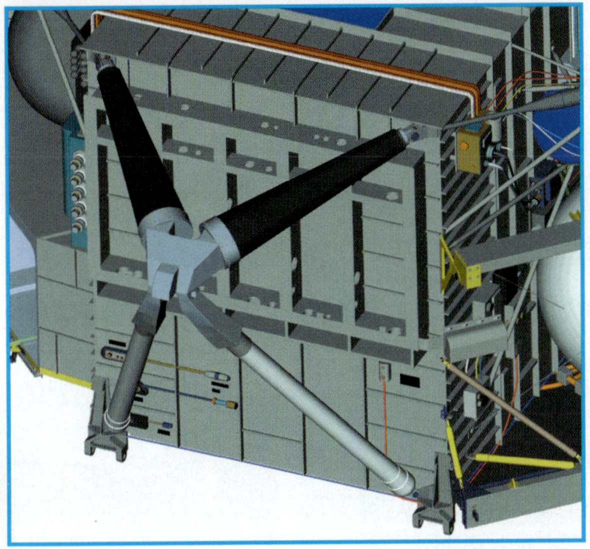

REAR OF LM STRUCTURE, RELOCATION OF BATTERIES TO MAINTAIN CENTER OF GRAVITY

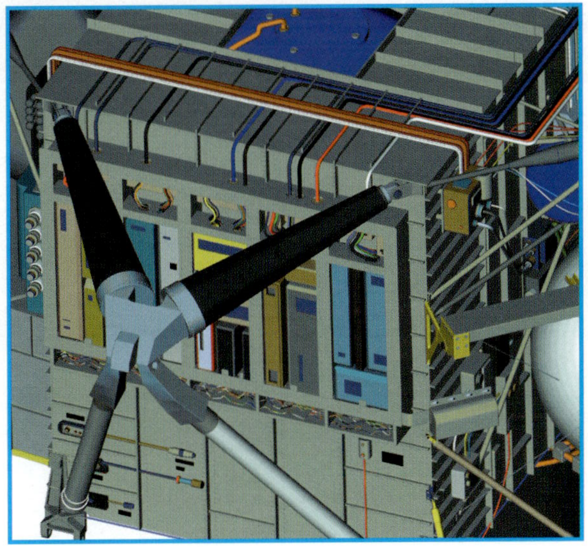

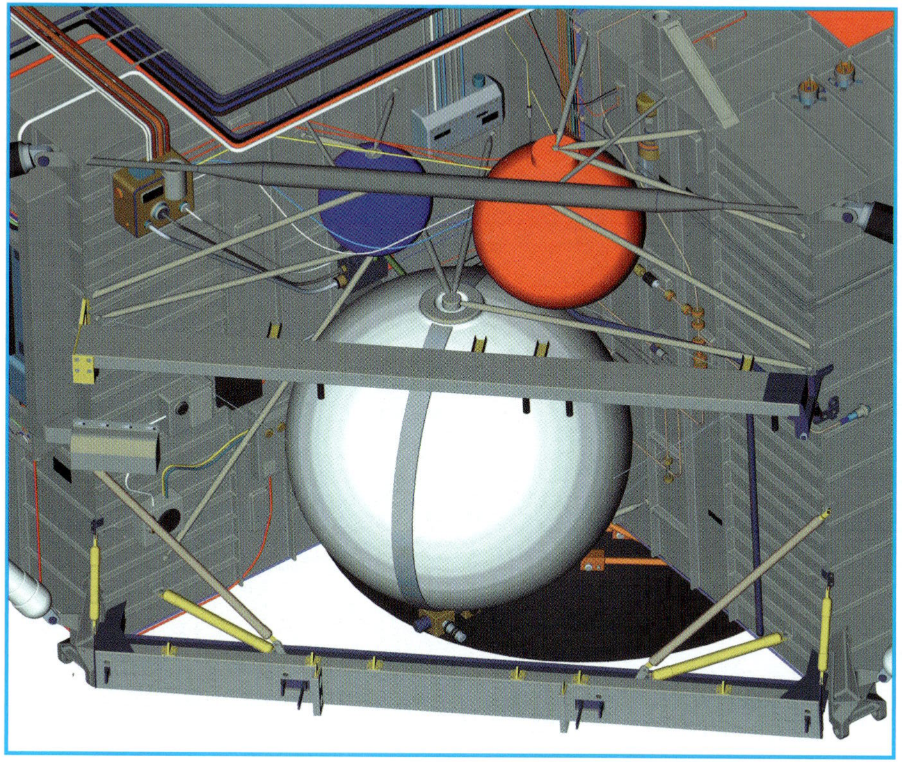

QUAD 3 – ADDITION OF PALLET SUPPORT STRUCTURE

QUAD 4 – ADDITION OF WATER, OXYGEN TANKS, AND WASTE CONTAINMENT CONTAINER

J-SERIES EXTENDED STAY MESA

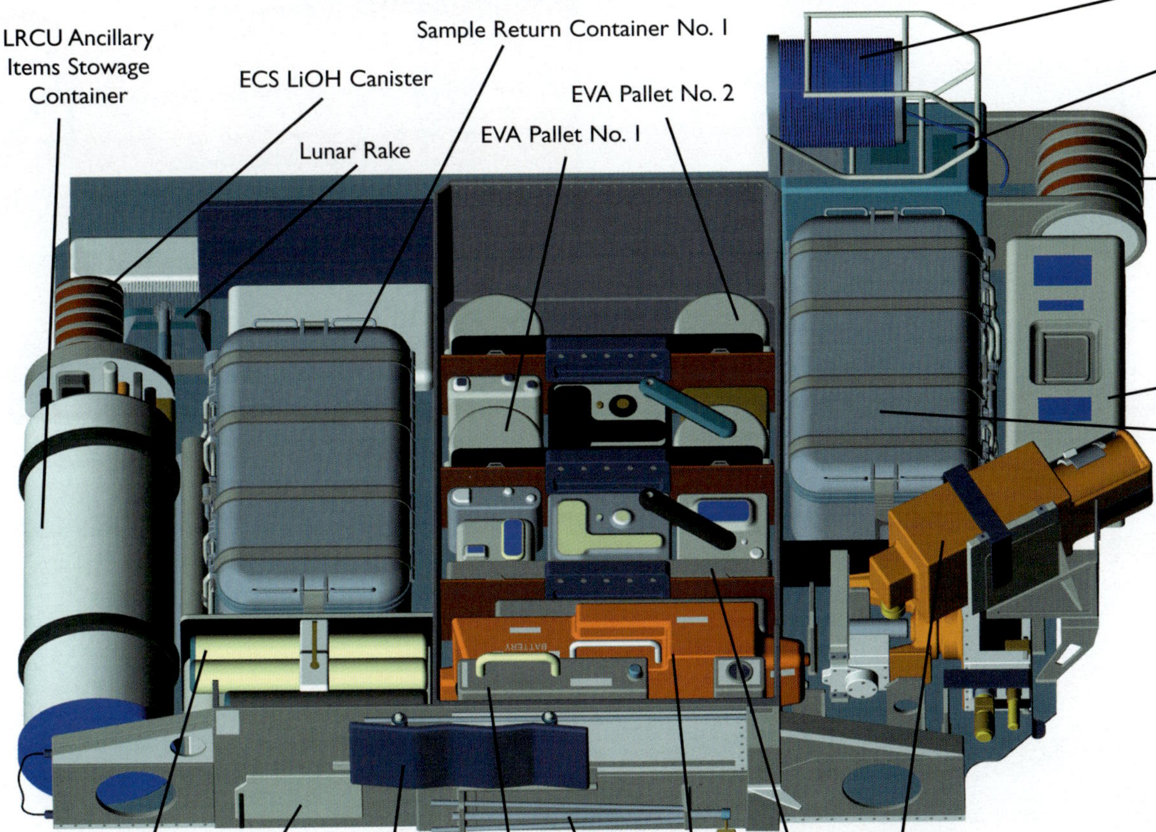

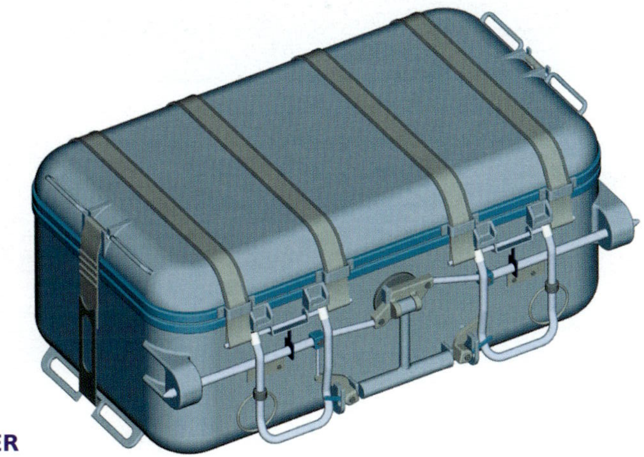

LUNAR SAMPLE RETURN CONTAINER

– Scott P. Sullivan –

J-SERIES QUAD 3 LUNAR ROVER PALLET DEPLOYMENT

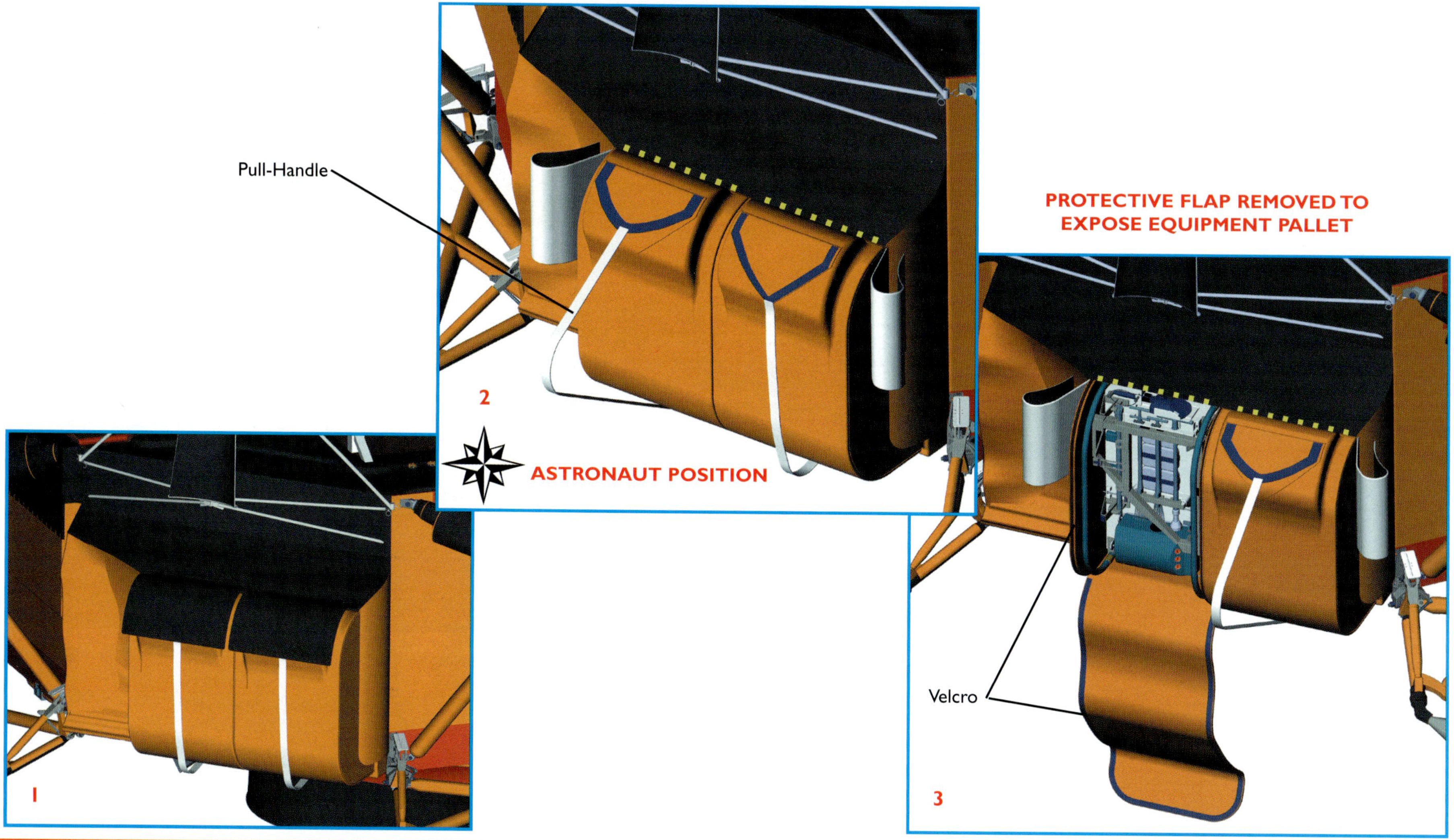

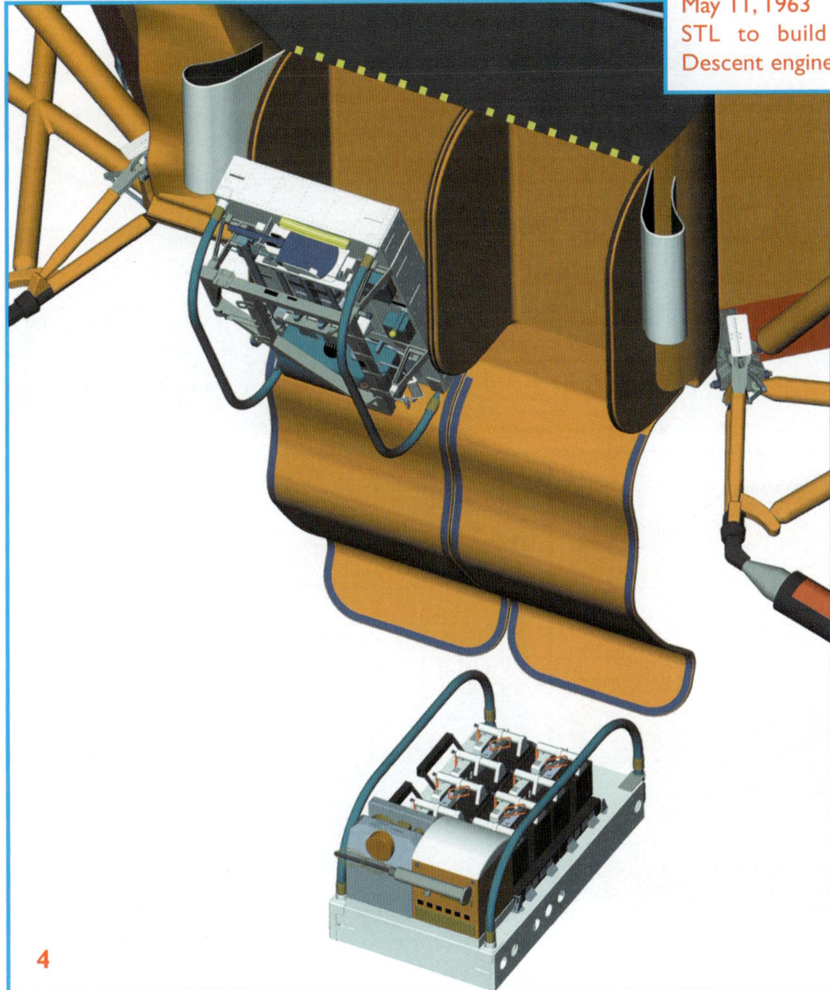
4

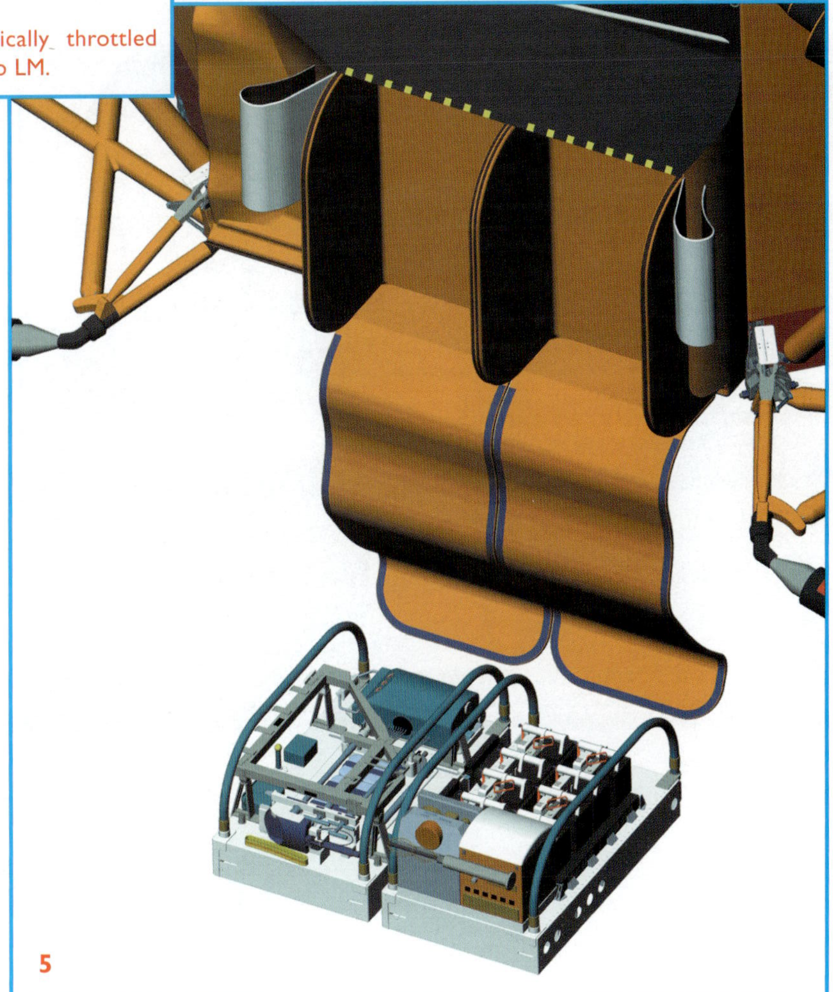
5

Lunar Module Chronology

May 11, 1963
STL to build the mechanically throttled Descent engine for the Apollo LM.

J-SERIES QUAD 4 EXTENDED STAY MESA DEPLOYMENT

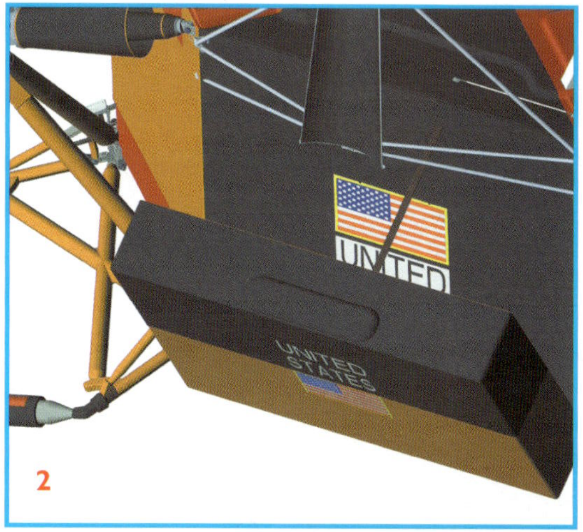

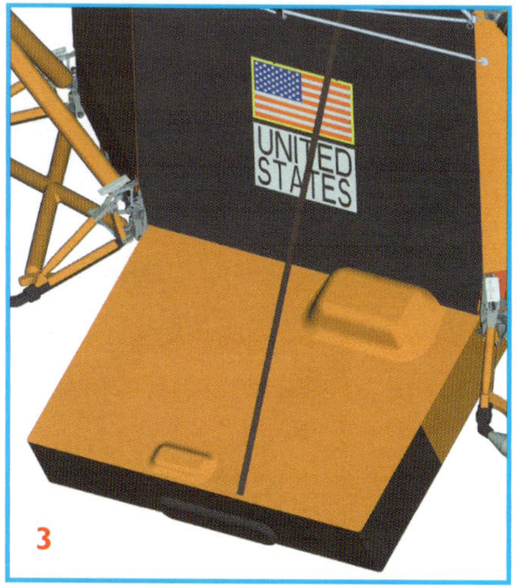

REMOVAL OF THERMAL COVERINGS BY ASTRONAUTS

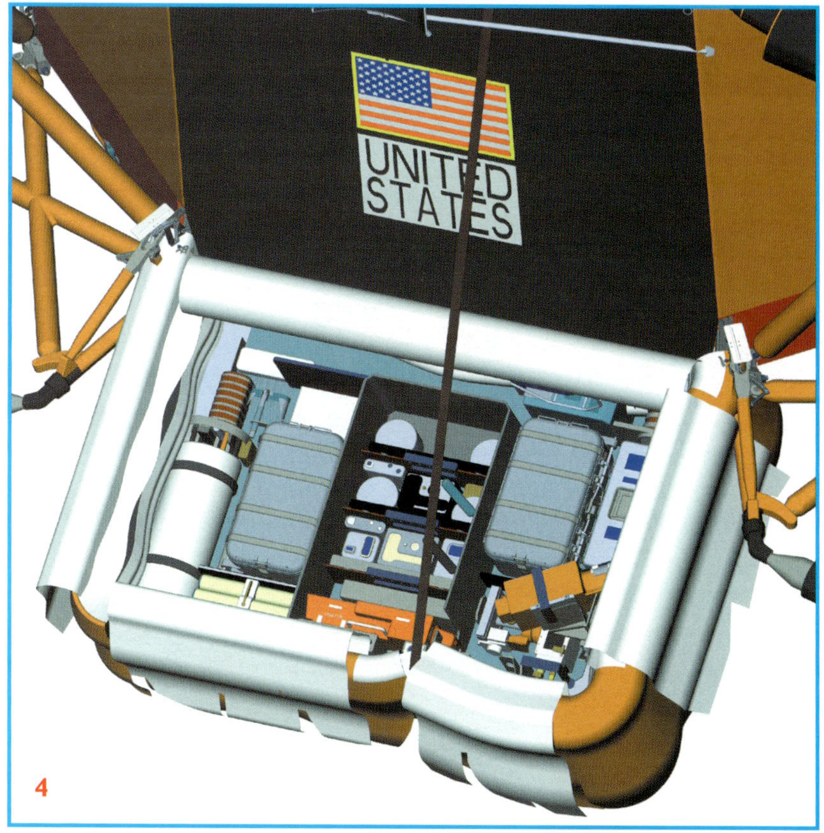

THE DESCENT STAGE FROM THE GROUND UP

The descent stage of the Apollo Lunar Module consisted of machined parts and chemically milled panels / stiffeners that were mechanically fastened. Compartments formed by this structural arrangement housed the descent engine and tanks for propellant, helium, oxygen and water. Fabrication of the descent stage began with the joining of the machined "picture frames" and the chem-milled panel / stiffener assemblies to form the engine compartment. After the outrigger bulkhead assemblies were attached to the engine compartment with machined cap strips, the eight remaining panel / stiffener assemblies were added.

1

PANELS / STIFFENERS

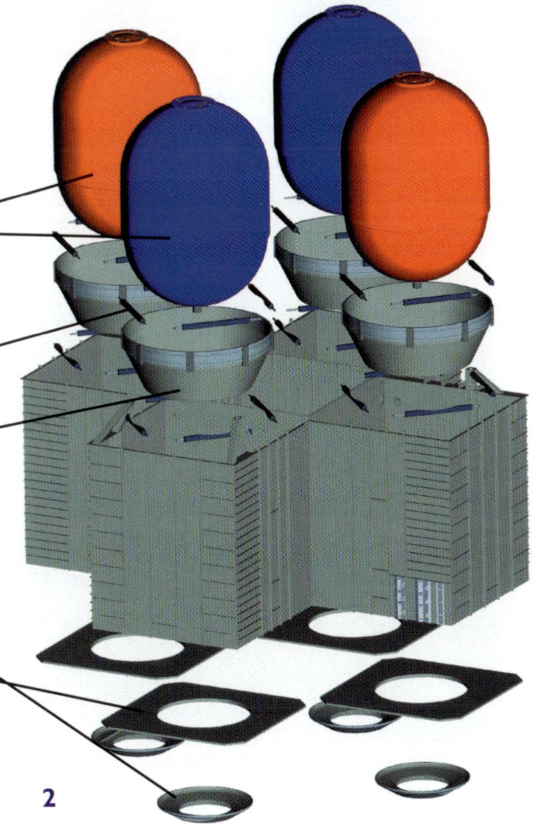

2

FIRST STRUCTURAL PANELS LOADED INTO THE STRUCTURAL ASSEMBLY FIXTURE

CHECK-OUT OF TANK SKIRT

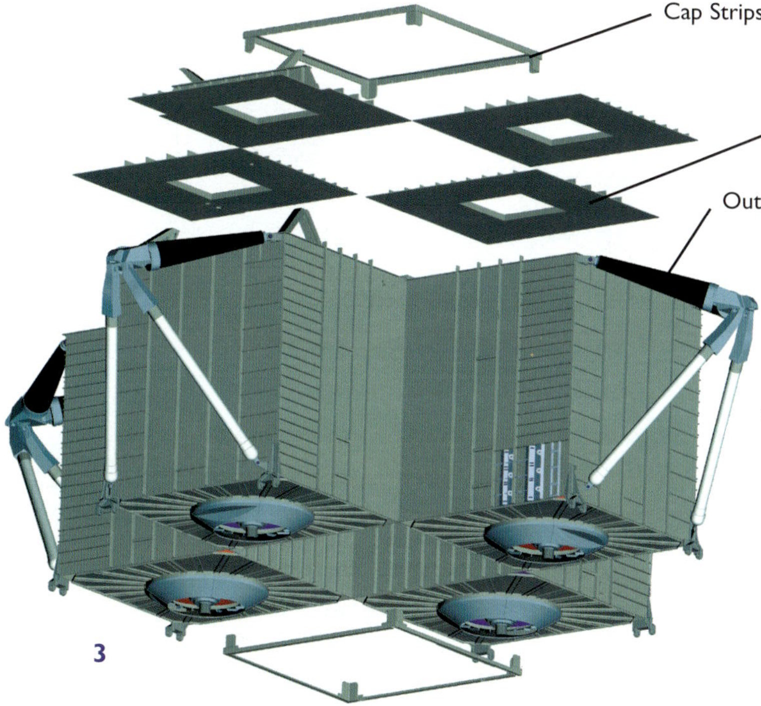

3

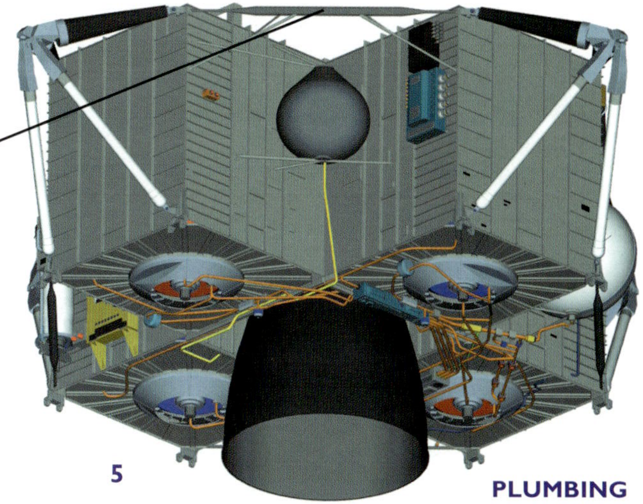

5

PLUMBING

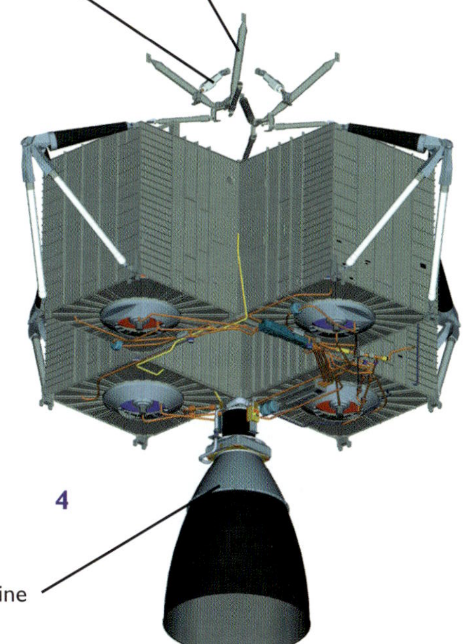

4

DESCENT STAGE IN STRUCTURAL SUPPORT FIXTURE PRIOR TO INSTALLATION OF DETAIL PARTS, COVER ASSEMBLIES, AND STRUTS

FUEL TANKS AND BRACKETRY INSTALLED

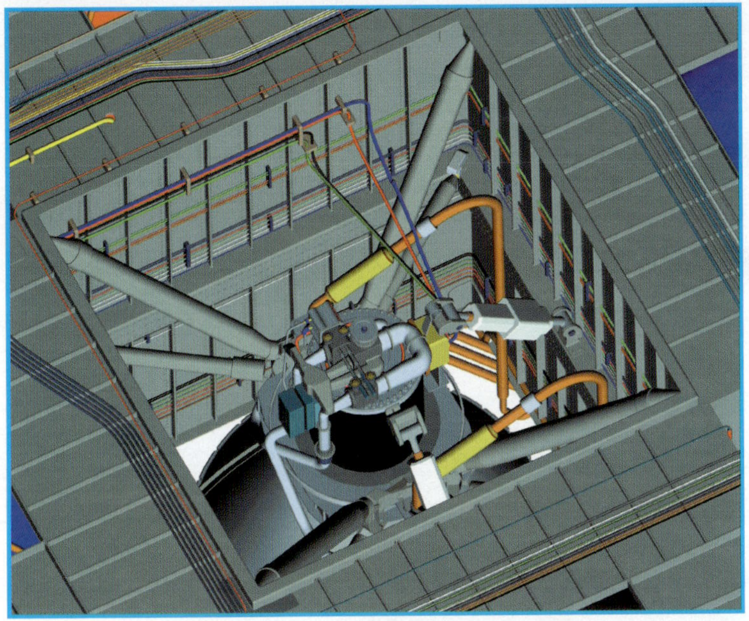

6

DESCENT ENGINE ASSEMBLED

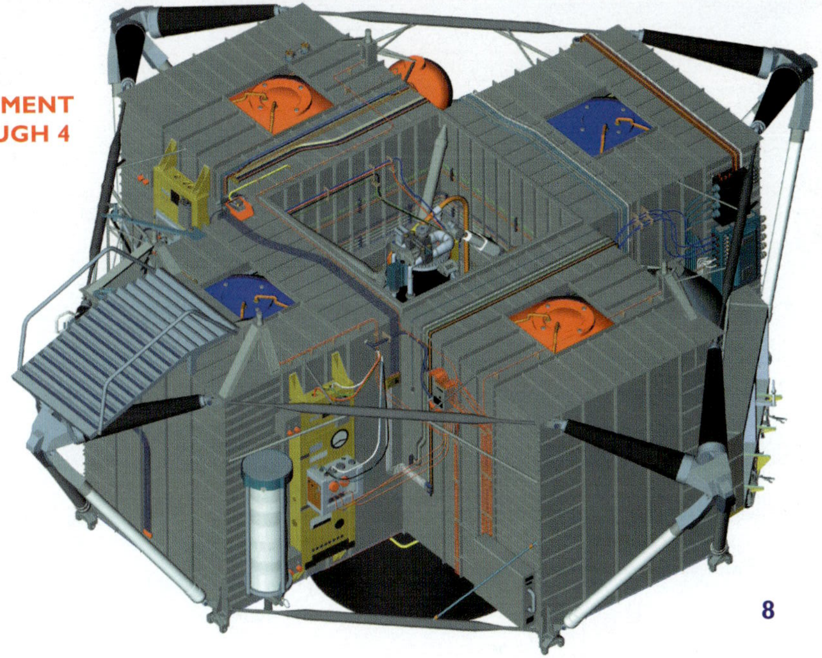

8

VARIOUS EQUIPMENT
QUADS 1 THROUGH 4

ELECTRICAL COMPONENTS
(CIRCUIT INTERRUPTERS)

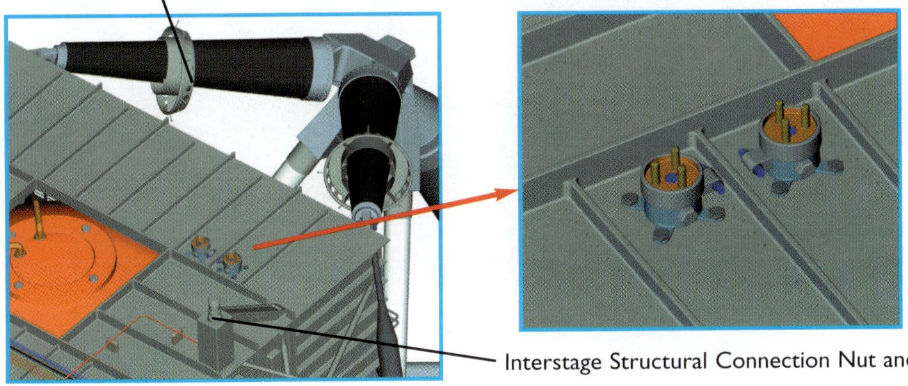

7

ELECTRICAL WIRING

LANDING GEAR ASSEMBLIES

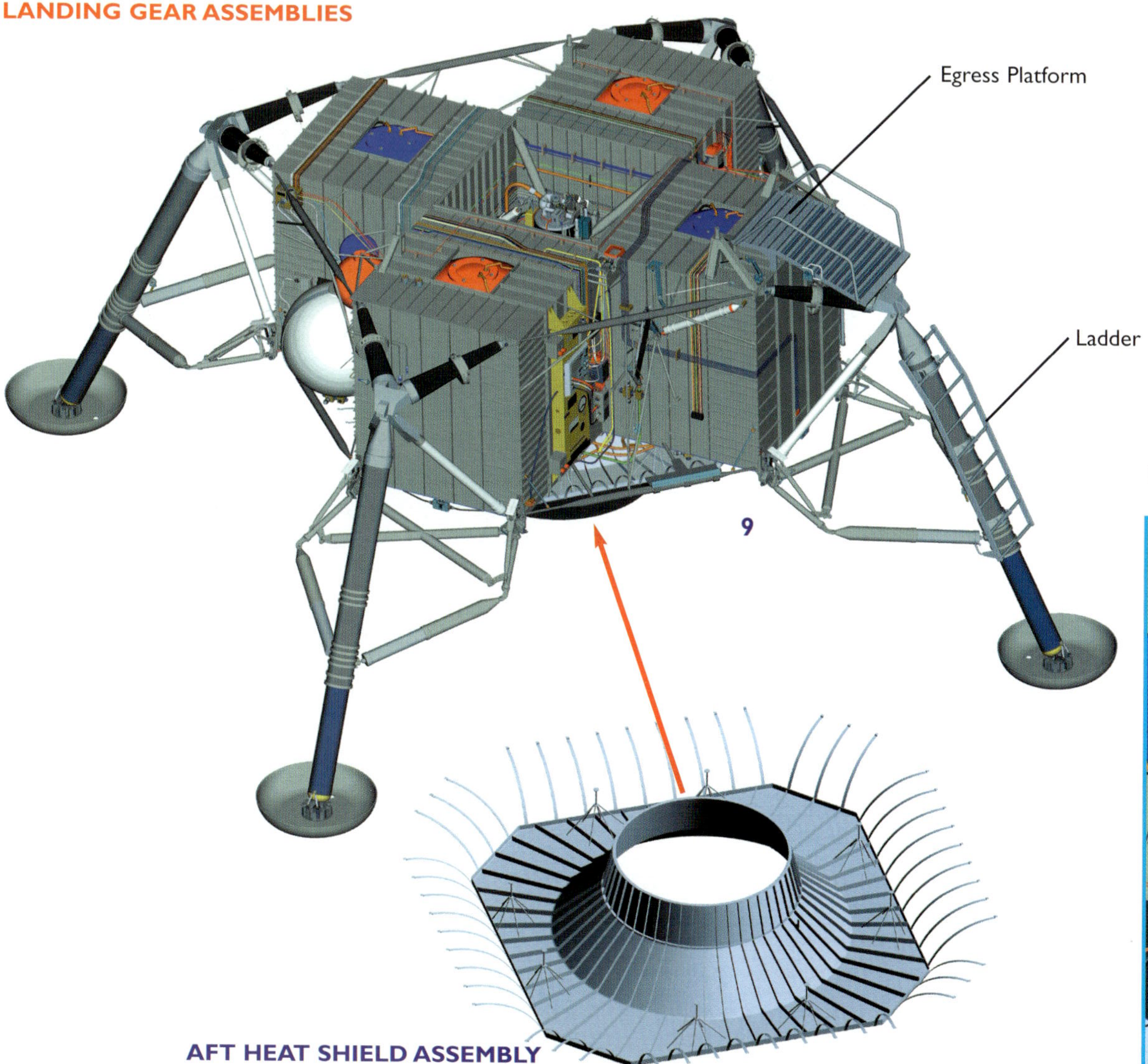

AFT HEAT SHIELD ASSEMBLY

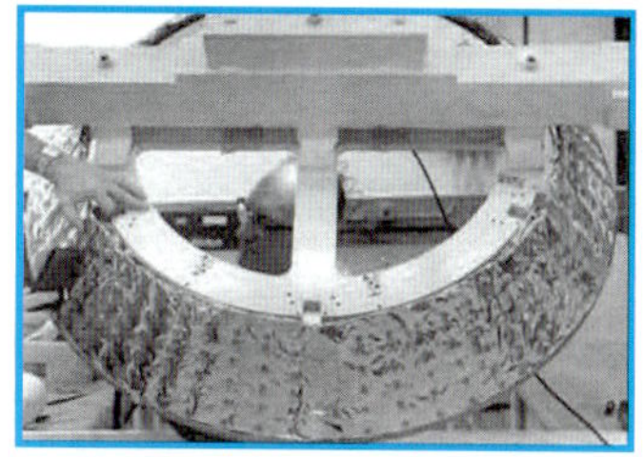

DESCENT ENGINE HEAT SHIELD BEING FABRICATED

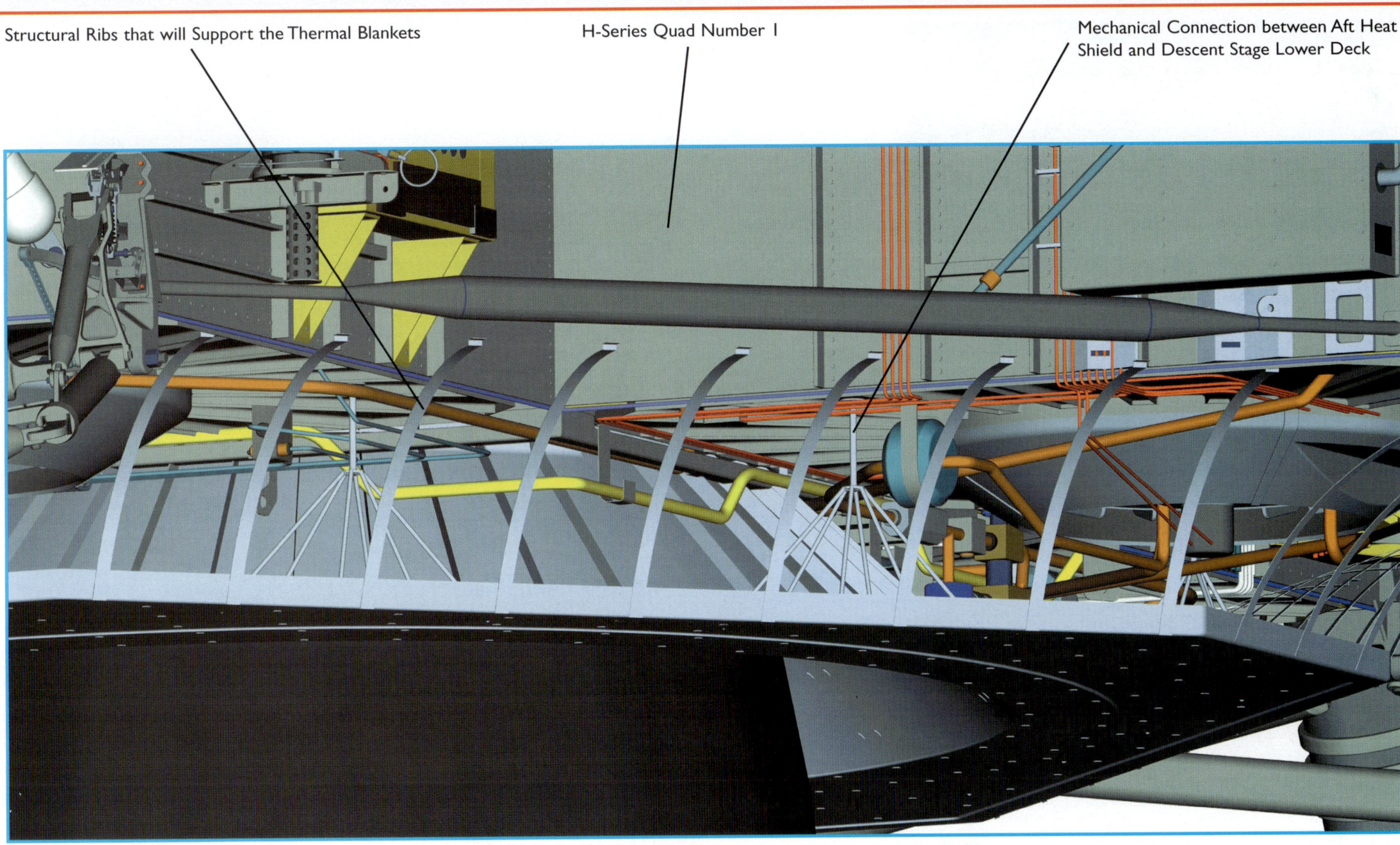

DETAIL VIEW OF AFT SECTION SHOWING THE AFT HEAT SHIELD, PLUMBING, WIRING AND ENGINE NOZZLE EXTENSION

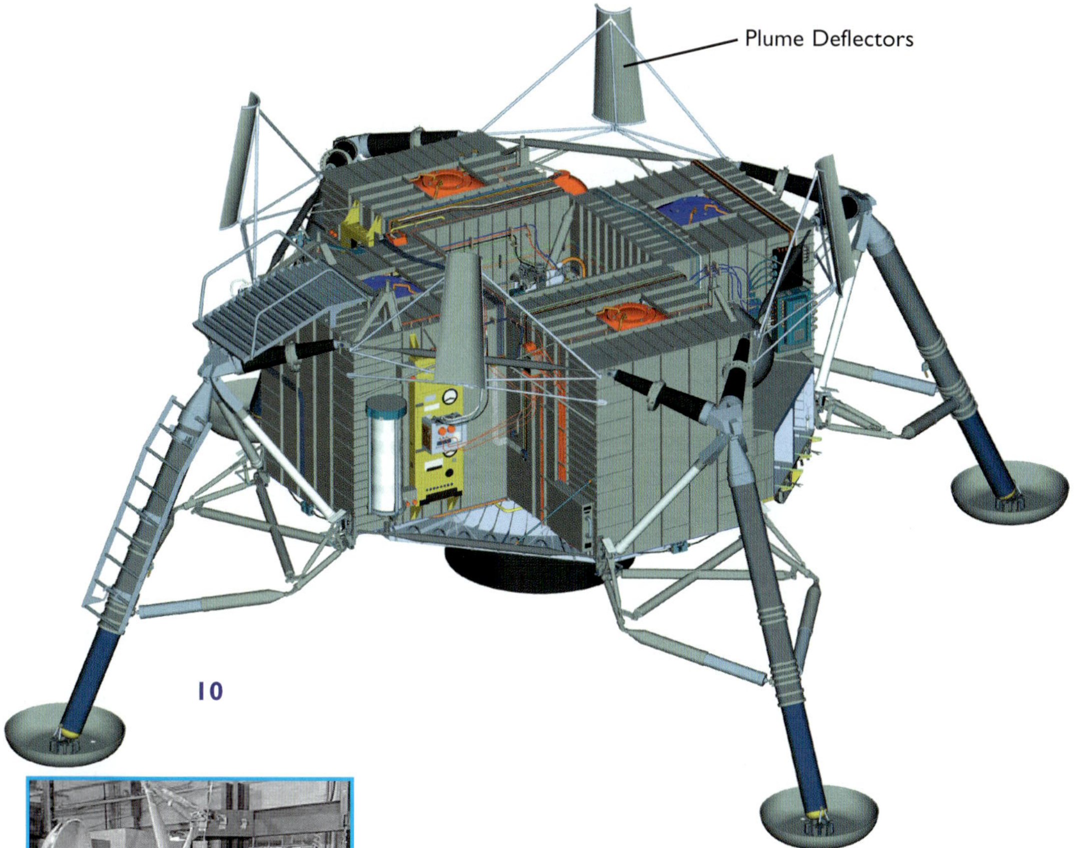

10

TESTING OF LANDING GEAR DEPLOYMENT SYSTEM

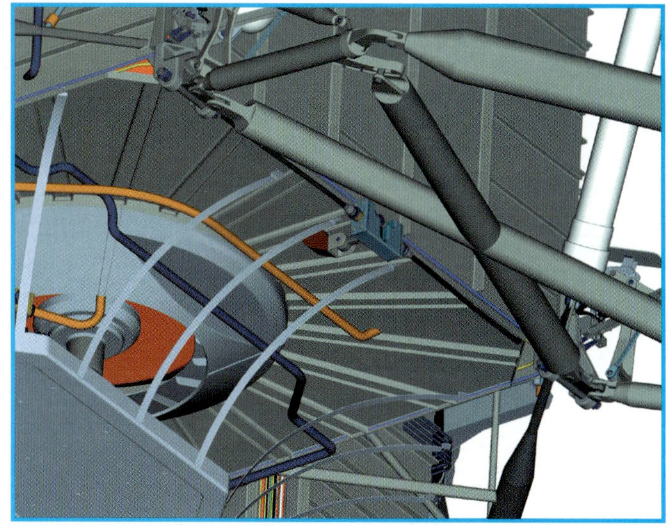

11

LANDING GEAR UPLOCK AND WIRING

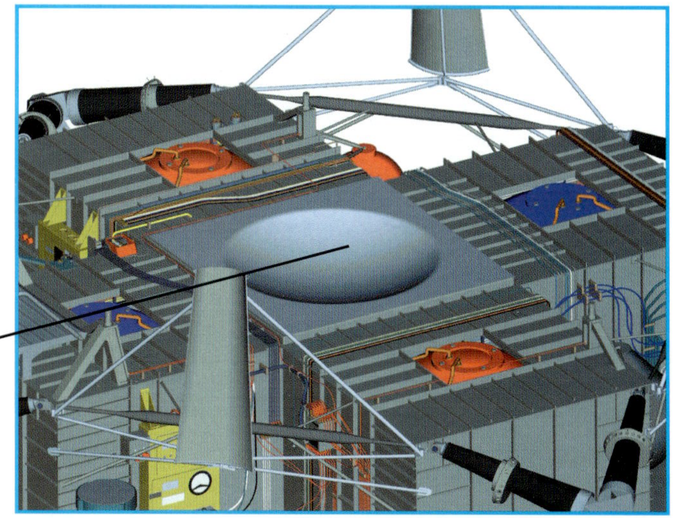

12

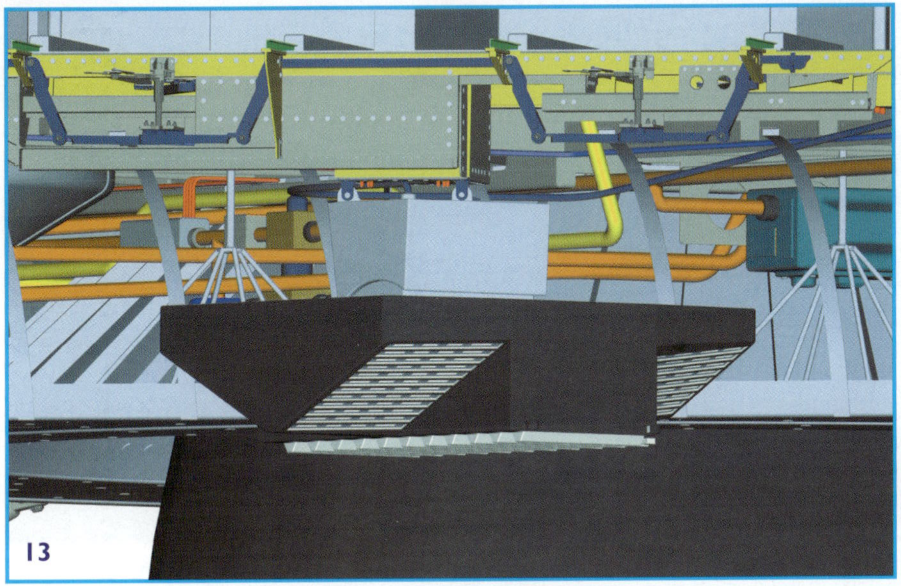

LANDING RADAR ASSEMBLY

Lunar Module Chronology

May 23, 1961
Major redesign of the Apollo guidance computer.

Lunar Module Chronology

May 31, 1961
First estimates of reliability for the Apollo Lunar Module.

LANDING RADAR PROTECTIVE SHIELD

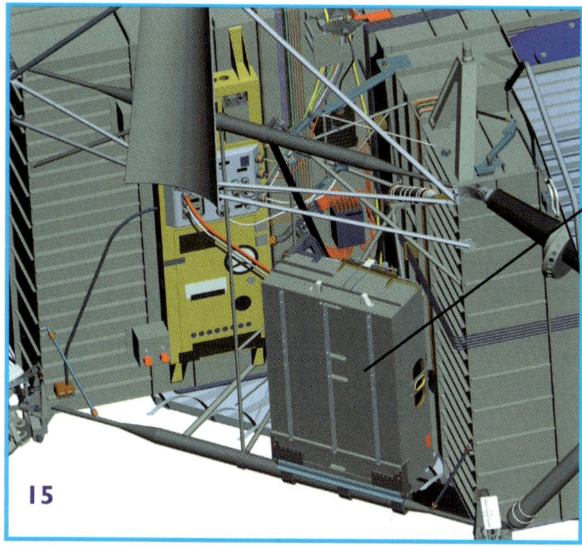

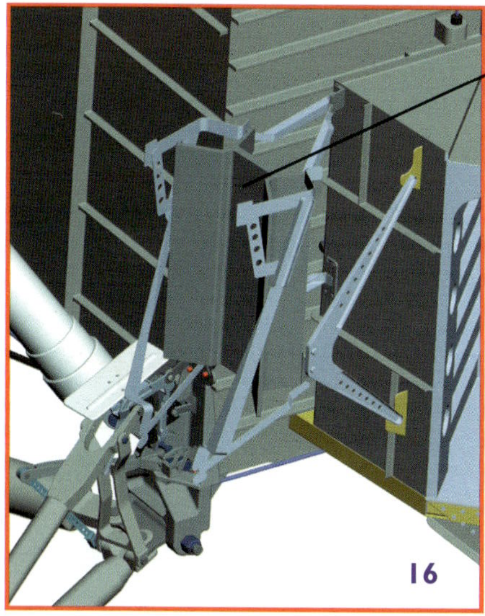

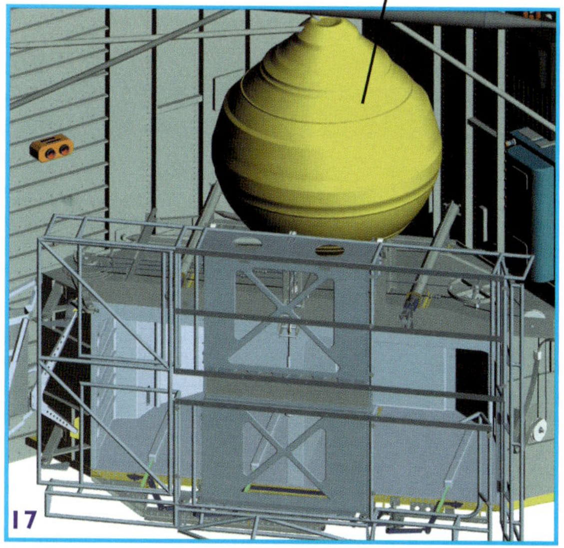

MESA

RTG Cask Support Structure

ALSEP Assembly and Water Tank Thermal Insulation

Lunar Module Chronology

July 3, 1963
STL gets go-ahead to develop a parallel descent engine for the LM.

AFT THERMAL INSULATION

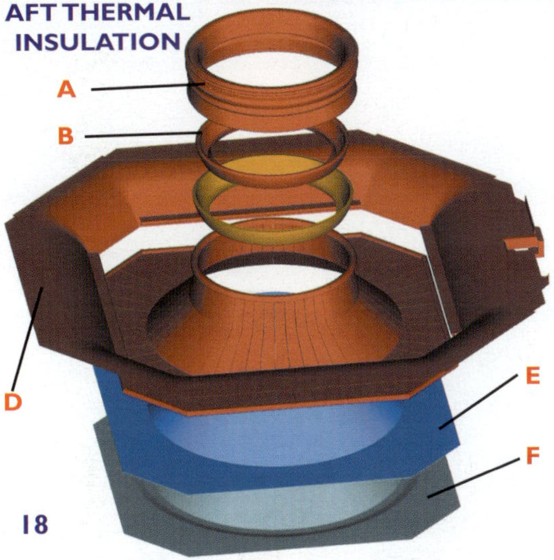

18

UPPER THERMAL INSULATION

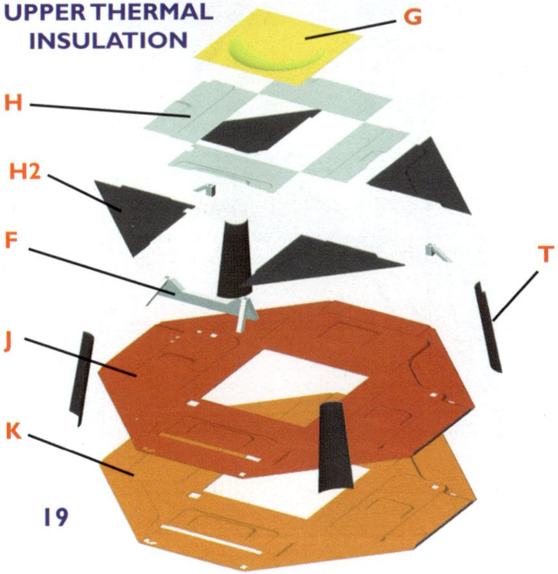

19

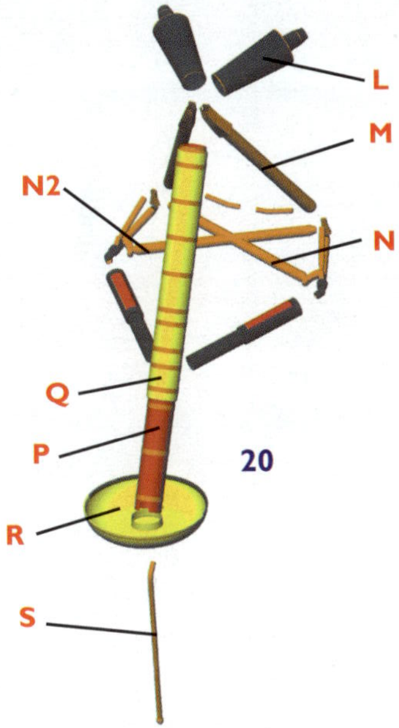

20

LANDING GEAR THERMAL INSULATION

THE TYPICAL THERMAL BLANKET WAS BENEATH THE EXTERNAL THERMAL LAYER SHOWN ON THE LANDING GEAR

(A) Bellows (Aluminized Kapton)
(B) Engine Heat Shield (Min-K)
(C) Columbium
(D) Thermal Blanket
(E) Nickel-Fiberfax Insulation
(F) Teflon Coated Beata Fabric
(G) Ascent Engine Blast Shield (Aluminized Kapton)
(H) Nickel Foil
(H2) Black Pyro Mark Paint
(J) Upper Deck Thermal Layer (5 mils Aluminized Kapton)
(K) Thermal Blanket
(L) Outrigger Thermal Covering
(Black Pyro Mark Paint over Aluminized Kapton)
(M) Outrigger Thermal Covering
(5 mils Aluminized Kapton)
(N) Downlock Assembly Thermal Coverings
(2 mils Aluminized Kapton)
(N2) Cross-Over Truss Assembly Thermal Covering
(2 mils Aluminized Kapton)
(P) Secondary Strut Thermal Covering
(Black Pyro Mark Paint over Aluminized Kapton)
(Q) Primary Strut Thermal Covering
(½ mil Aluminized Kapton)
(Q2) Primary Strut Thermal Covering
(5 mils Aluminized Kapton)
(R) Landing Pad Thermal Covering
(½ mil Aluminized Kapton)
(S) Probe Thermal Covering
(2 mils Aluminized Kapton)
(T) Plume Deflector Thermal Build-Up
From Outside to Inside:
1. Inconel (1.25 mils)
2. Inconel Mesh (varying thicknesses)
3. H-Film (5 mils)

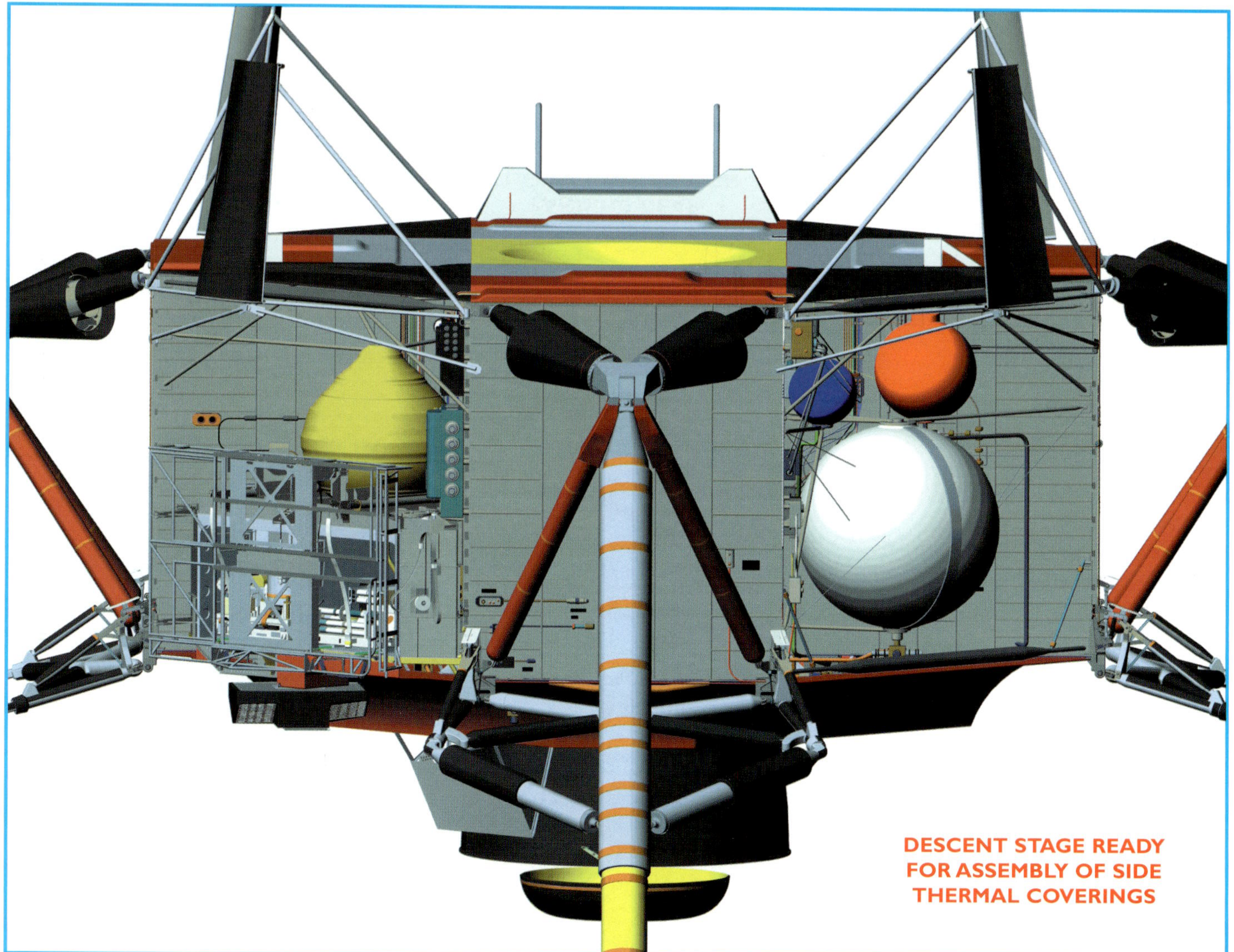

DESCENT STAGE READY
FOR ASSEMBLY OF SIDE
THERMAL COVERINGS

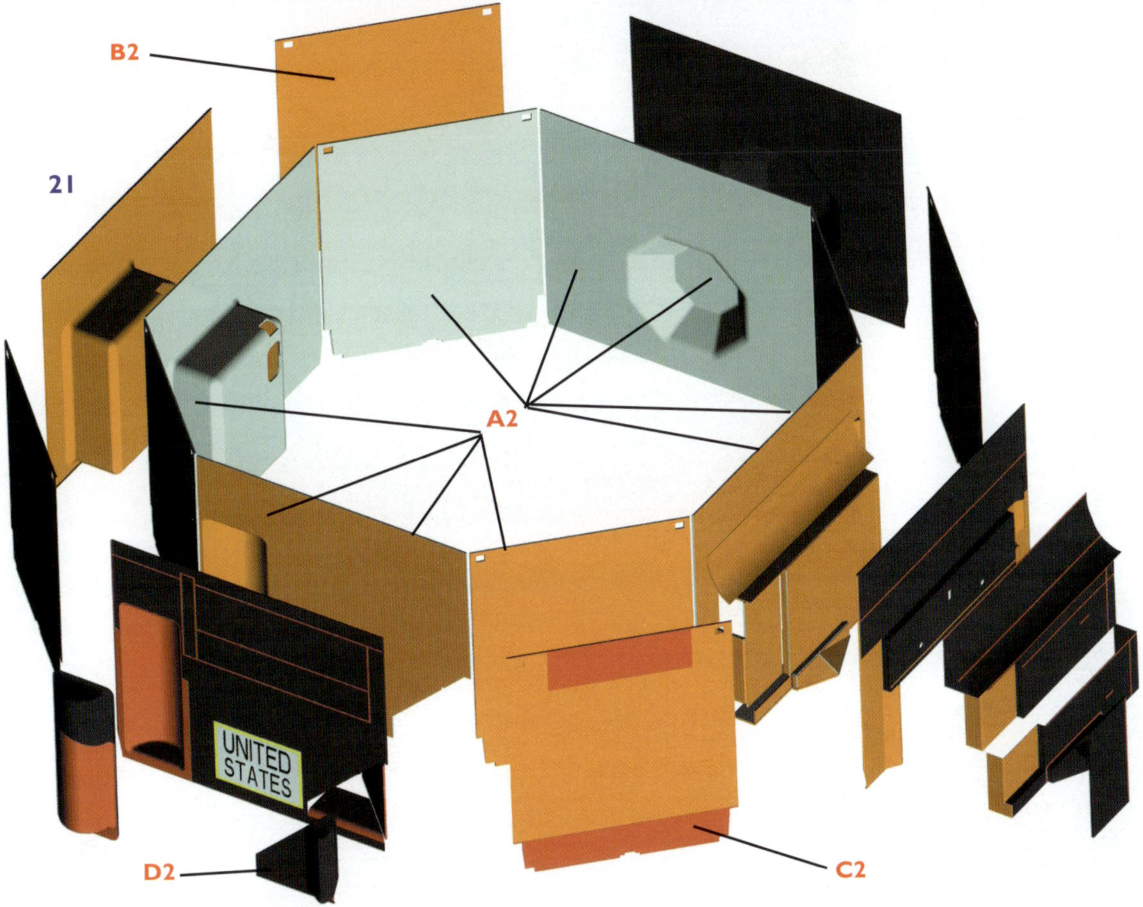

SIDE THERMAL INSULATION

(A2) Thermal Blanket (Typ)
(B2) Exterior Thermal Layer (2 mils of Aluminized Kapton) All Gold Colored External Panels
(C2) Exterior Thermal Layer (5 mils of Aluminized Kapton) All Brownish Surfaces
(D2) Black Pyro Mark Paint

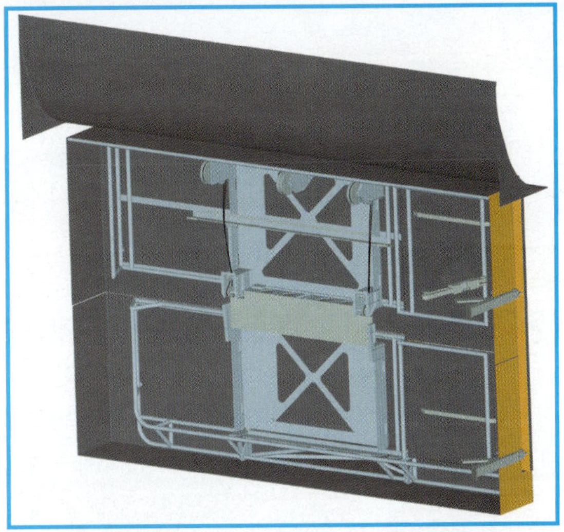

QUAD 2 ALSEP BAY AND DOOR DETAIL VIEWS

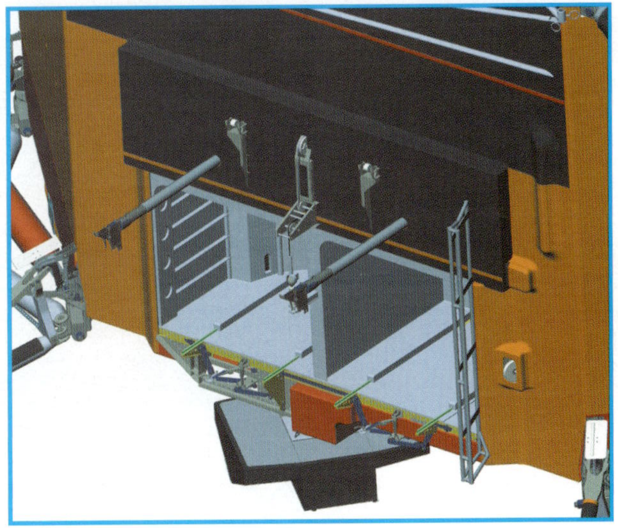

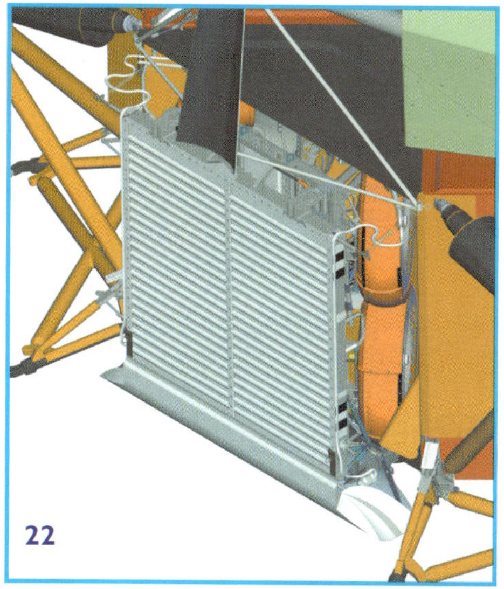
22

QUAD 1 – LRV

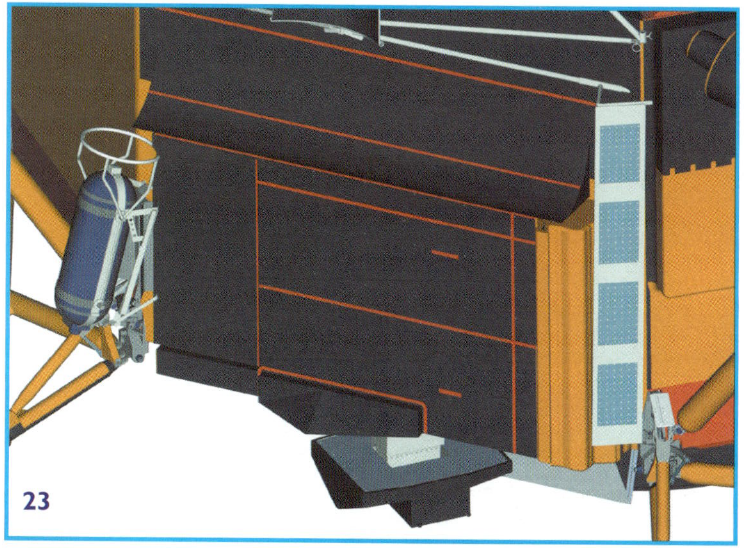
23

QUAD 2 – ALSEP, RTG CASK, AND SOLAR COLLECTION STRIP

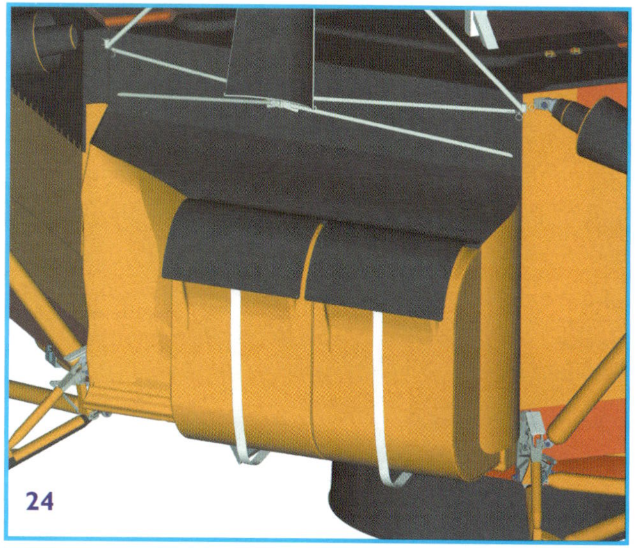
24

QUAD 3 – LRV EQUIPMENT PALLETS

25

QUAD 4 – EXTENDED STAY MESA

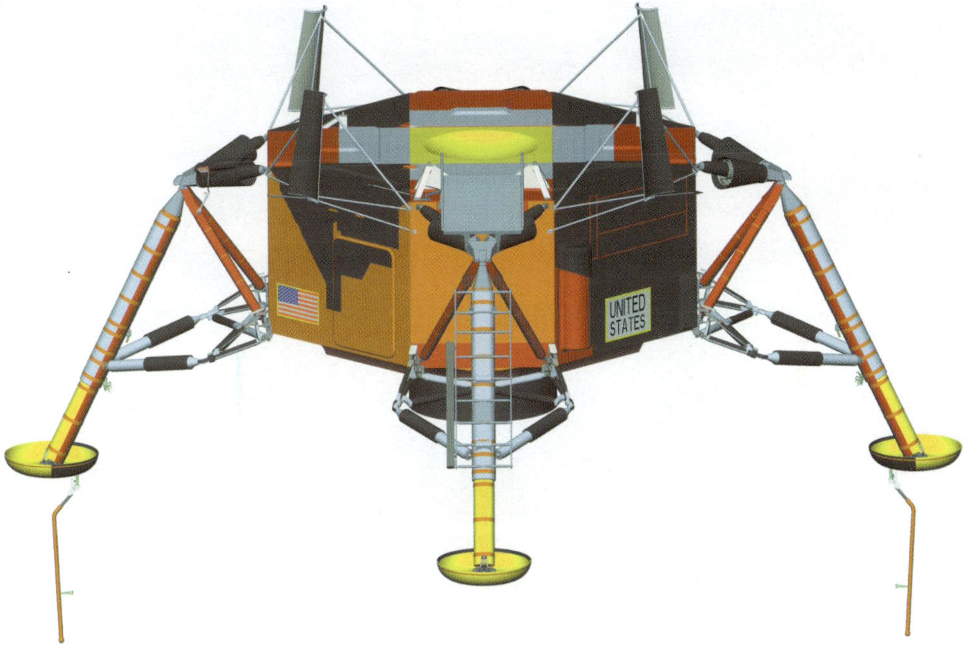

THE H-SERIES DESCENT STAGE

"Well, some people seem to have the idea that space vehicles are made by pushing buttons and using automated manufacturing techniques, and all that. Actually, it is quite different – quite the opposite, in fact. There is no real production involved with spacecraft. They are hand made. Its almost like the old cottage industry, where an individual craftsman painstakingly molds, or files, or does something with great skill. It is much more like that."

Tom Kelly from "The Grumman Story" by R. Thruselsen

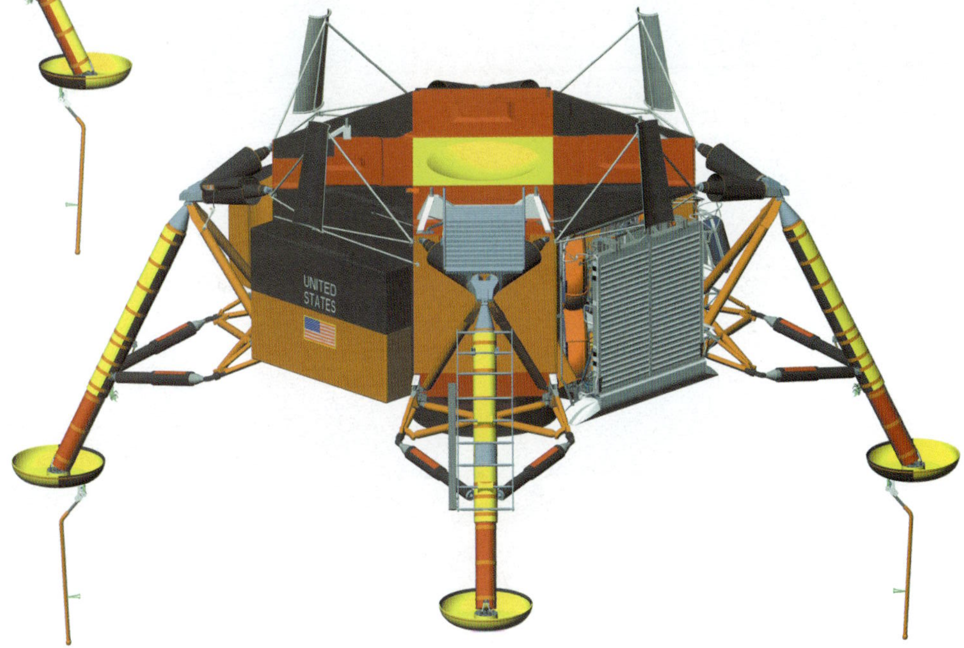

THE J-SERIES DESCENT STAGE

"The challenge was in the application of known principles and techniques. The other challenge was in the quality control. Even with all the money we were paying for this stuff, it was still very difficult to get 100 percent quality control. There was a real premium on hand-craftsmanship on the LM – as there was, indeed, on the entire Apollo spacecraft."

Tom Kelly from "The Grumman Story" by R. Thruselsen

THE LUNAR MODULE ASCENT STAGE OVERVIEW

The ascent stage (which sat atop the descent stage) of the Apollo Lunar Module was the manned portion of the spacecraft, and the only portion of the spacecraft that was pressurized and contained an atmosphere. It inluded a crew compartment, a hypergolic ascent engine, an aft equipment bay and tank section, and 16 reaction control engines. The astronauts used the crew compartment as an operations center during their lunar stay. Lunar descent, lunar landing, lunar launch, and spacecraft rendezvous and docking with the Command and Service Modules were also conducted from this compartment.

H-SERIES FRONT

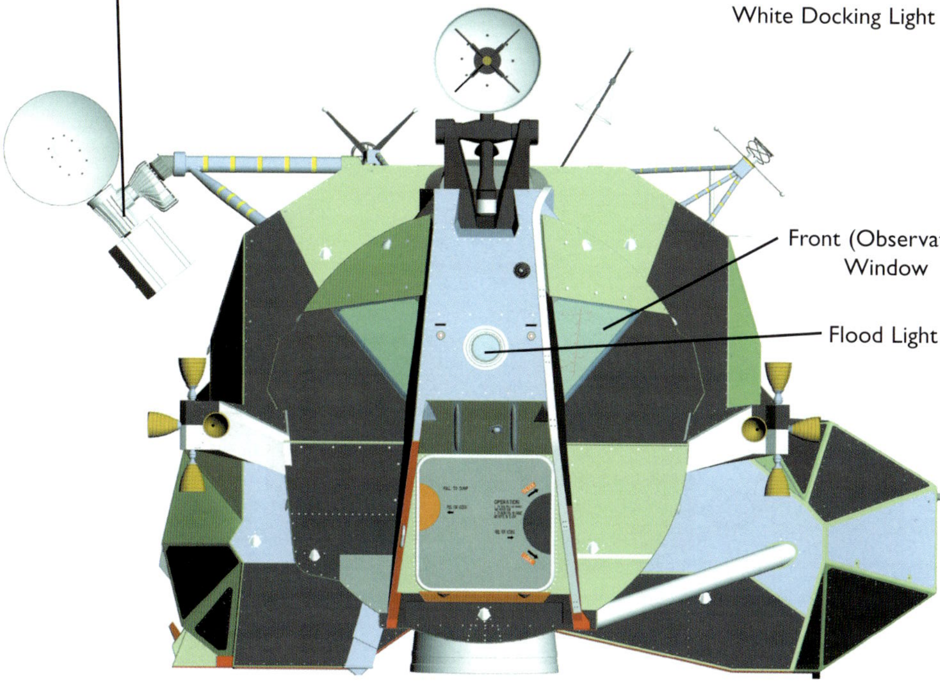

J-SERIES FRONT

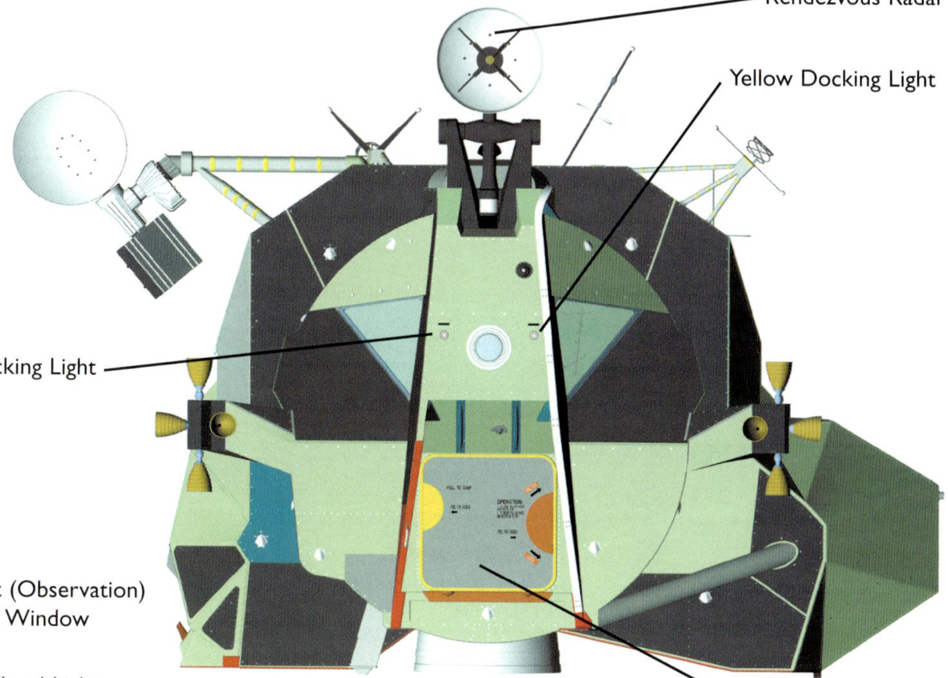

MAJOR SUBSYSTEMS:

(A) Guidance and Navigation
(B) Crew Provisions / Displays
(C) Environmental Control
(D) Electro-Explosive Devices
(E) Instrumentation
(F) Electrical Power
(G) Propulsion
(H) Reaction Control
(J) Communications

H-SERIES BACK

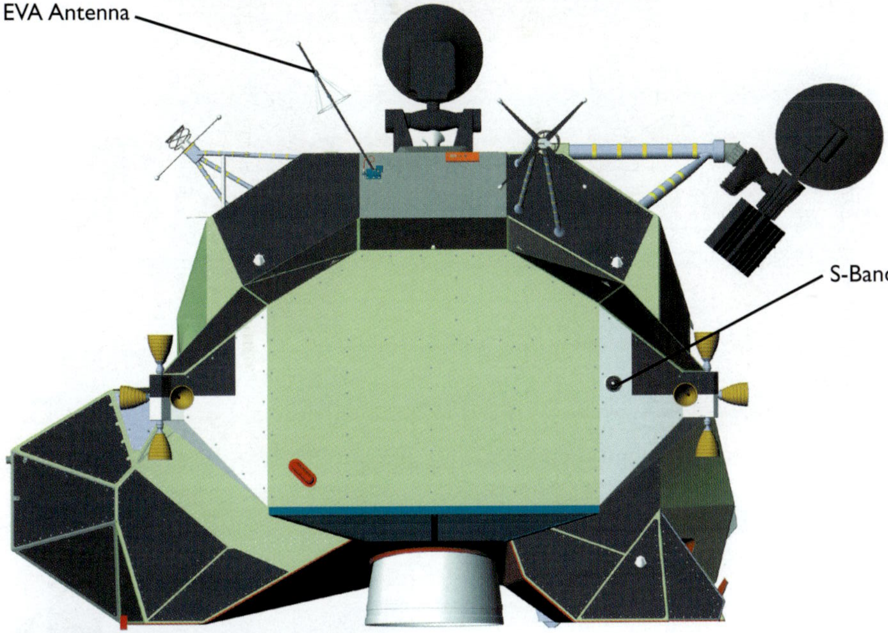

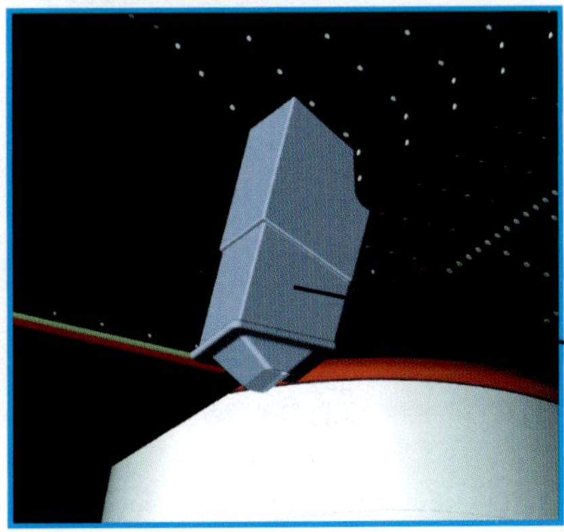

Detail View of Interstage Umbilical Shaft

WEIGHT DRY: 4,500 POUNDS
WEIGHT FUELED: 9,700 POUNDS

HEIGHT: 12 FEET 4 INCHES
DIAMETER: 14 FEET 1 INCH

J-SERIES BACK

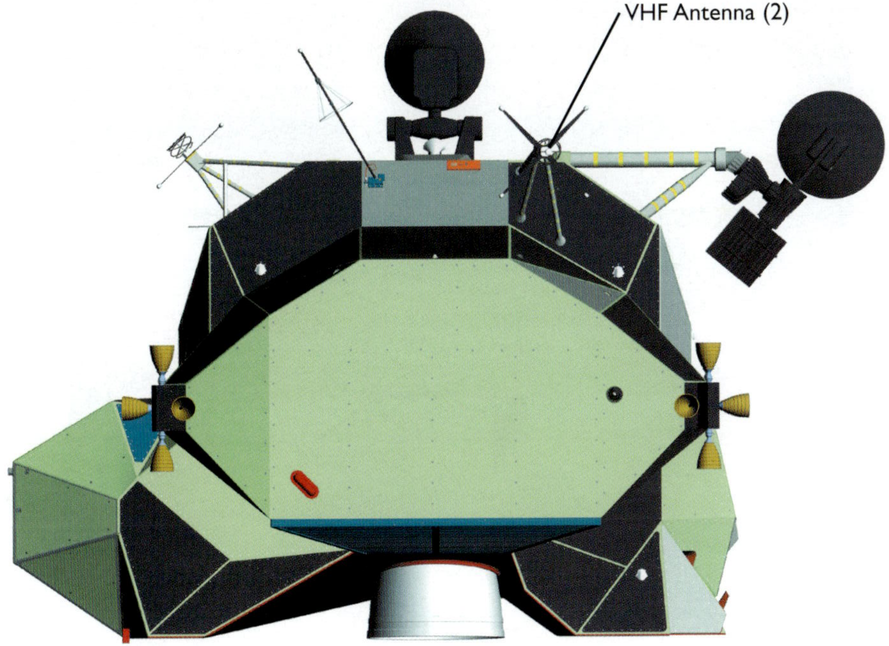

The ascent stage was the control center for the Lunar Module; it was comprised of three main areas: the crew compartment, the midsection, and the equipment bay.

The ascent stage was designed to provide a controlled environment for the astronauts while separated from the CSM; provide required visibility for lunar landing, stay, and ascent for rendezvous and docking with the CSM; provide for astronaut equipment and transfer between the LM and CSM, and between the LM and the lunar surface; and to protect the astronauts and equipment from micrometeoroid penetration.

J-SERIES LEFT

H-SERIES LEFT

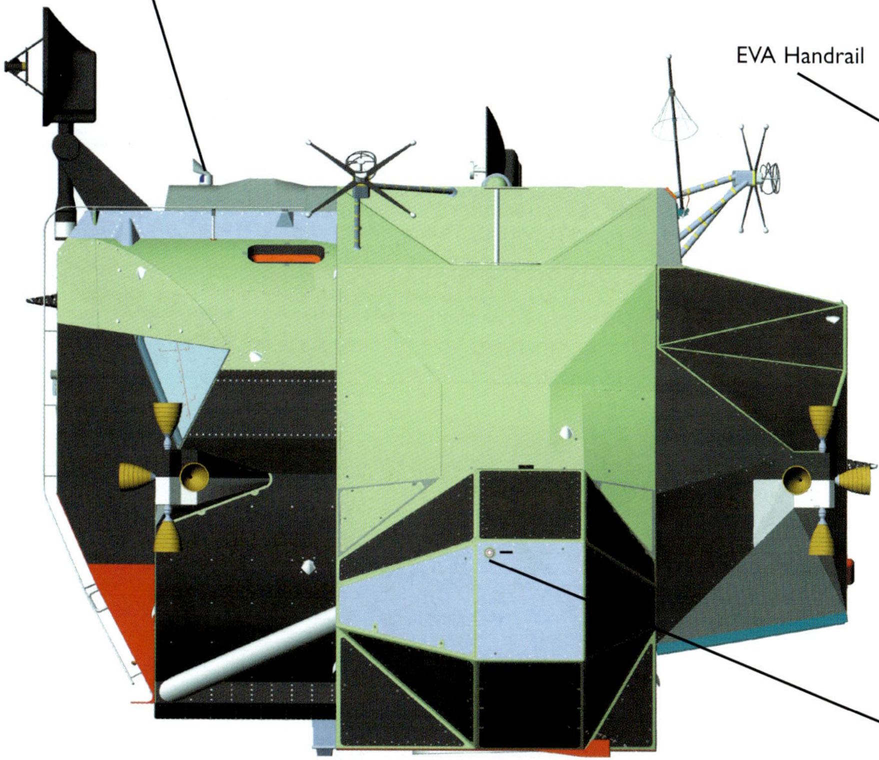

Lunar Module Chronology

July 19, 1963
Marquardt begins development of the LM reaction control engine thruster.

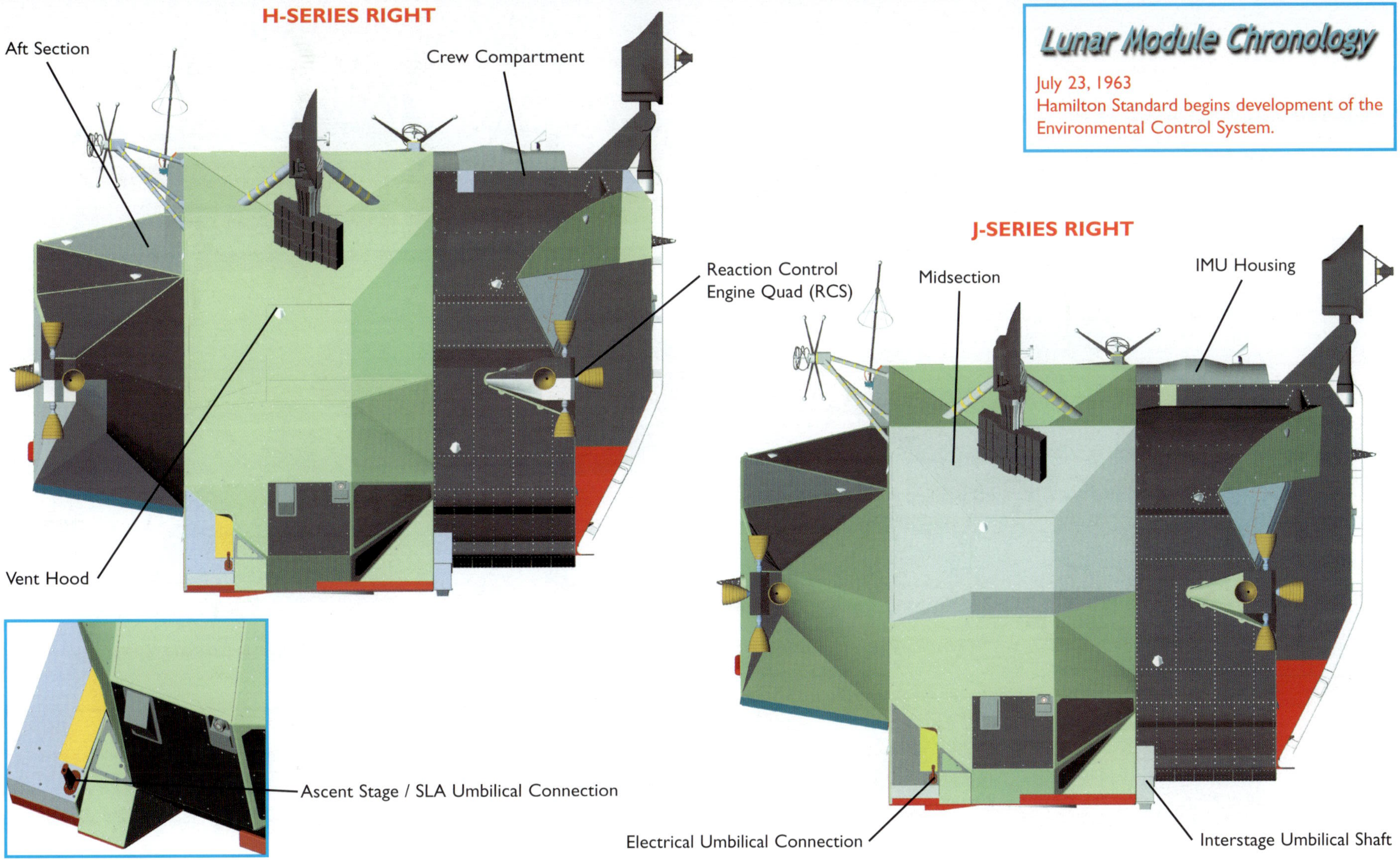

Lunar Module Chronology

July 23, 1963
Hamilton Standard begins development of the Environmental Control System.

Lunar Module Chronology

August 2, 1963
Grumman tasked to design LM with thruster capability with CSM attached.

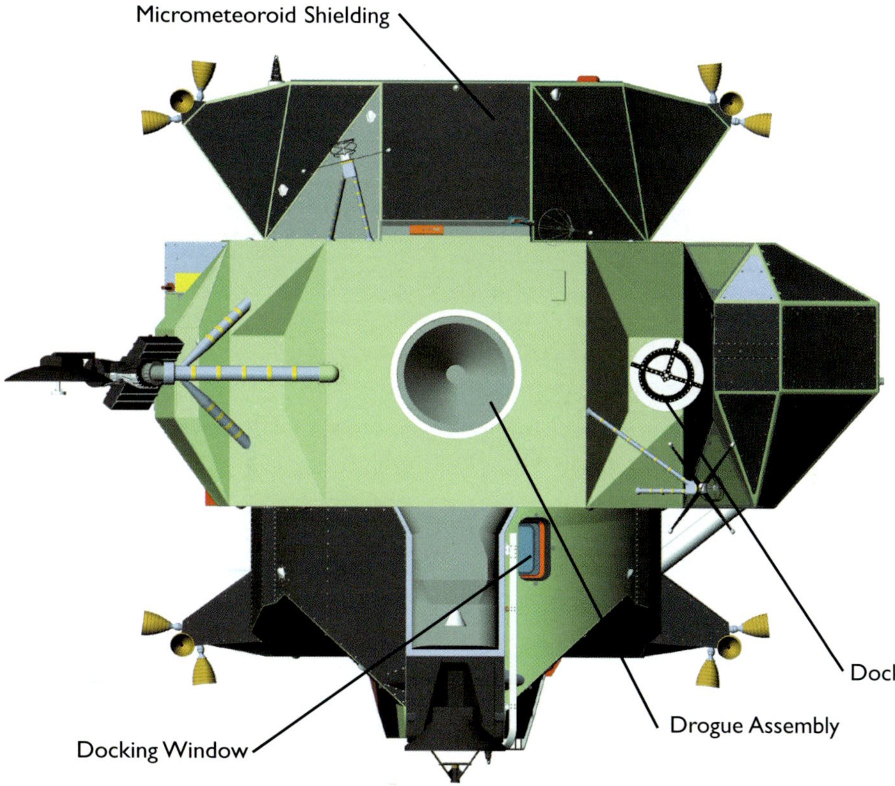

H-SERIES TOP

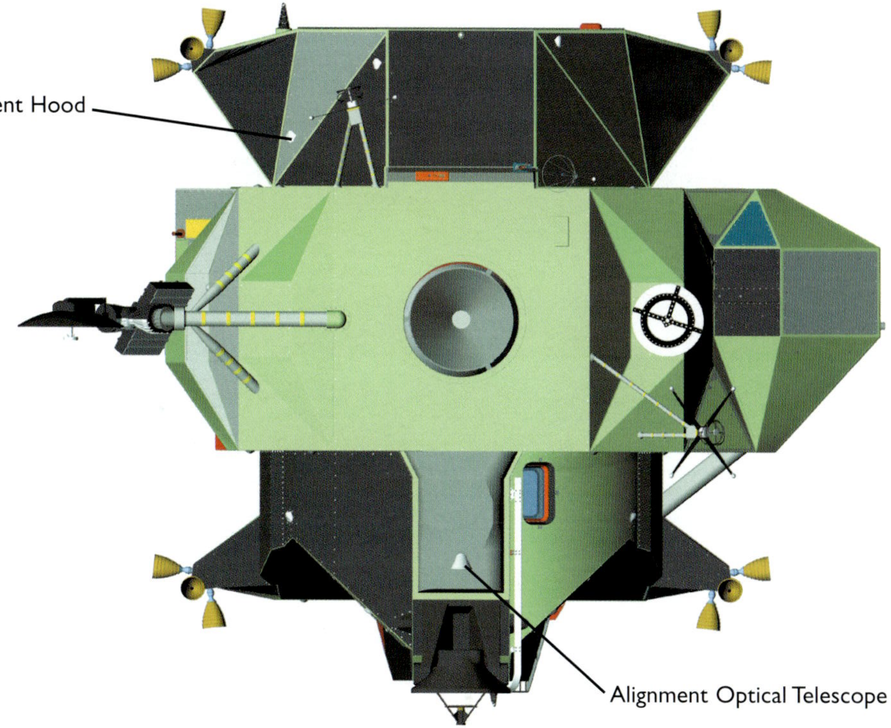

J-SERIES TOP

Lunar Module Chronology

August 31, 1963
Full-scale cardboard model of Lunar Module created by Grumman engineers.

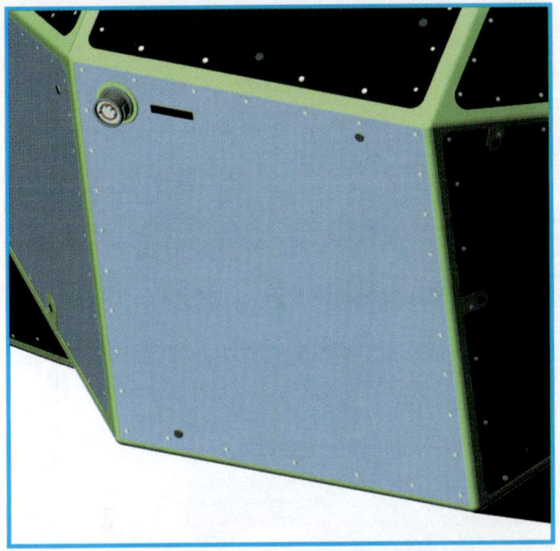

DETAIL VIEW OF RED DOCKING LIGHT

DETAIL VIEW OF FLOOD LIGHT

BOTTOM (VIRTUALLY THE SAME FOR H-SERIES AND J-SERIES)

Lunar Module Chronology

September 16, 1963
Inspection of first ascent stage mock-up.

Lunar Module Chronology

November 8, 1963
Go-ahead to RCA to develop the LM radar.

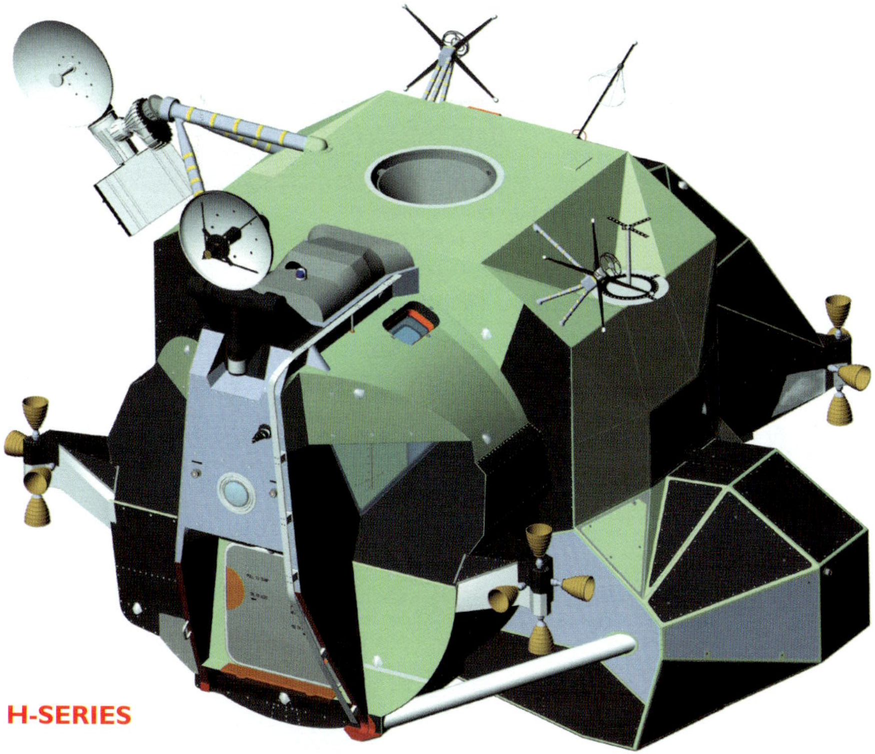

H-SERIES

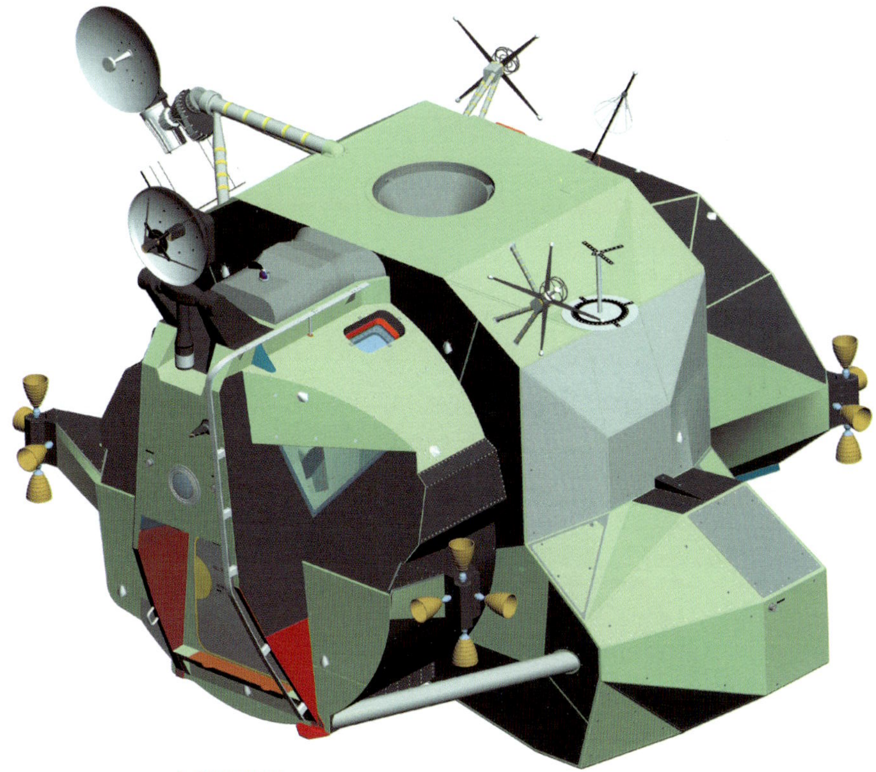

J-SERIES

Lunar Module Chronology

January 16, 1964
First full-throttle firing of STL Apollo Lunar Module descent engine.

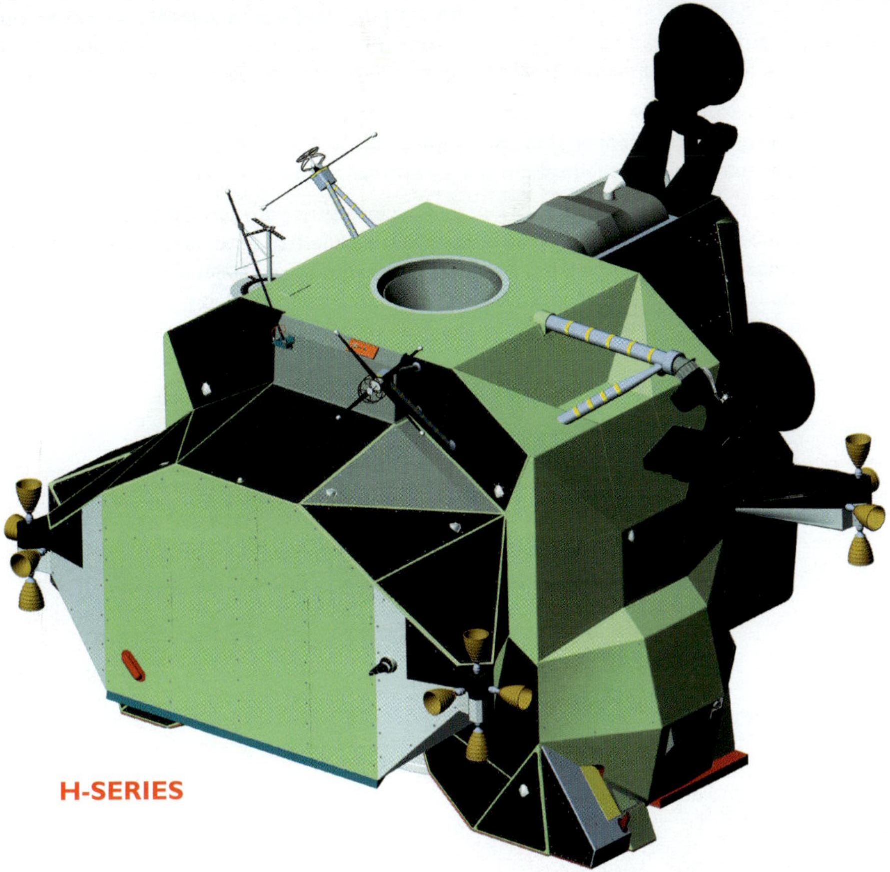

H-SERIES

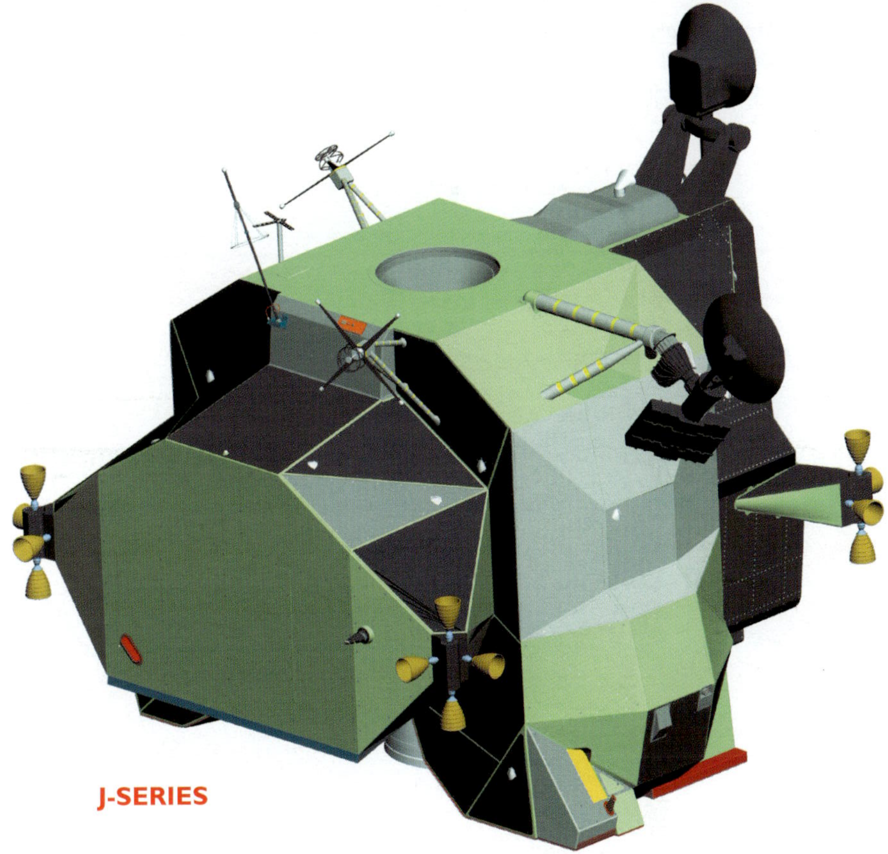

J-SERIES

Lunar Module Chronology

February 13, 1964
Studies conducted on post-Apollo lunar exploration.

Lunar Module Chronology

February 11, 1964
Apollo Lunar Module to use inertial reference system.

REACTION CONTROL SYSTEM (RCS) ENGINE

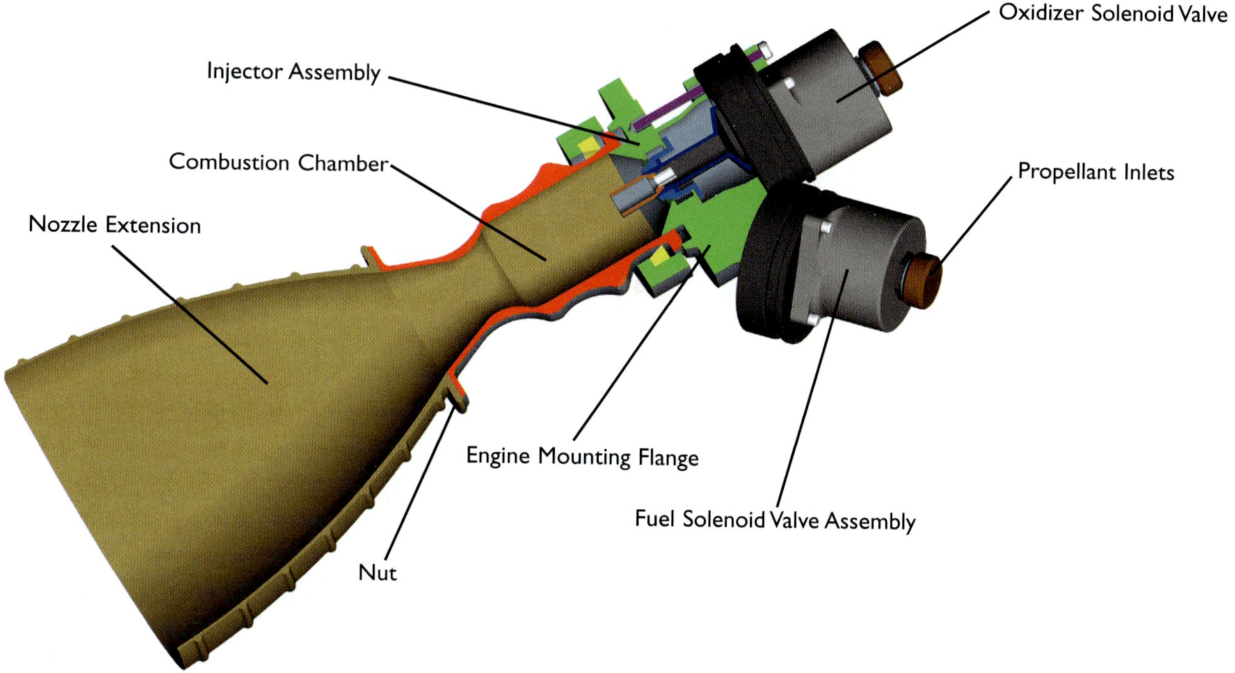

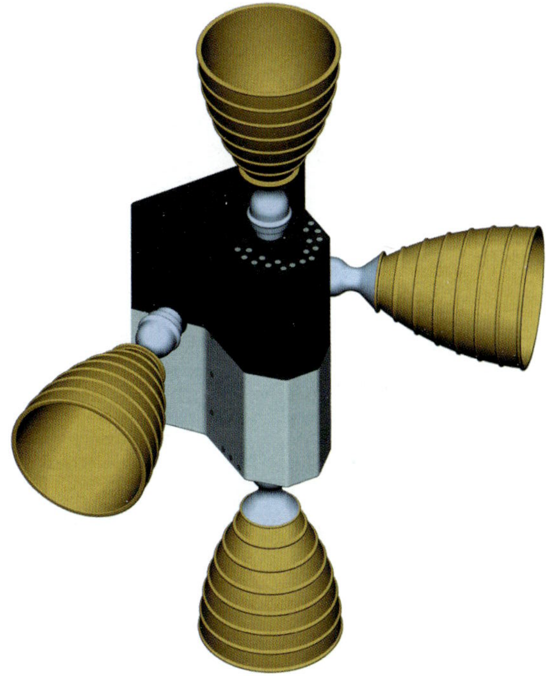

RCS QUAD ASSEMBLY

The reaction control subsystem (RCS) provided thrust impulses that stabilized the LM during the descent and ascent trajectory, and controlled attitude and translation movement of the LM about and along its three axes during hover, landing, and docking maneuvers. The RCS also provided the thrust required to separate the LM from the CSM, and the +X axis acceleration (ullage maneuver) required to settle main propulsion system propellant before a descent or ascent engine start. The RCS accomplished its tasks during coasting periods or while the descent or ascent engine was firing. The RCS operated in response to automatic control commands from the Guidance, Navigation and Control Subsystem (GN&CS) or manual commands from the astronauts.

The 16 thrust chamber assemblies (thrusters) and the propellant and helium sections that comprise the RCS were located on the ascent stage. The propellants used in the RCS were identical with those used in the main propulsion system. The fuel, Aerozine 50, is a mixture of approximately 50% each of hydrazine and unsymmetrical dimethyl-hydrazine. The oxidizer was nitrogen textroxide. The injection ratio of oxidizer to fuel was approximately 2 to 1. The propellants were hypergolic, meaning that they ignite and burn on contact.

The thrusters were small rocket engines, each capable of delivering 100 pounds of thrust. They were arranged in clusters of four, mounted on four outriggers equally spaced around the ascent stage. In each cluster, two thrusters were mounted parallel to the LM X-axis, facing in opposite directions; the other two were spaced 90 degrees apart perpendicular to the X-axis and parallel to the Y-axis and Z-axis.

COMMUNICATIONS SUBSYSTEM ANTENNA LOCATIONS

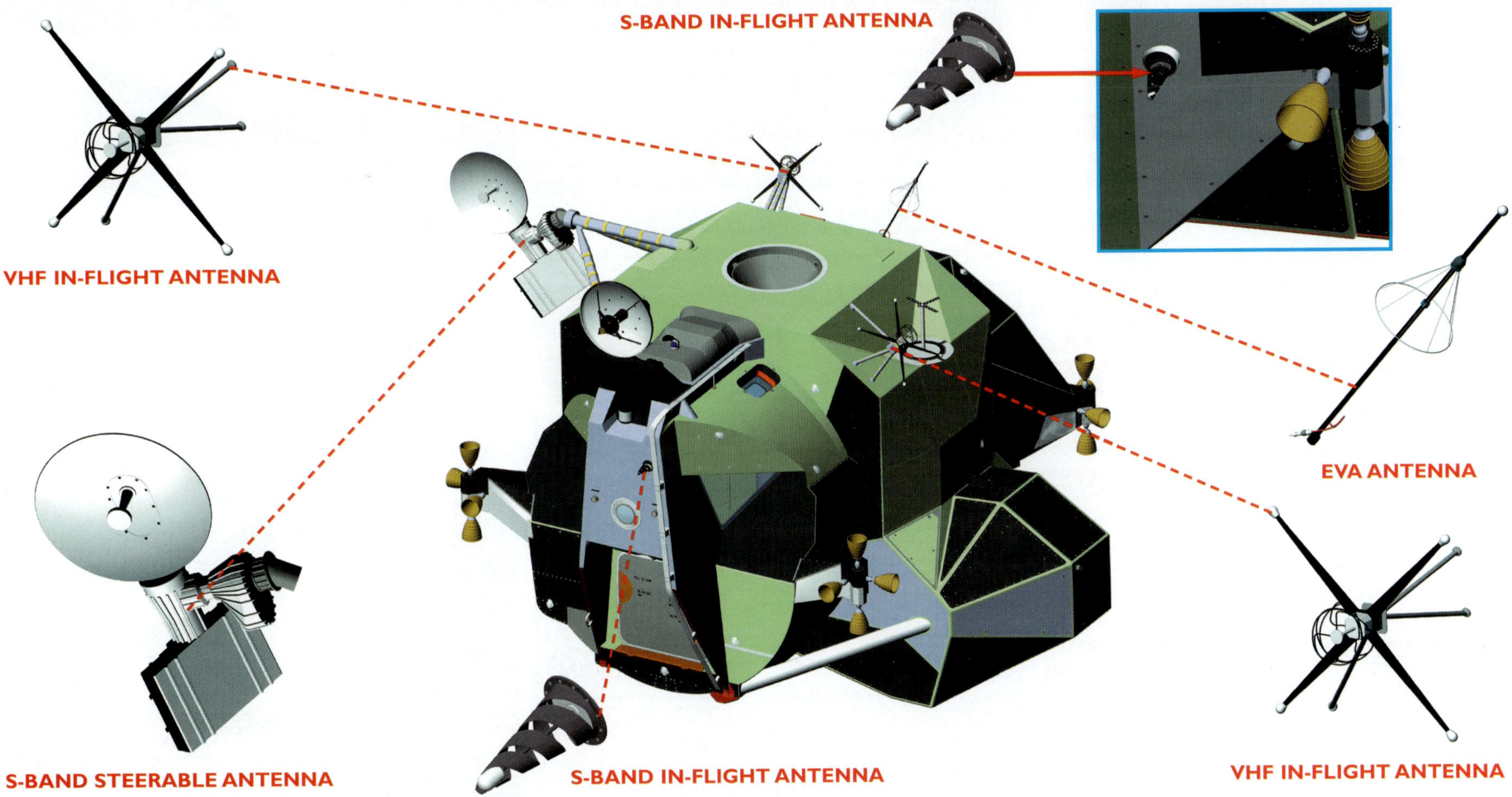

The communications subsystem (CS) provided in-flight and lunar surface communications links between the LM and CSM, the LM and mission control, and the LM and the extravehicular astronaut (EVA). When both astronauts were outside the LM, the LM relayed communications between the astronauts and mission control. When the astronauts were on the LRV, the Lunar Communications Relay Unit (LCRU), mounted on the front of the LRV, was the communications relay. The CS consisted of the S-band and VHF equipment.

S-BAND STEERABLE AND VHF ANTENNAS

The S-band transceiver assembly provided deep space communications between the LM and Mission Control. S-band communications consisted of voice and pseudo-random noise ranging transmission from Mission Control to the LM; and voice, pseudo-random noise ranging turnaround, biomed, and subsystem data transmission from the LM to Mission Control. The assembly consisted of two identical phase-locked receivers, two phase modulators with driver and multiplier chains, and a frequency modulator. The receivers and phase modulators provided the ranging, voice, emergency-keying, and telemetry transmit-receive functions.

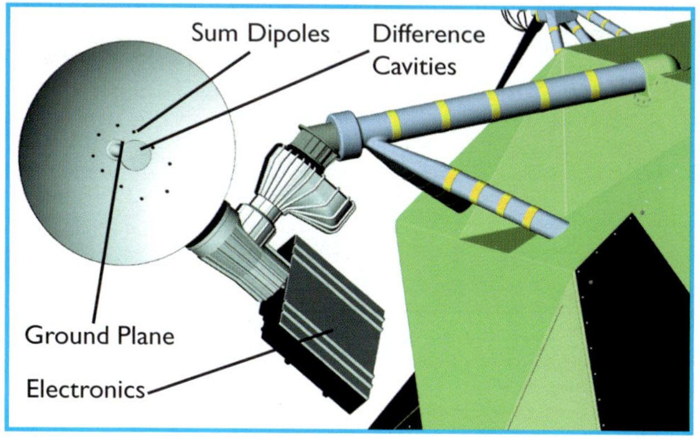

S-BAND STEERABLE ANTENNA

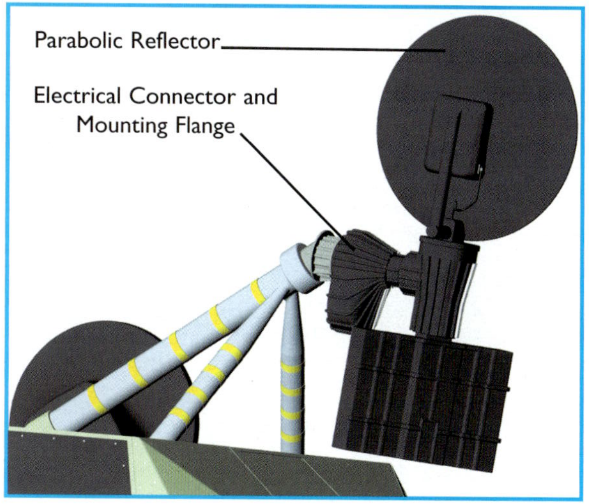

The S-band steerable antenna was a 26-inch diameter parabolic reflector with a point surface feed that consisted of a pair of cross-sleeved dipoles over a ground plane. The prime purpose of this antenna was to provide deep-space voice and telemetry communications, and deep-space tracking and ranging. One side of the antenna was white in color and the other side was black; this was to ensure that neither side became too hot or too cold, considering that there was a 250 degree difference between light and shadow in space and on the Moon.

The two S-band in-flight antennas were omnidirectional; one was forward and one aft on the LM. The antennas were circularly polarized radiators that collectively covered 90% of the sphere at -3 db or better. These antennas were primarily S-band antennas for the LM when in flight.

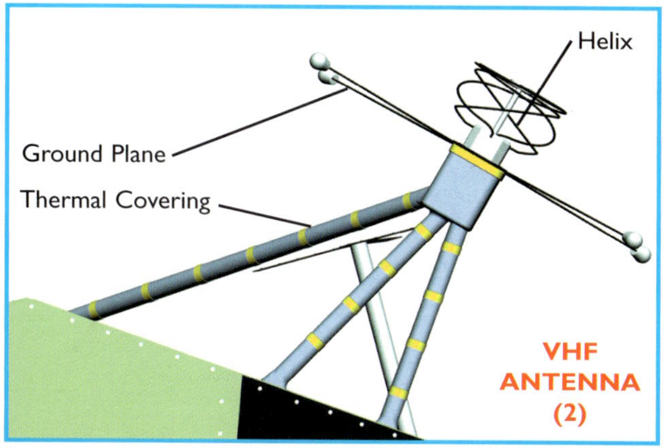

VHF ANTENNA (2)

The two VHF in-flight antennas were omnidirectional, right-hand circularly polarized antennas whose primary function was to provide voice communications between the LM and CSM; provide CSM-LM-CSM ranging when used with the ranging tone transfer assembly (RTTA) and during transmission blackout with Mission Control; and to provide low-orbit telemetry transmissions from the LM to the CSM.

The VHF EVA antenna was an omnidirectional conical antenna which was used for LM-EVA communications when the LM was on the lunar surface. It was mounted at the rear of the ascent stage and up-righted (unstowed) by an astronaut by turning a handle inside the LM, which actuated a simple pulley mechanism that turned a shaft attached to antenna.

EVA AND RENDEZVOUS ANTENNAS

EVA ANTENNA STOWED

EVA ANTENNA RAISED AND DEPLOYED

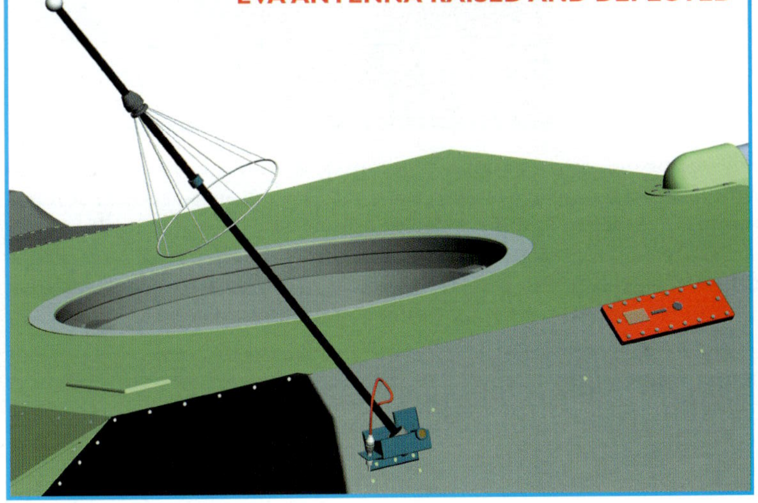

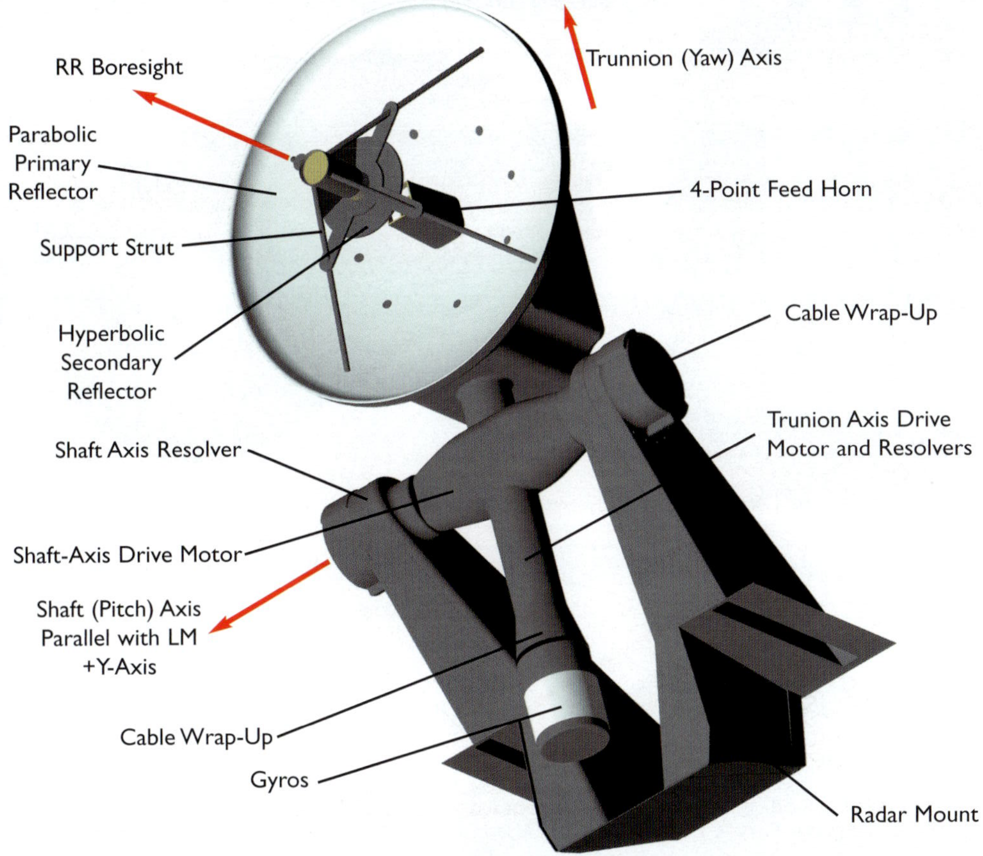

The rendezvous radar, operated in conjunction with a CSM transponder, acquired and tracked the CSM before and during rendezvous and docking. The radar tracked the CSM during the descent phase of the mission to supply tracking data for any kind of required abort maneuver, and during the ascent phase of the mission to supply data for rendezvous and docking. When the radar tracked the CSM, continuous measurements of range, range rate, angle, and angle rate (with respect to the LM) were provided simultaneously to the primary guidance and navigation section and to the LM cabin displays. This allowed rendezvous to be performed automatically under computer control, or manually by the astronauts.

DOCKING WINDOW

The Lunar Module had two triangular shaped windows used by the astronauts for observation (located in the front of the LM, one in front of each flight station) during the lunar landing and launch, and during the lunar stay. The docking window was set in the curved overhead portion of the crew compartment and provided visibility during docking maneuvers.

Each of the three windows consisted of two separate panes, vented to the space environment. The outer pane was made of Vycor glass with a thermal coating (59 layers of blue-red metallic oxide) on the outboard surface, which reduced the mirror effect of the windows and increased their normal light-transmission efficiency, and an anti-reflective coating on the inboard surface. The inner pane was made of chemically tempered, high-strength structural glass. It was sealed with a Raco seal (the docking window inner pane had a dual seal) and had a defog coating on the outboard surface and an anti-reflective coating on the inboard surface. Both panes were bolted to the window frame through retainers.

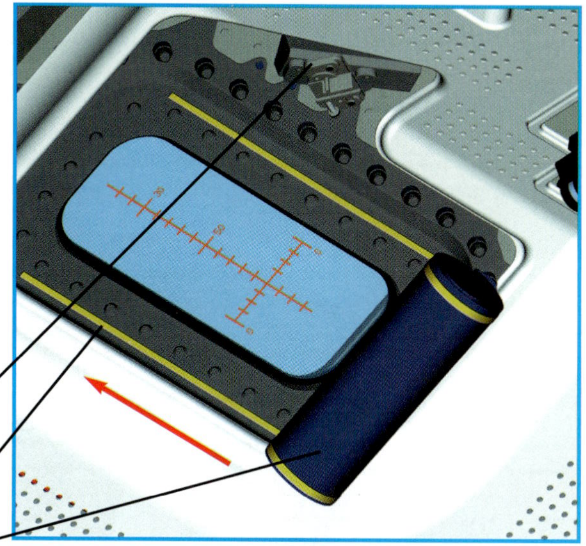

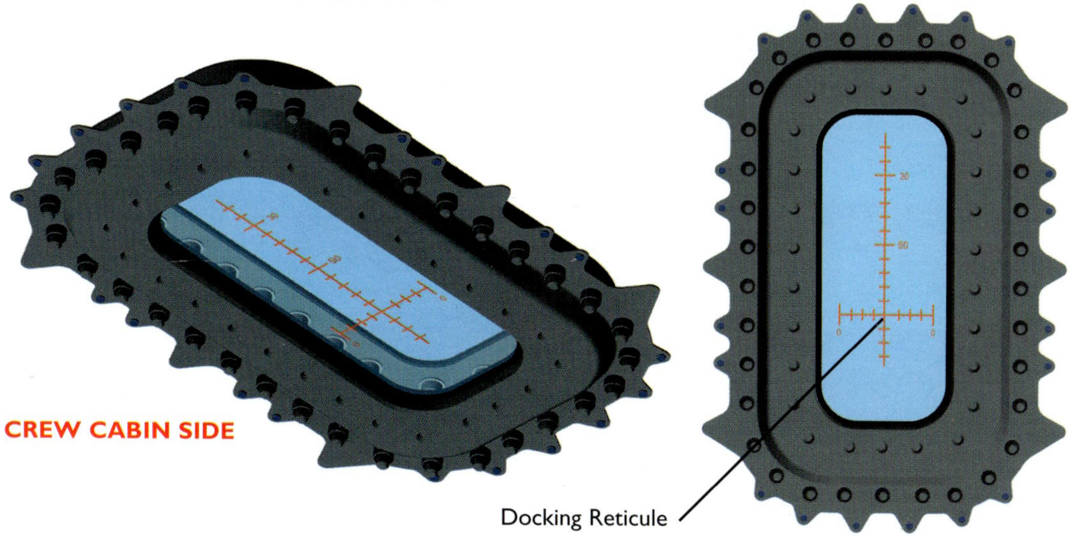

Lunar Module Chronology

February 16, 1964
Grumman subcontracts for engines for the LM.

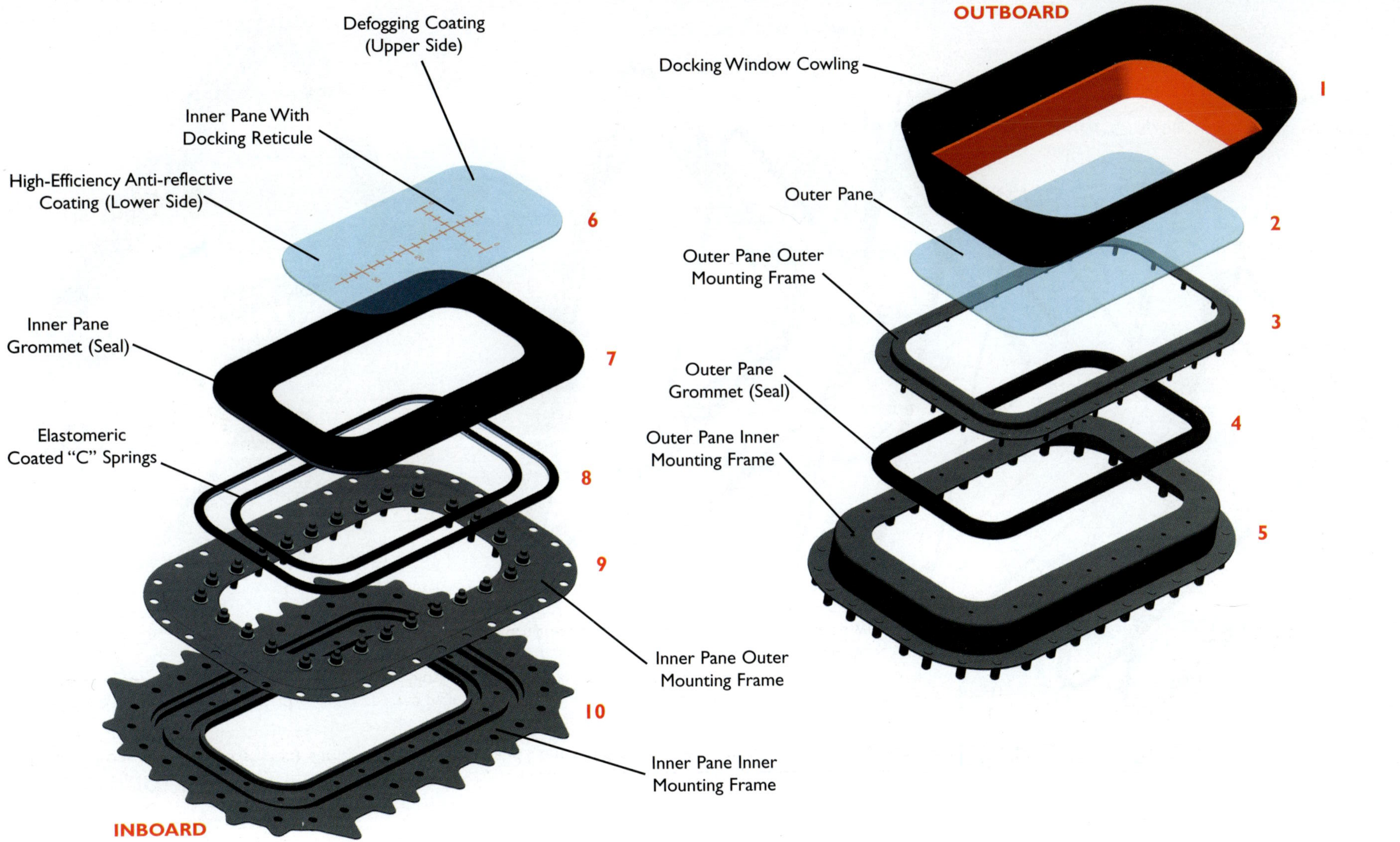
OUTBOARD
Docking Window Cowling
1
Outer Pane
2
Outer Pane Outer Mounting Frame
3
Outer Pane Grommet (Seal)
4
Outer Pane Inner Mounting Frame
5
Defogging Coating (Upper Side)
Inner Pane With Docking Reticule
High-Efficiency Anti-reflective Coating (Lower Side)
6
Inner Pane Grommet (Seal)
7
Elastomeric Coated "C" Springs
8
Inner Pane Outer Mounting Frame
9
Inner Pane Inner Mounting Frame
10
INBOARD

FRONT WINDOWS

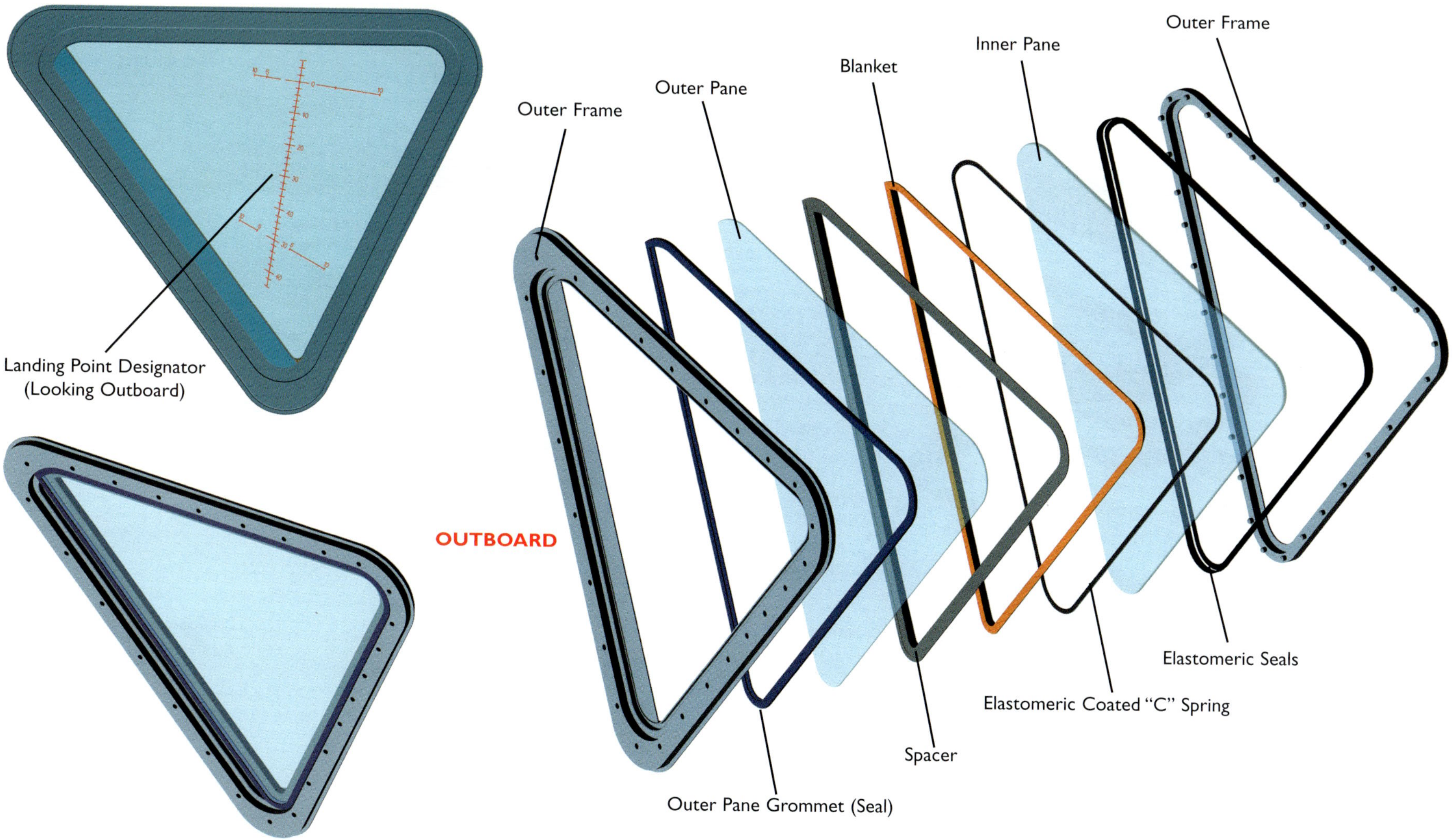

All three windows were heated to prevent fogging. The two triangular windows had approximately two square feet of viewing area; they were canted down to the side to permit adequate peripheral and downward visibility. The docking window, above the commander's flight station, had approximately 80 square inches of viewing area.

A window shade, with an anti-reflective coating on its outboard side, was provided for each window. Normally, the shade was rolled up at the window edge. A glare shield mounted between each of the front windows and the control and flight display panels reduced the window reflection of internal panel lighting.

The crewman optical alignment sight (COAS) provided the commander with gross range cues and closing rate cues during the docking maneuver. The COAS provided the commander with a fixed line-of-sight attitude reference image, which appeared to be the same distance away as the target. The COAS was a collimating instrument. It weighed 1.5 pounds, was 8 inches long and operated from a 28-volt DC power source. The COAS consisted of a lamp with an intensity control, a reticule, a barrel-shaped housing and mounting track, and a combiner and power receptacle.

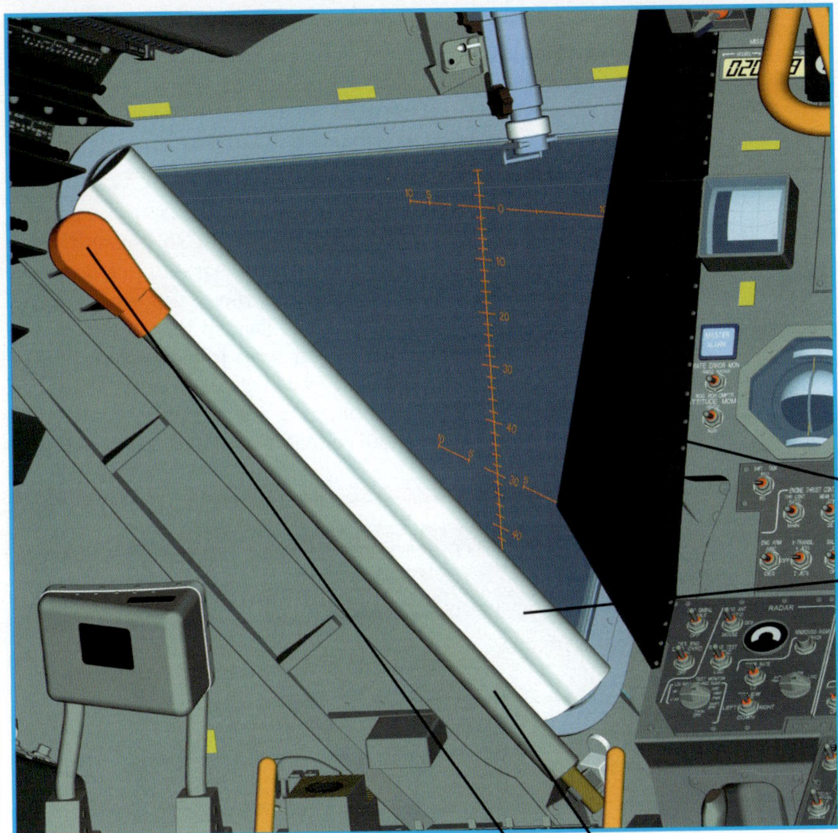

The crash bar could be pivoted about its orange head and attached to the window frame at the opposite end to protect the windows from falling gear or crewmembers.

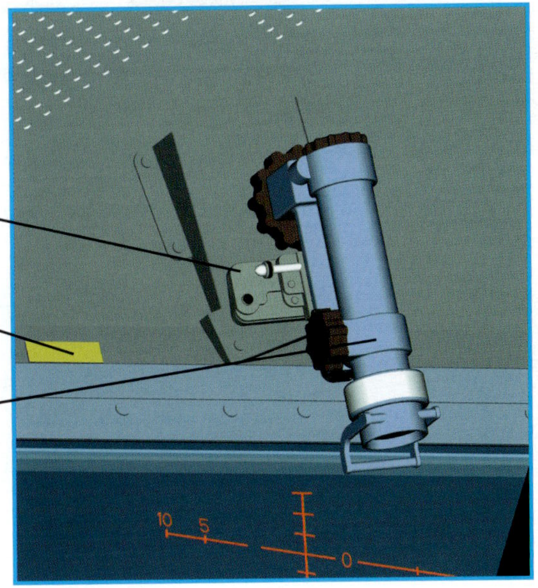

Lunar Module Chronology

March 12, 1964
North American, the prime contractor for the Apollo CSM, was directed to study the feasibility of using LM propulsion as a backup to the Service Module propulsion.

ALIGNMENT OPTICAL TELESCOPE (AOT)

ALIGNMENT OPTICAL TELESCOPE ATOP THE LM ASCENT STAGE

The alignment optical telescope, mounted on the navigation base to provide mechanical alignment and a common reference between the telescope and the inertial measurement unit, was a unity-power, periscope-type device with a 60-degree conical field of view. It was operated manually by the astronauts, and had a movable shaft axis parallel to the LM X-axis and a line of sight approximately 45 degrees from the X-axis in the Y-Z plane. The telescope line of sight was fixed in elevation and movable in azimuth to six detent positions. These detent positions were selected by turning a detent selector knob on the telescope; they were located at 60-degree intervals.

The optical subsection consisted of the alignment optical telescope and a computer control and reticule dimmer assembly. The telescope was 36 inches long and used by the astronauts to take angular measurements of celestial objects. The astronauts (using the computer control and reticule dimmer assembly) transferred sightings taken with the telescope to the computer. This assembly also controlled the brightness of the telescope reticule pattern.

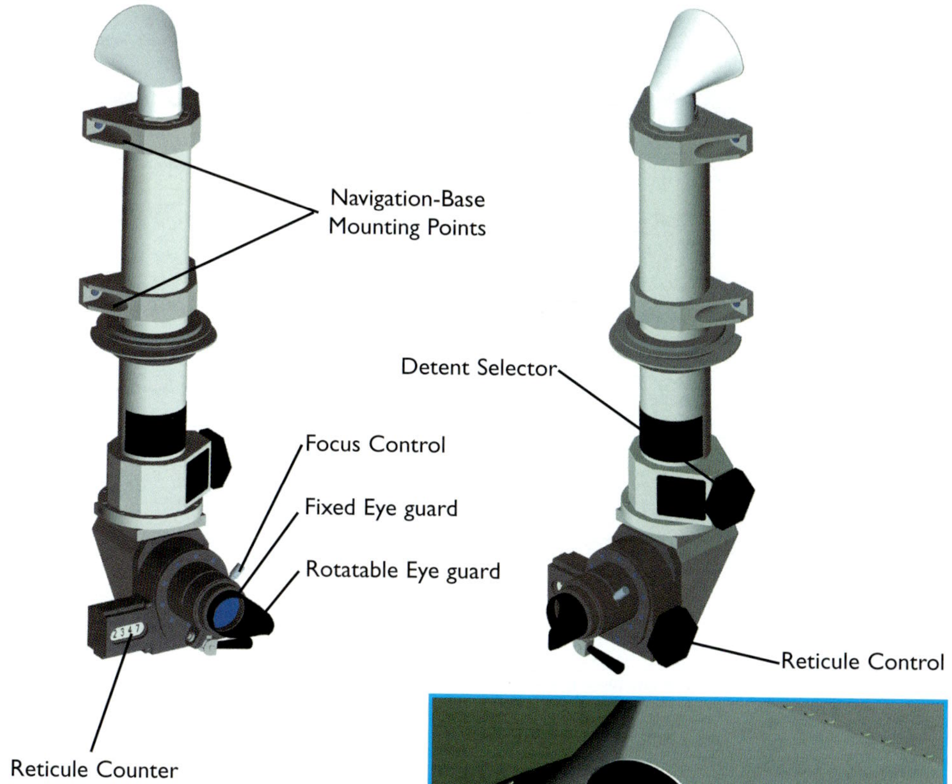

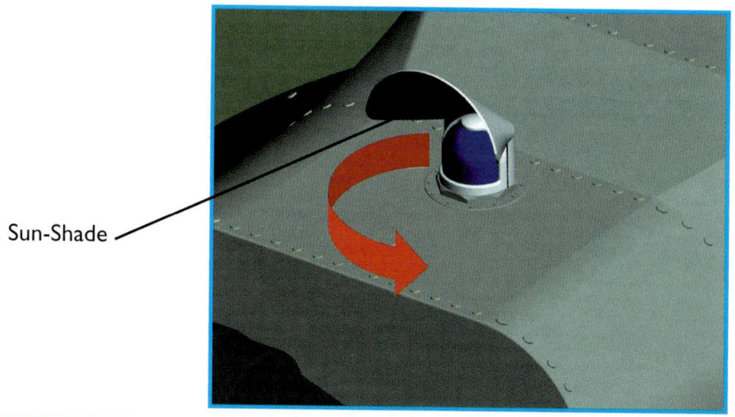

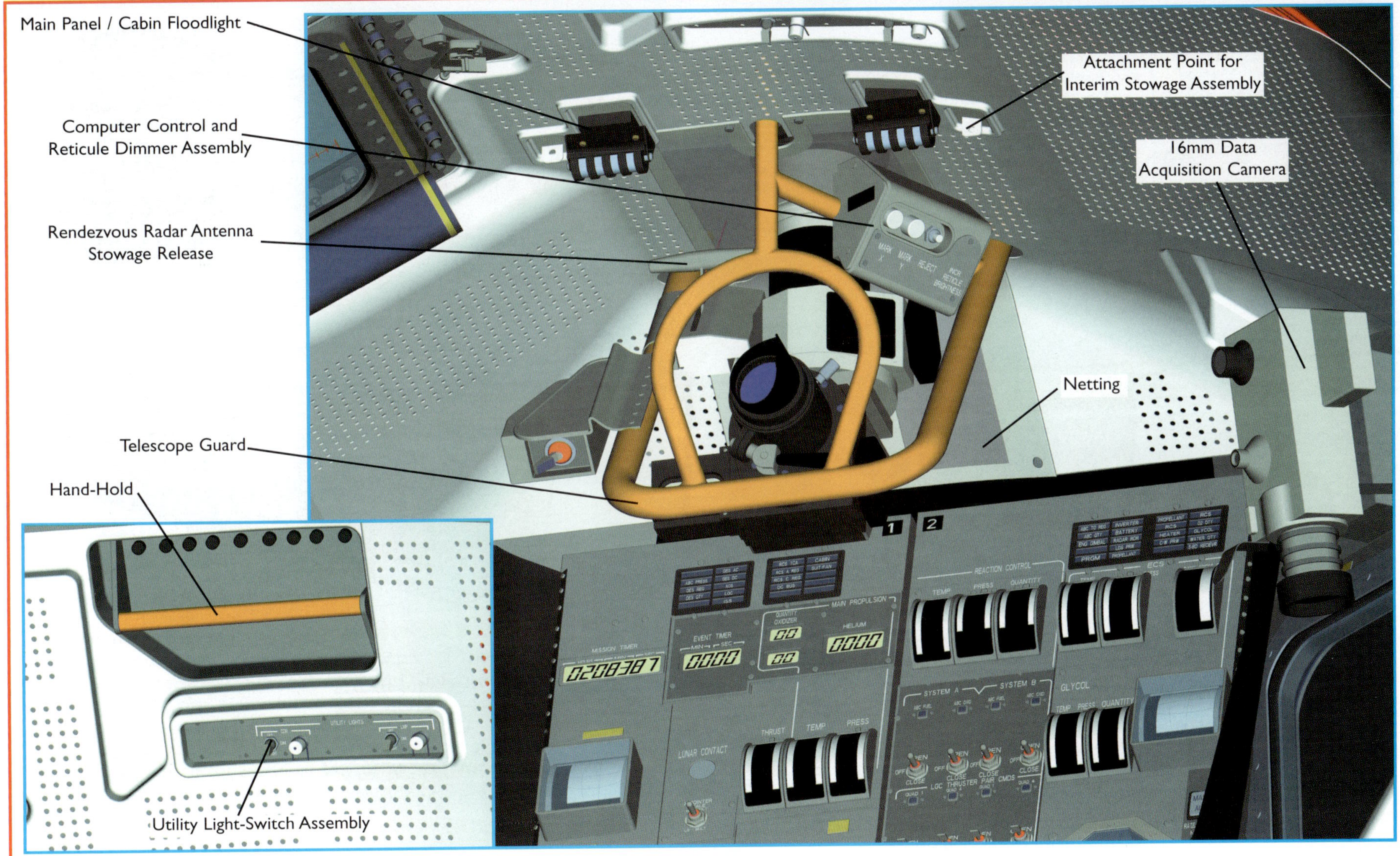
Main Panel / Cabin Floodlight
Computer Control and Reticule Dimmer Assembly
Rendezvous Radar Antenna Stowage Release
Telescope Guard
Hand-Hold
Utility Light-Switch Assembly
Attachment Point for Interim Stowage Assembly
16mm Data Acquisition Camera
Netting

FRONT HATCH

The forward hatch was in the front face assembly, just below the lower display panels. The hatch was used for transfer of astronauts and equipment between the LM and lunar surface, or for in-flight extravehicular activity (EVA) while docked with the CM. The hatch was approximately 32 inches square; it was hinged to swing inboard when open. A cam latch assembly forced a lip around the outer circumference of the hatch into a preloaded electrometric silicone compound seal secured to the LM structure.

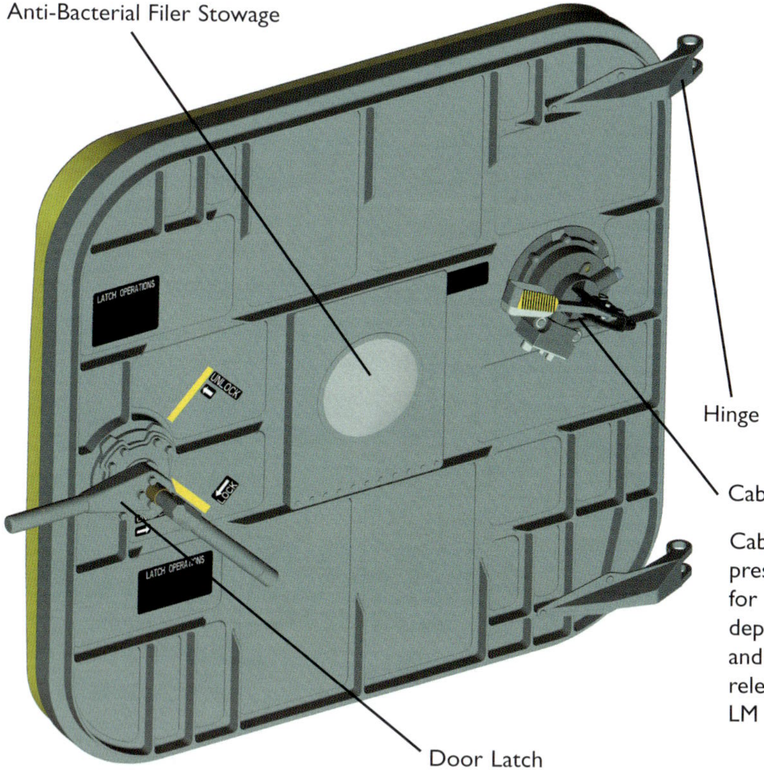

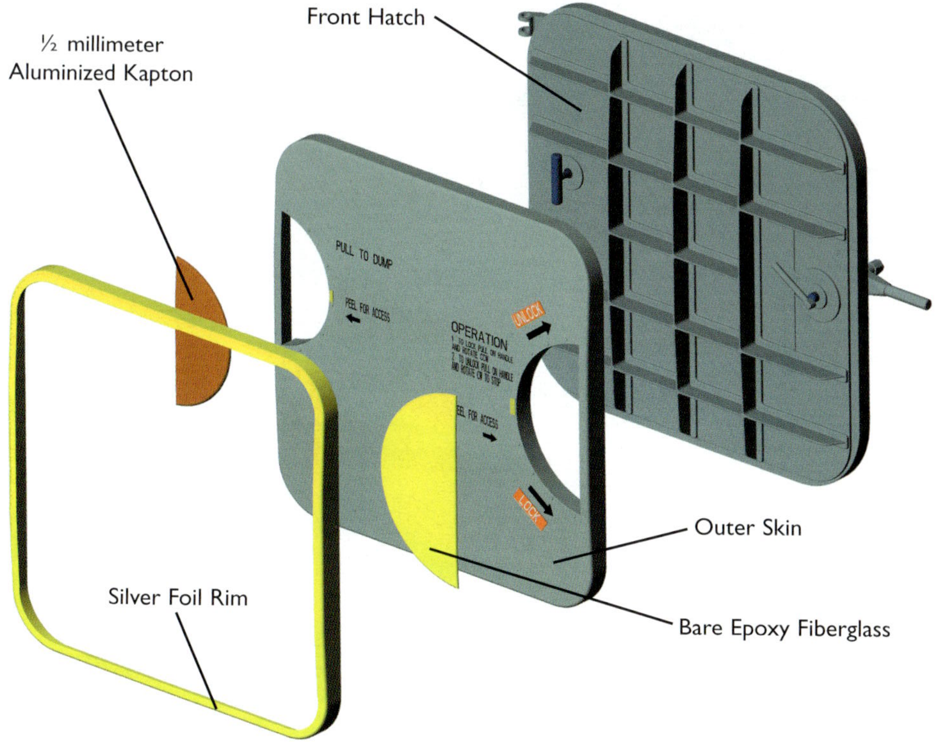

Cabin pressurization forced the hatch lip further into the seal, ensuring a pressure-tight contact. A handle was provided on both sides of the hatch, for latch operation. To open the hatch, the cabin must be completely depressurized; rotating the latch handle opened the hatch. The cabin relief and dump valve could also be operated from outside the LM. Quick-release pins in the latch plate and hinges could be pulled from inside the LM to open the hatch in an emergency.

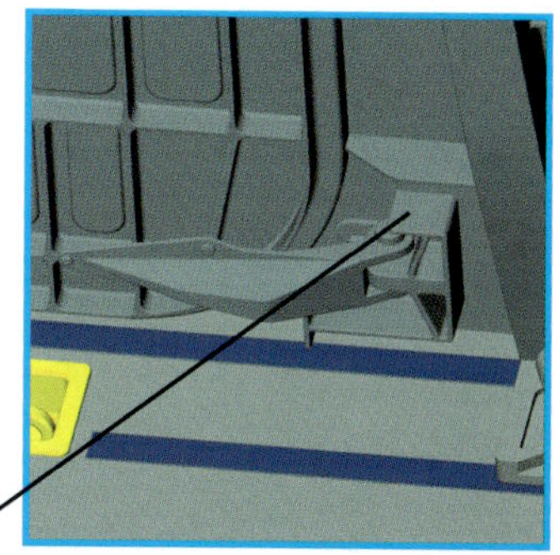

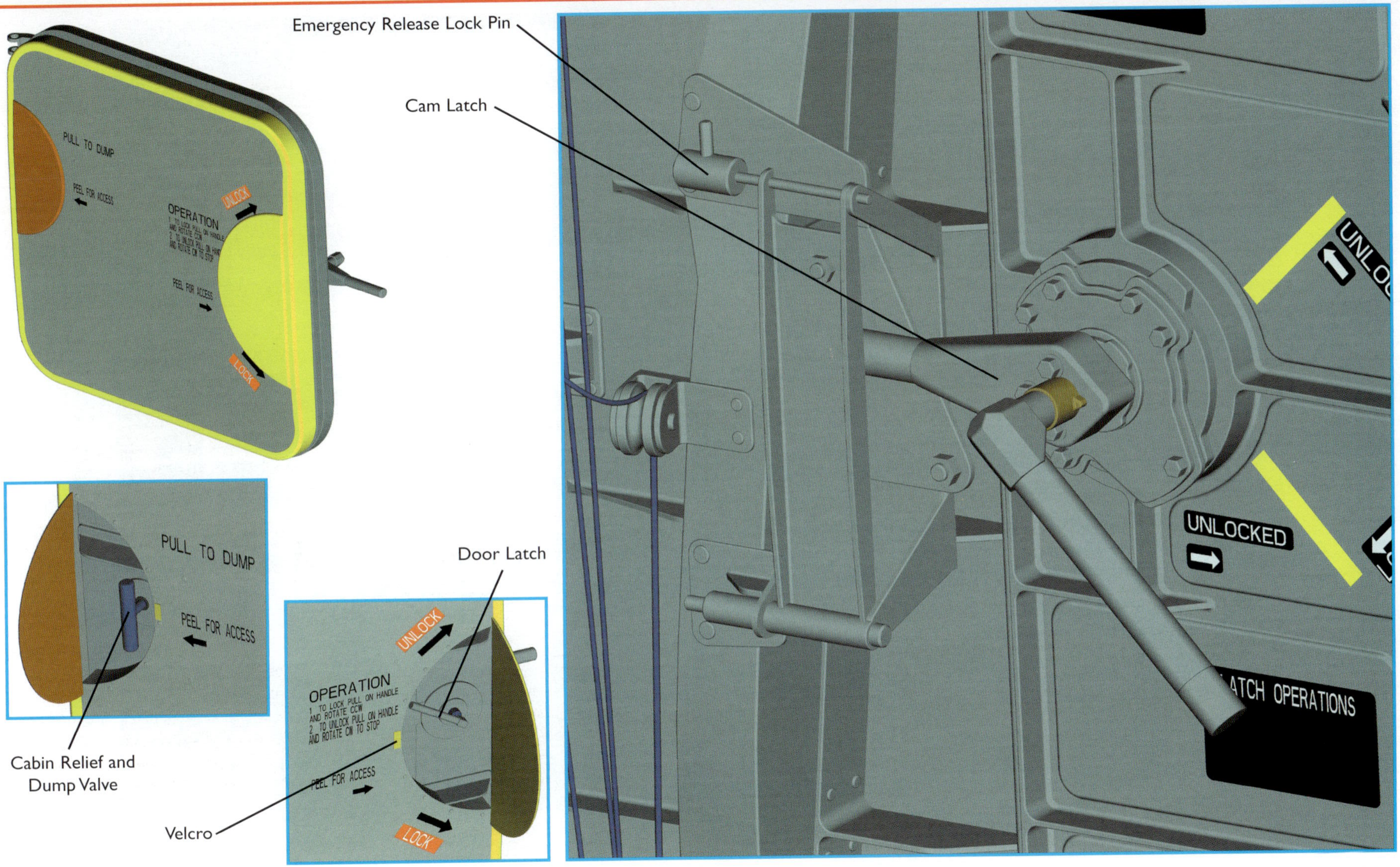

Emergency Release Lock Pin
Cam Latch
UNLOCKED
LATCH OPERATIONS
PULL TO DUMP
PEEL FOR ACCESS
OPERATION
1 TO LOCK PULL ON HANDLE AND ROTATE CCW
2 TO UNLOCK PULL ON HANDLE AND ROTATE CW TO STOP
UNLOCK
LOCK
Door Latch
Cabin Relief and Dump Valve
Velcro

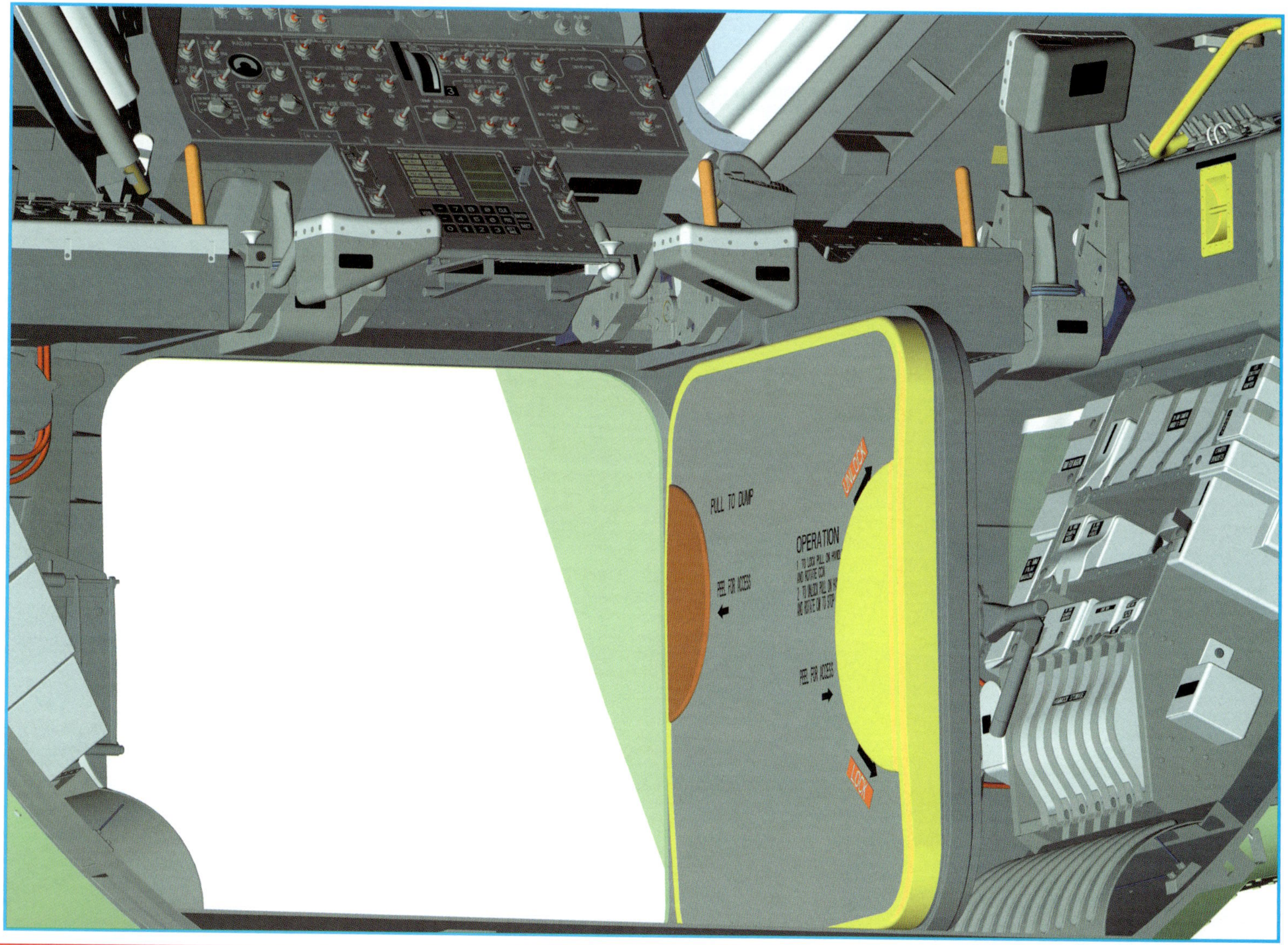
PULL TO DUMP
PEEL FOR ACCESS
OPERATION
PEEL FOR ACCESS
UNLOCK
LOCK

CREW COMPARTMENT

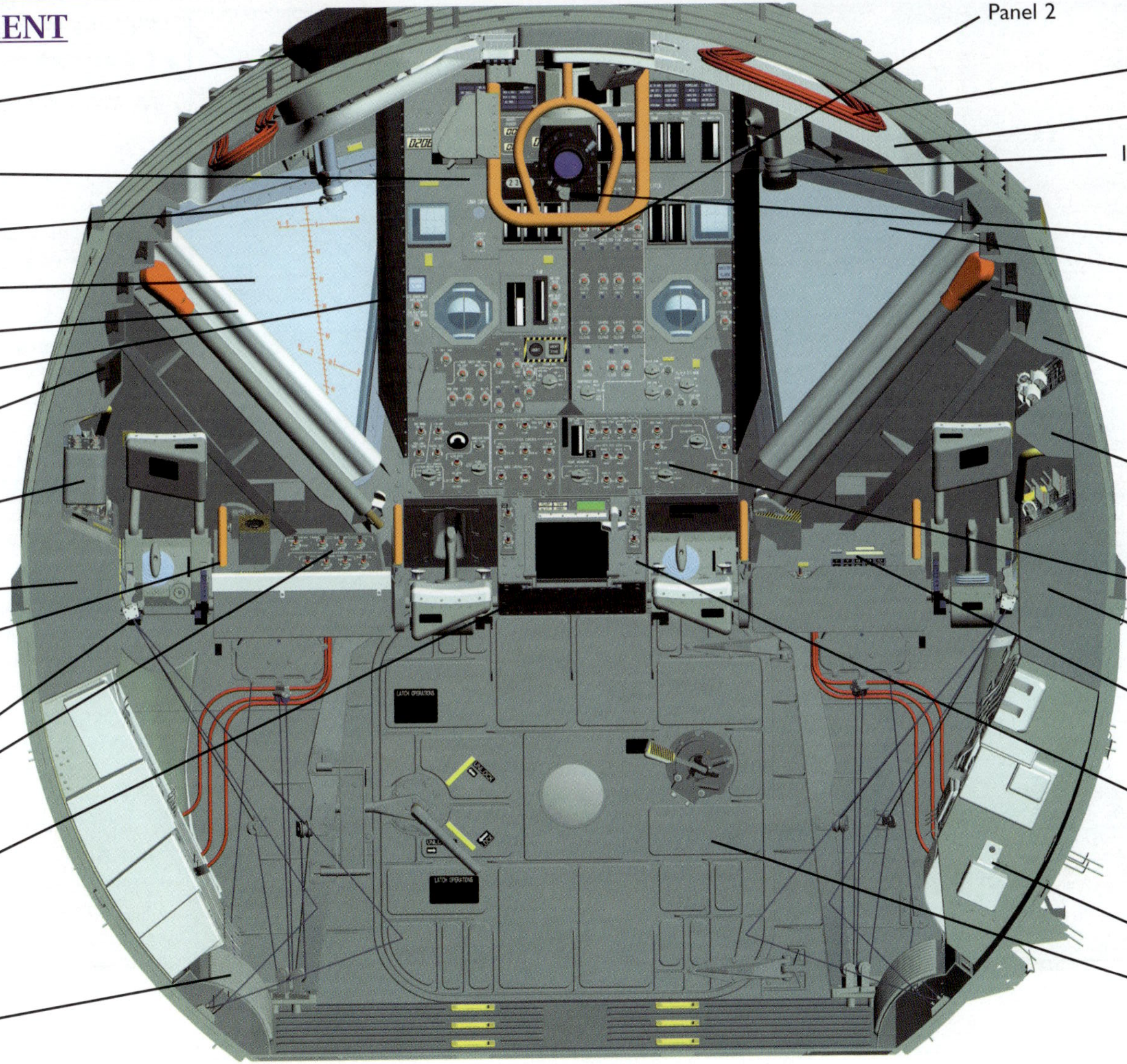

FULL STOWAGE COMPLIMENT FOR FORWARD CREW CABIN

The crew cabin and midsection that made up the cabin had an overall volume of 235 cubic feet. The basic structure was primarily aluminum alloy; titanium was used for fittings and fasteners. Aircraft-type construction methods were used. Skin and web panels were chemically milled to reduce weight, and mechanical fasteners joined the major structural assemblies with epoxy as a sealant. Structural members were fusion welded wherever possible to minimize cabin air pressurization leaks. The basic structure included supports for thrust control engine clusters and various antennas. Thermal insulation and a micrometeoroid shield enveloped the entire basic structure.

The crew compartment was the front area of the ascent stage; it was cylindrical (92 inches in diameter and 42 inches deep). The commander's flight station was to the left and the LM pilot's to the right of the front hatch. The flight station centerlines were 44 inches apart. For maximum downward visibility the upper portion of the crew compartment was constructed to extend forward of the lower portion. The area had control and display panels, body restraints, landing aids, a front window for each astronaut, a docking window above the commander's flight station and other accessory equipment. Each flight station had an attitude controller, a thrust / translation controller, and adjustable armrests for each controller. The front hatch was also in the front face assembly.

Interim Stowage Assembly

Utility Light Assembly

Lunar Orbital Map and Star Chart

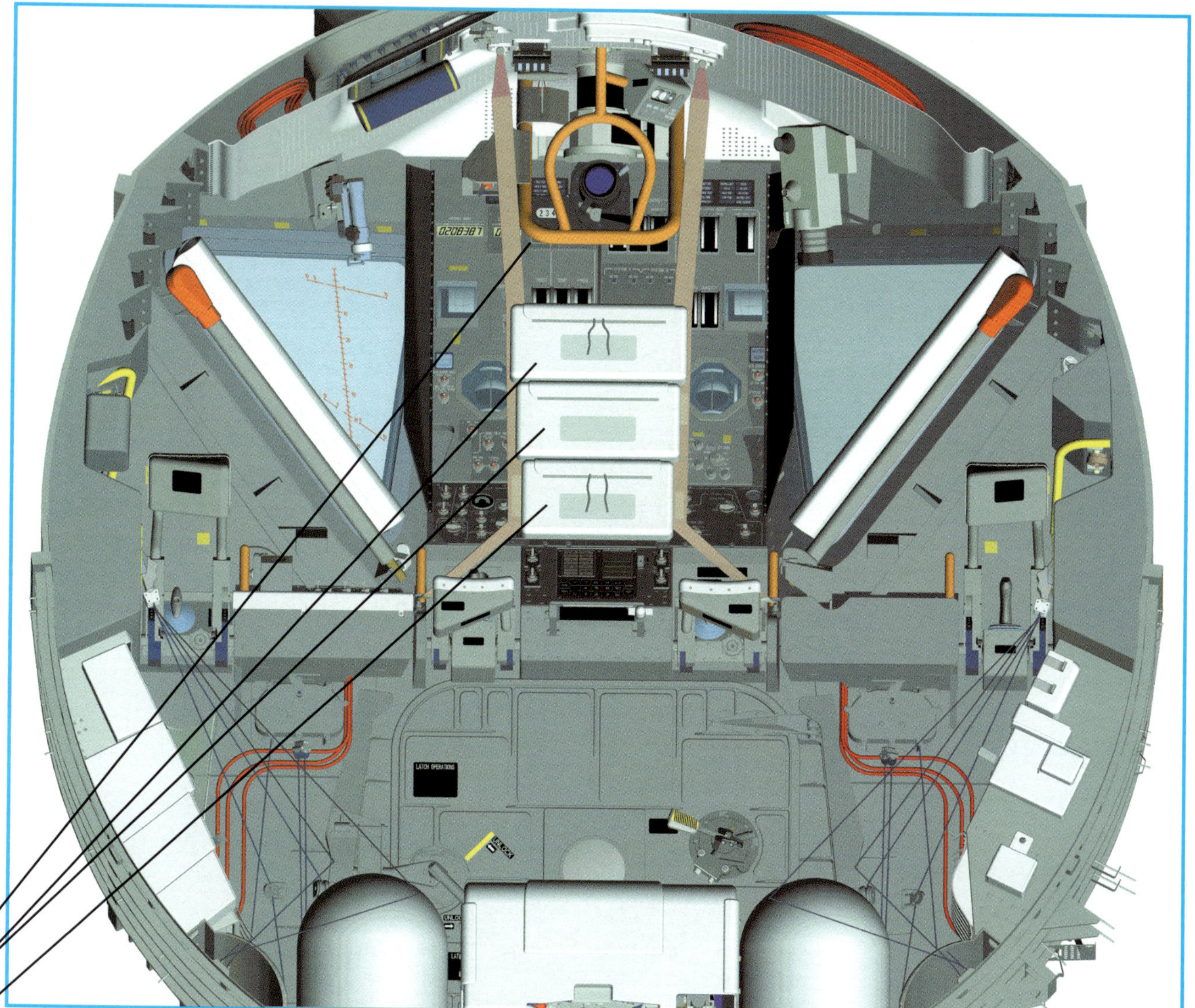

CREW EQUIPMENT

Underneath the control and display panels, to the right and left of the LM pilot and commander's flight stations respectively, were compartment cabinets. Equipment used during the mission was stowed in these cabinets. This equipment included food, personal hygiene items, EVA waist tethers, a camera, camera lens filters, a spare bulb, and a multipurpose special tool having a modified Allen-head. The forward equipment stowage also included a spare Environmental Control System (ECS) lithium hydroxide canister, waste collection containers, a PLSS lithium hydroxide cartridge, and a PLSS condensate container.

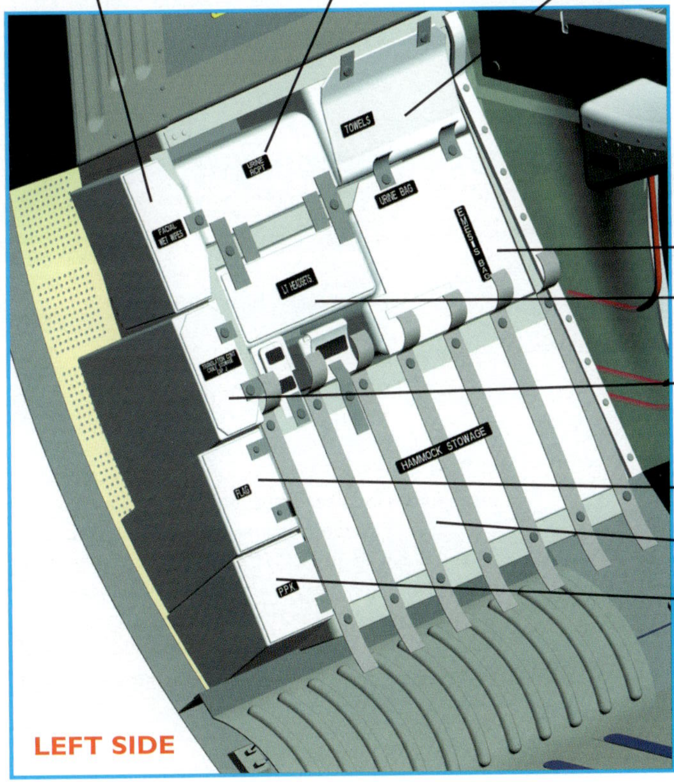

A portable life support system (PLSS) donning station was behind the optical alignment telescope station. The crew compartment deck (flight station floor) measured approximately 36 by 55 inches. Non-flammable Velcro pile strips were bonded to the deck's upper surface to provide restraining force for the astronauts while at their flight stations. Velcro was also on the soles of the astronaut's in-flight boots. Hand grips aligned with the forward hatch and recessed in the deck aided with ingress and egress.

The control and display panels contained all devices necessary to control, monitor and observe subsystems performance. The arrangement of the panels permitted either astronaut to fly the LM to the CSM. All panels were canted to facilitate viewing. Six of the panels were in front of the flight stations; the two upper panels – one inboard of each flight station – were at eye level. These panels were shock mounted to dampen vibrations. The next two lower panels were centered between the flight stations to enable sharing of the control functions. The remaining two panels were in front of each flight station, at waist height. The commander's panel contained lighting, mission timer, engine, and thrust chamber controls.

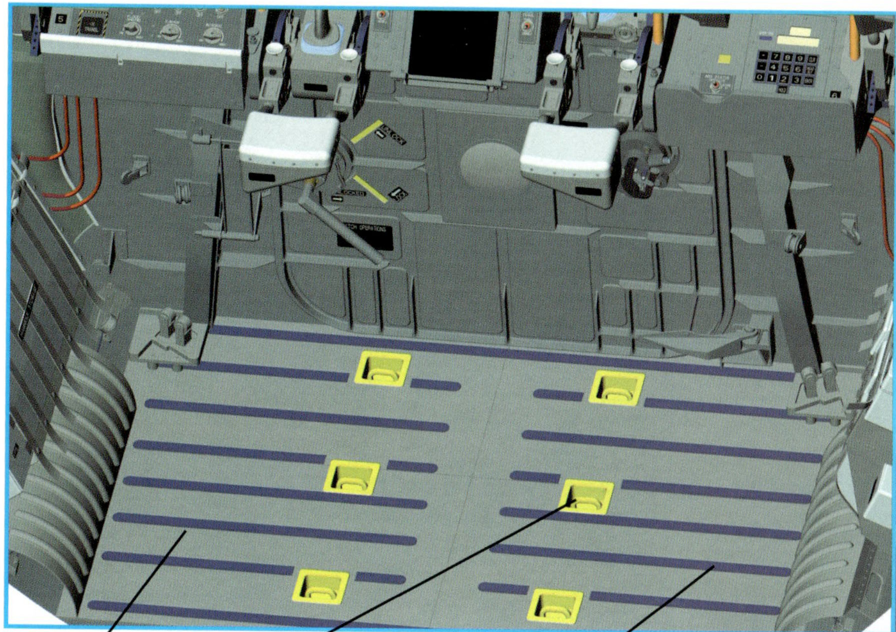

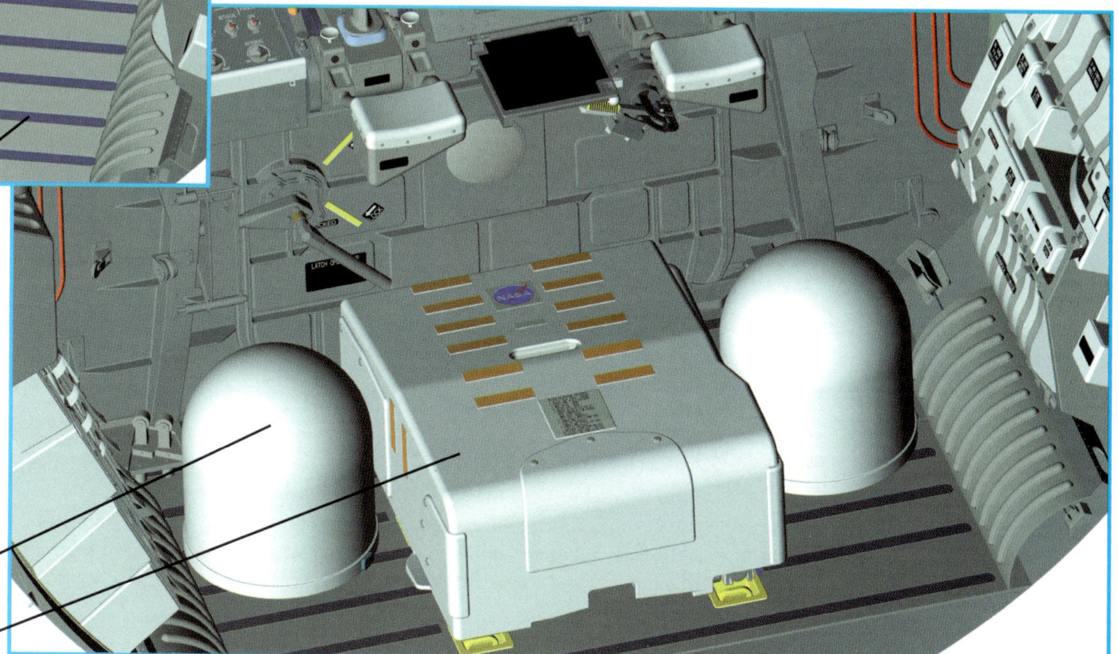

CONTROLS AND DISPLAYS

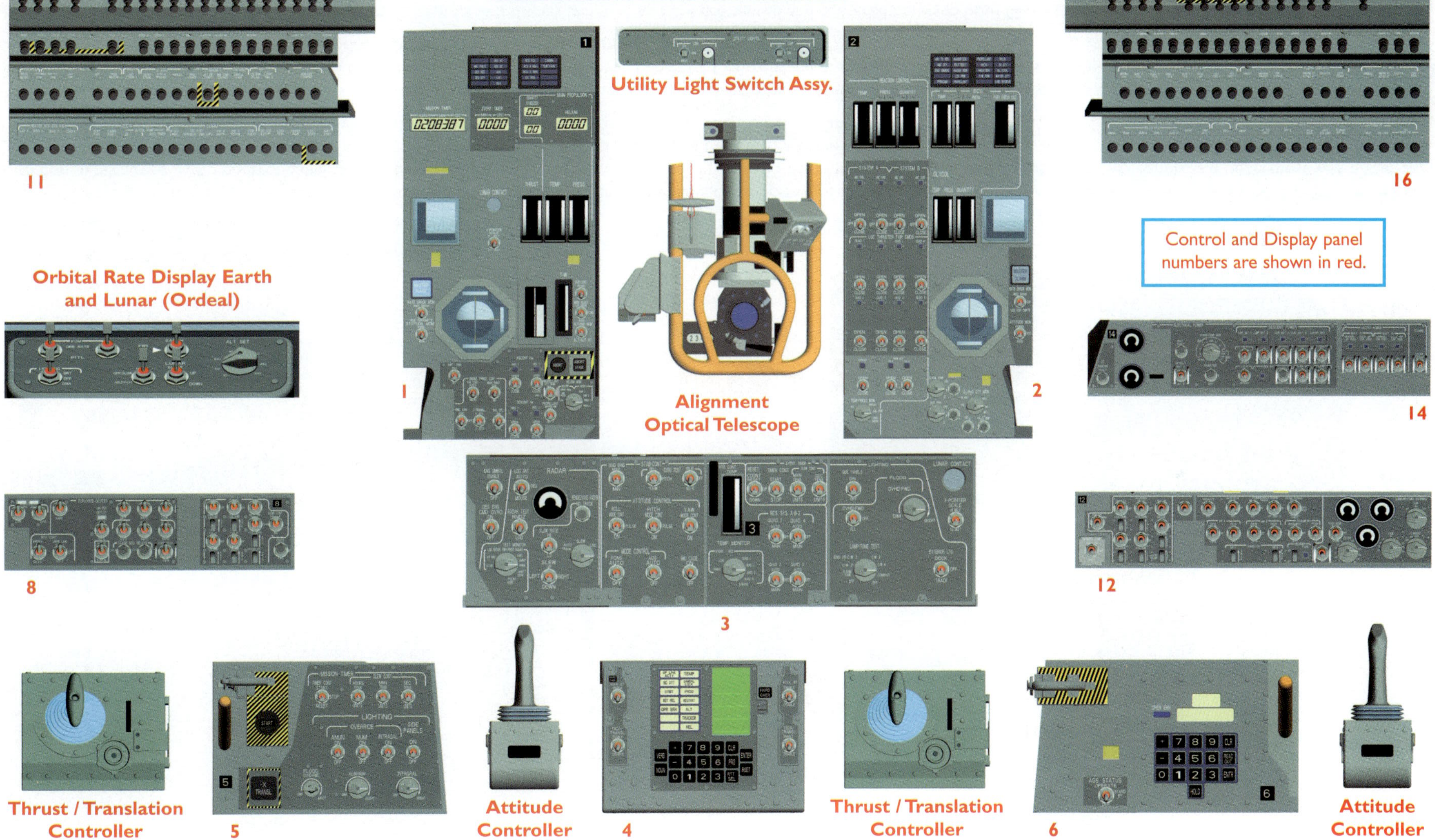

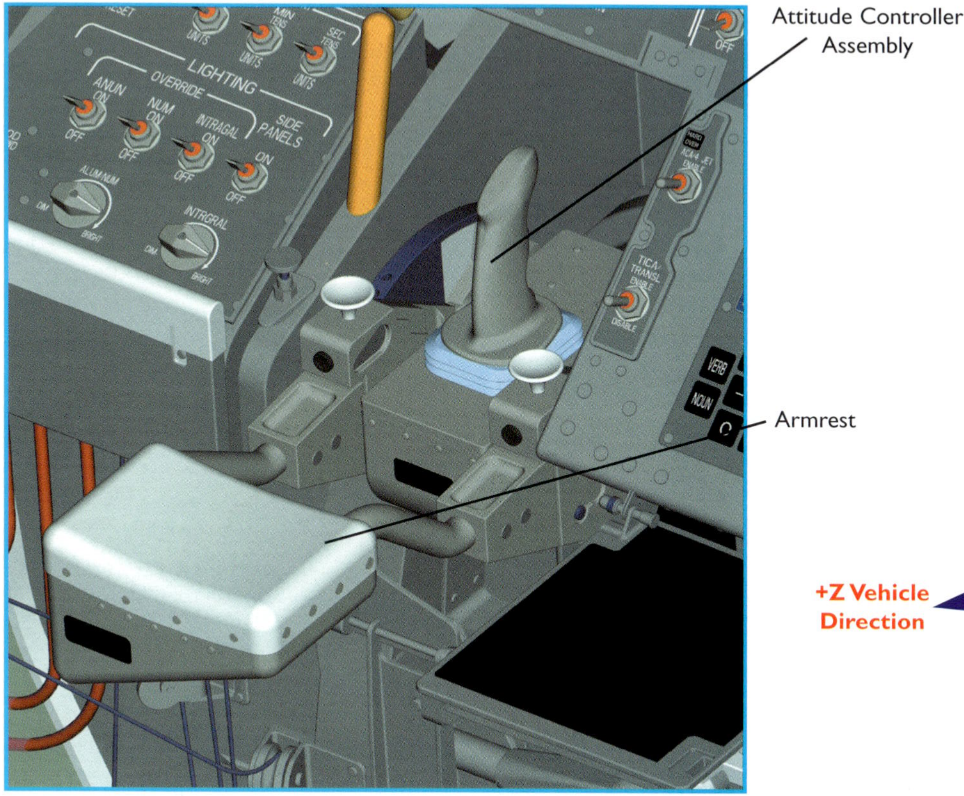

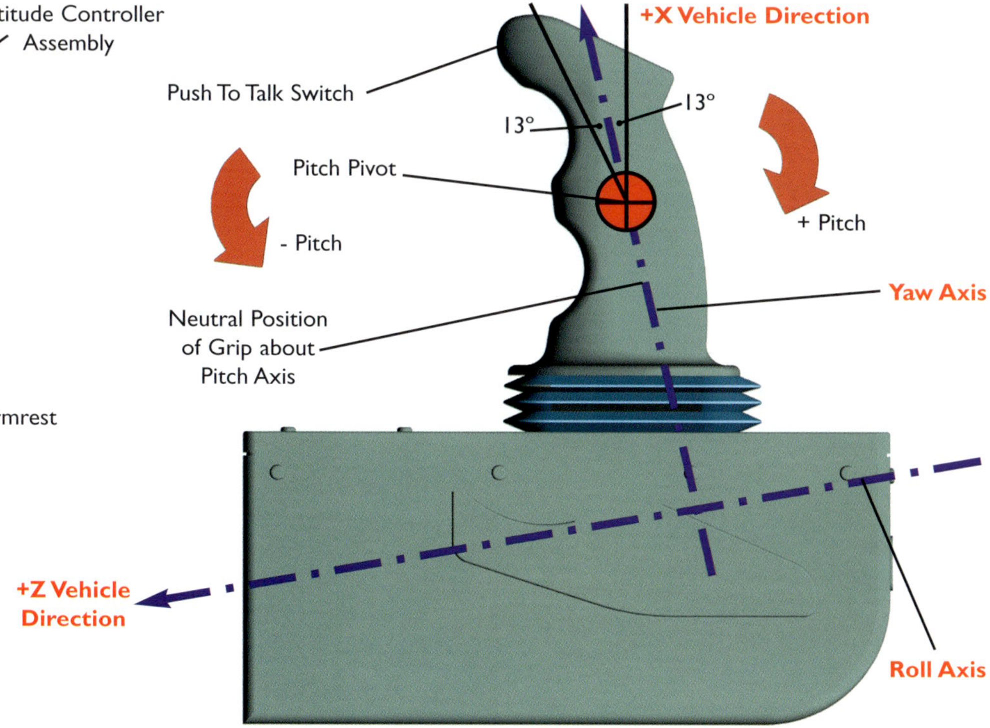

ATTITUDE CONTROLLER ASSEMBLY

The LM pilot's panel had abort guidance subsystem controls. To the left of the commander's station were three panels: a five-tier circuit breaker panel at the top, an explosive devices and communications audio control panel, and an Earth and lunar orbital rate and display panel. To the right of the LM pilot's station were two panels: the uppermost was a four-tier circuit breaker panel, the center contained controls and displays for electrical power, and the bottom panel contained communications controls and displays. The circuit breaker panels were canted to the line of sight so that the white band on each circuit breaker could be seen when the breakers were open.

In general, the controls and displays were in subsystem groupings, located in accordance with astronaut responsibilities. Certain controls were duplicated to satisfy mission and/or safety requirements; a system of interlocks prevented simultaneous operation of these controls. Controls and displays that enabled either astronaut to control the LM were centrally located so as to be accessible from both flight stations. Controls that could be operated inadvertently were appropriately guarded.

There were 12 control and display panels. The main control and display panels, 1 and 2, were canted and centered between both flight stations. Panels 3 and 4 were below these panels, within convenient reach of both astronauts. Panels 5, 8, and 11 were located for use by the commander. Panels 6, 12, 14, and 16 were located for use by the LM pilot.

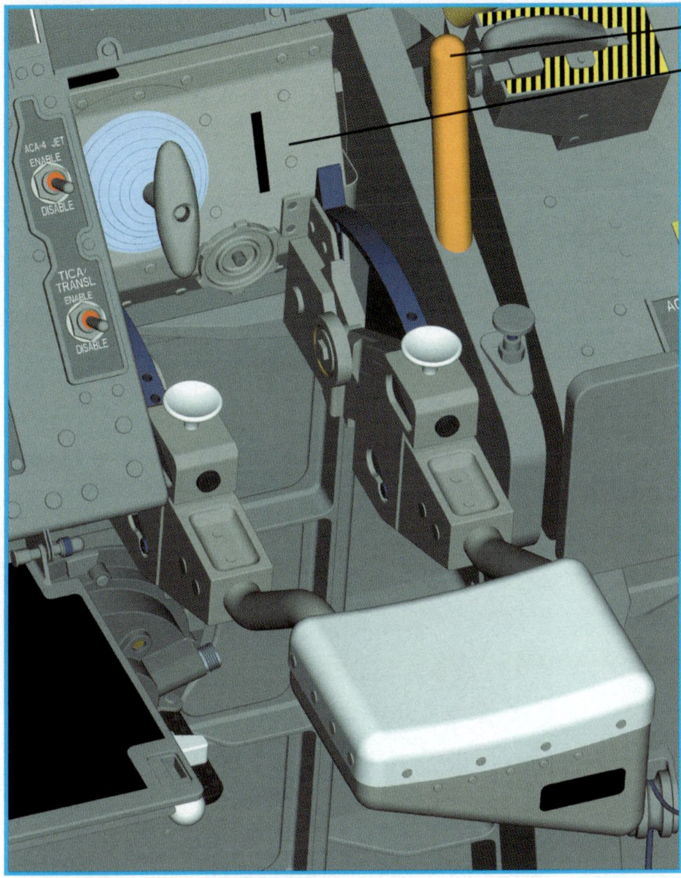

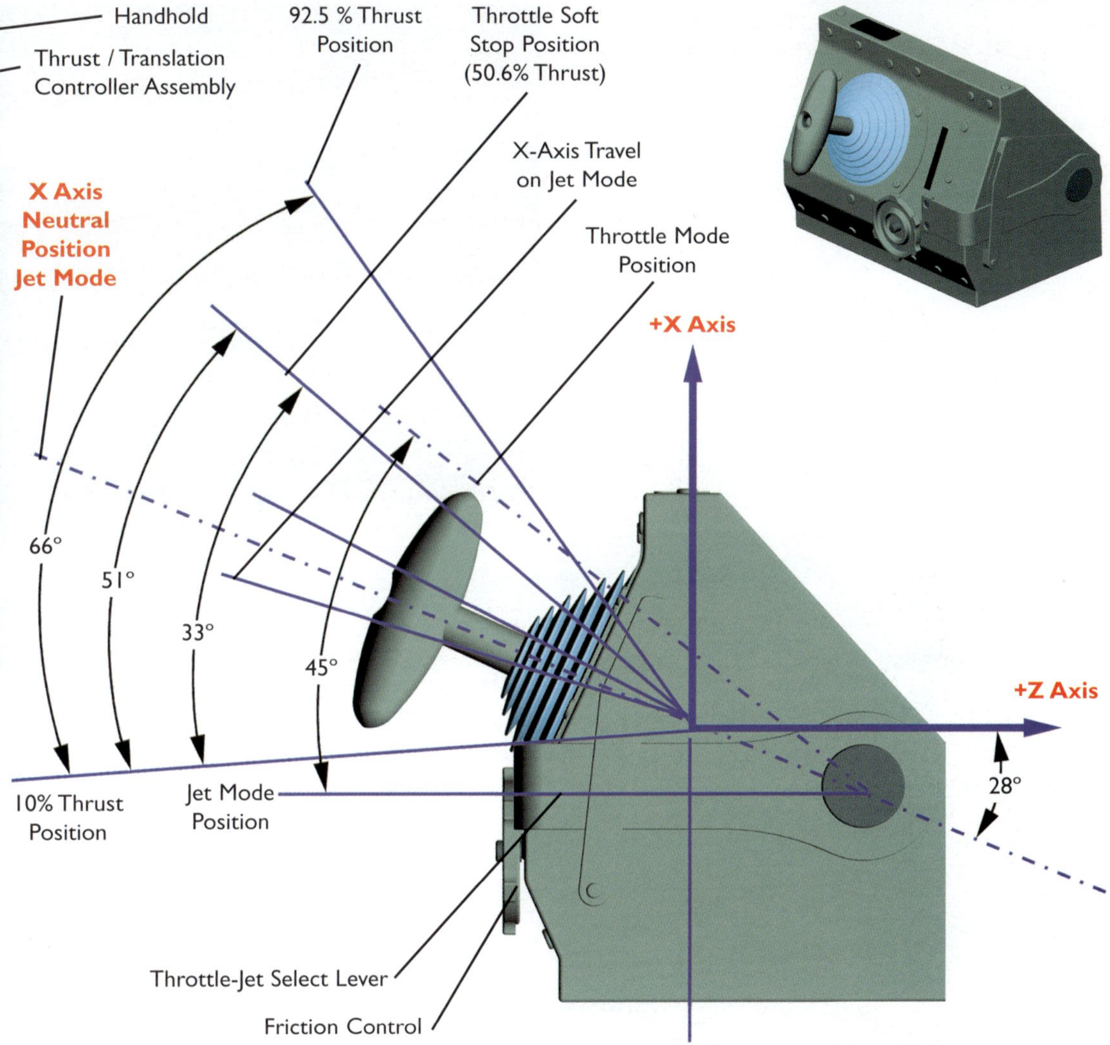

At the right of each flight station was a pistol-grip attitude controller assembly, used to control LM attitude changes. At the left of each station was a T-handle thrust / translation controller assembly. This assembly, an integrated translation and thrust controller, was used to command vehicle translations by firing thrusters in the Reaction Control Subsystem, and to throttle the descent engine.

THRUST / TRANSLATION CONTROLLER ASSEMBLY

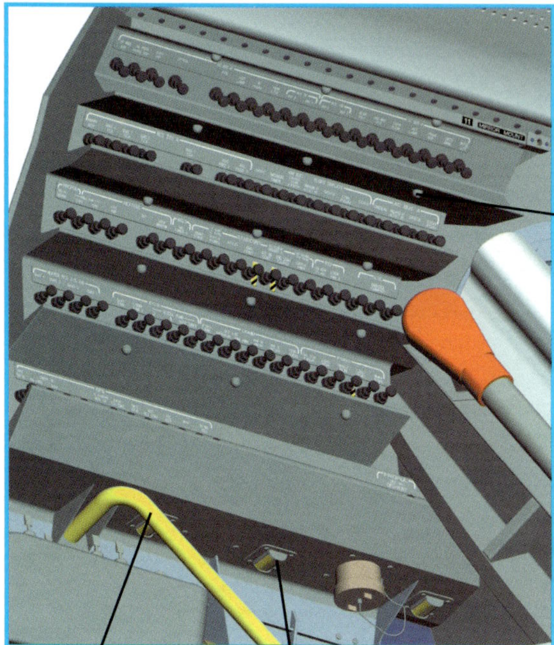

PANEL 11 – CIRCUIT BREAKERS

ORDEAL PANEL ORBITAL RATE DISPLAY (EARTH AND LUNAR)

PANEL 8 – EXPLOSIVE DEVICES, DESCENT PROPULSION AND AUDIO CONTROLS

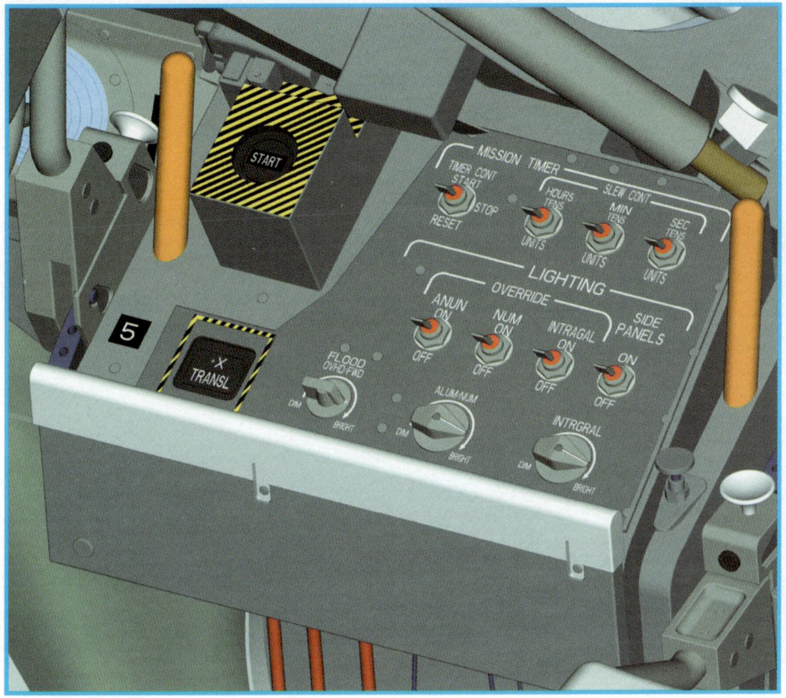

PANEL 5 – LIGHTING AND MISSION TIMER CONTROLS, ENGINE START AND STOP PUSHBUTTONS, AND X-TRANSLATION PUSHBUTTONS

The DEDA (Data Entry and Display Assembly), panel 6, was used by the astronauts to select the desired mode of operation, insert the desired targeting parameters, and monitor related data throughout the mission. Essentially, the DEDA consisted of a control panel to which electroluminescent displays and data entry push buttons were mounted and a logic enclosure that housed logic and input-output circuits.

PANEL 6 – ABORT GUIDANCE CONTROLS

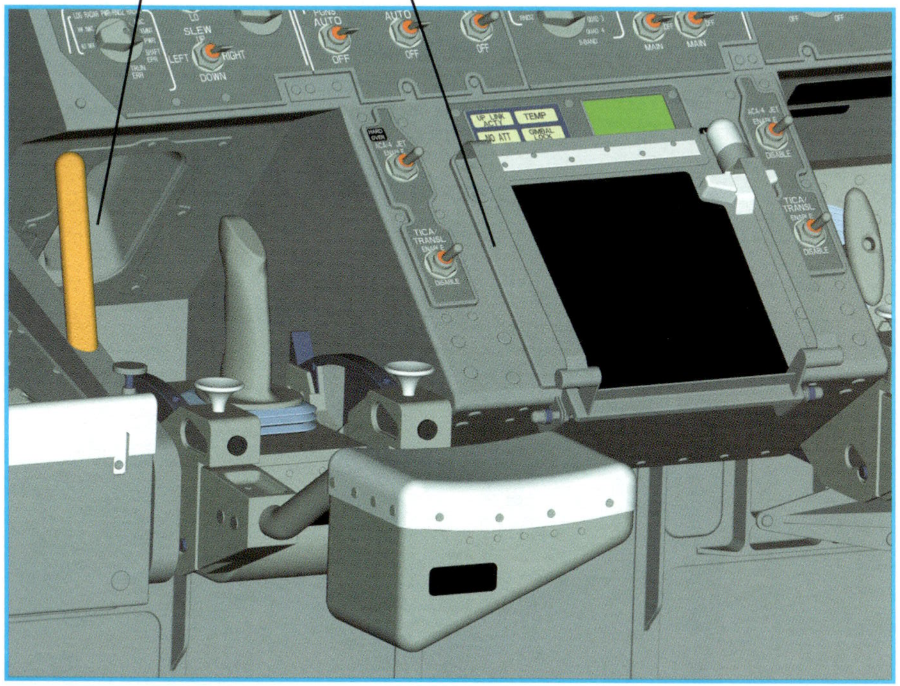

PANEL 4 – DISPLAY AND KEYBOARD ASSEMBLY – 2-WAY COMMUNICATIONS LINK BETWEEN ASTRONAUTS AND LM GUIDANCE COMPUTER

PANEL 14 – CONTROLS AND DISPLAYS FOR ELECTRICAL POWER DISTRIBUTION AND MONITORING

PANEL 12 – AUDIO, COMMUNICATIONS, AND COMMUNICATIONS ANTENNA CONTROLS AND DISPLAYS

Mirror Mount (Both Sides)

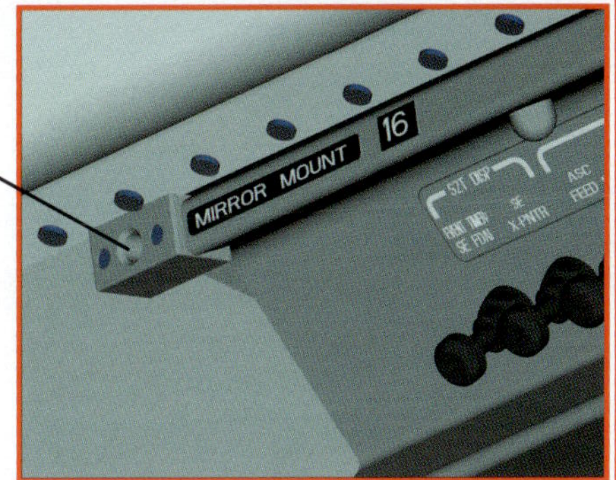

PANEL16 – CIRCUIT BREAKERS

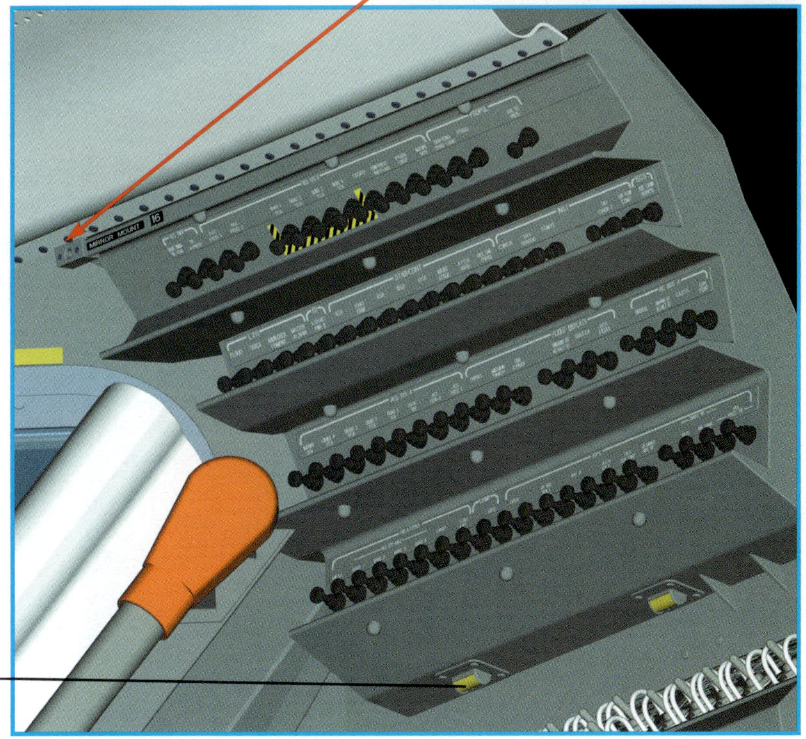

Panel Light

PANEL 1 – WARNING LIGHTS, DIGITAL COUNTERS, NAVIGATION INSTRUMENTS, ENGINE THRUST CONTROL SWITCHES, AND ENGINE FUEL AND ALTITUDE INDICATORS

PANEL 2 – CAUTION LIGHTS, REACTION CONTROL INDICATORS, ENVIRONMENTAL CONTROL INDICATORS, NAVIGATIONAL INSTRUMENTS, AND VARIOUS OTHER INDICATORS AND SWITCHES

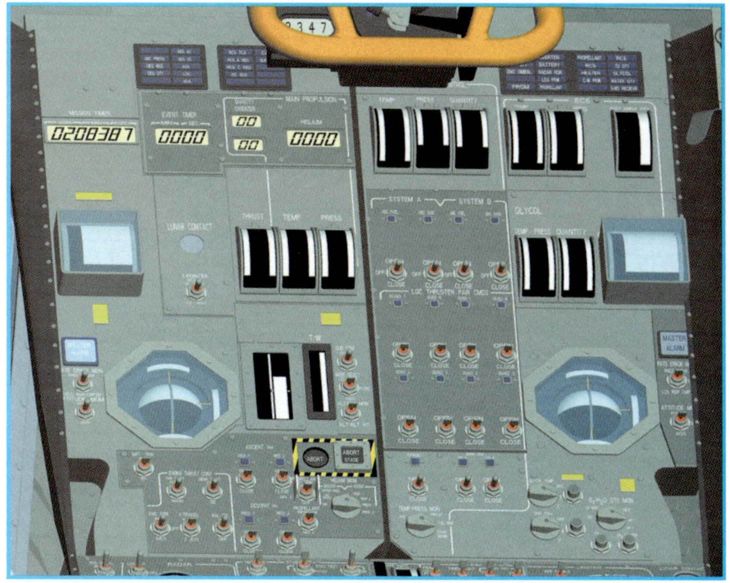

PANEL 3 – CONTROLS AND DISPLAYS FOR RADAR, STABILIZATION AND CONTROL, HEATER CONTROL, EVENT TIMER, AND LIGHTING

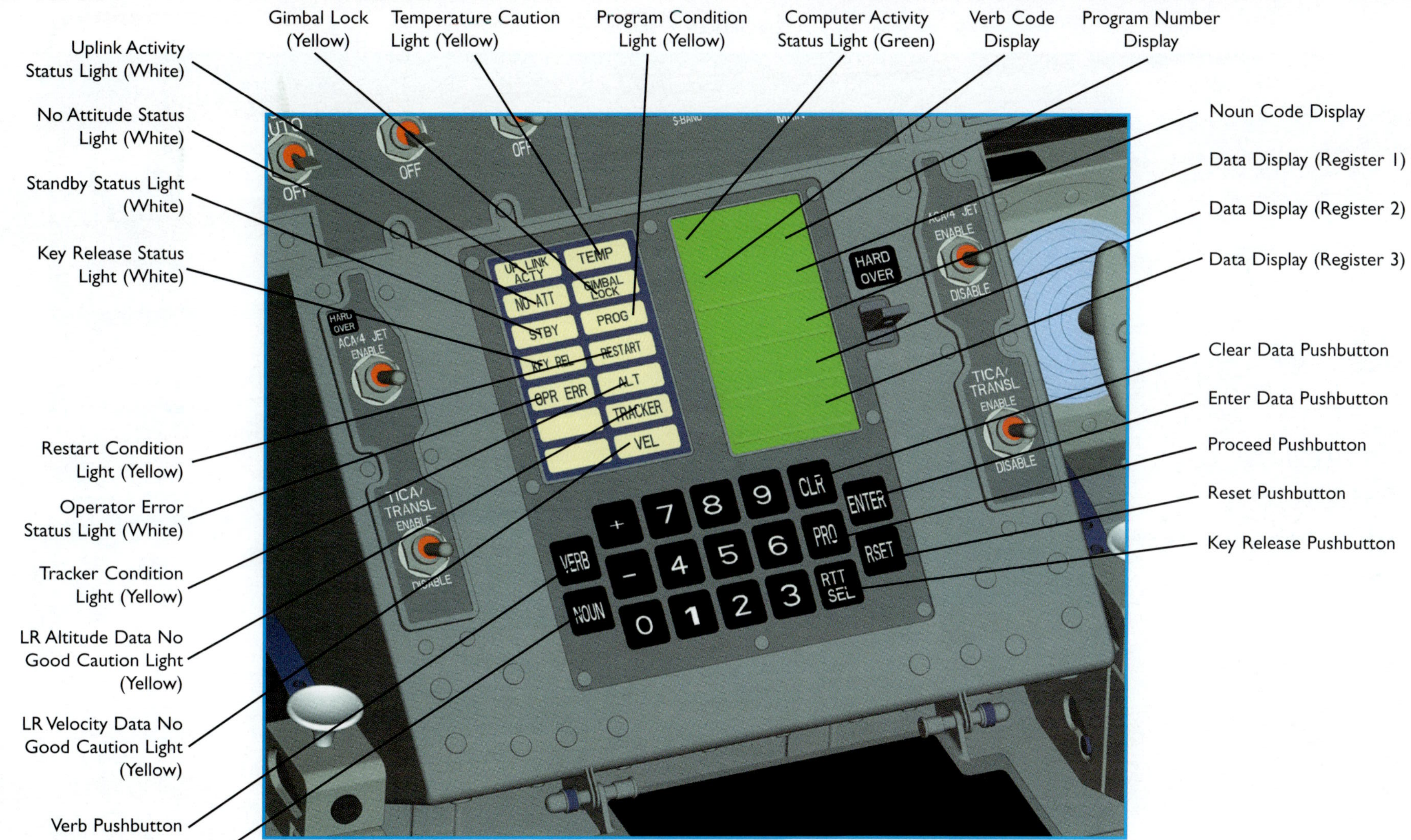

DISPLAY AND KEYBOARD ASSEMBLY DETAIL

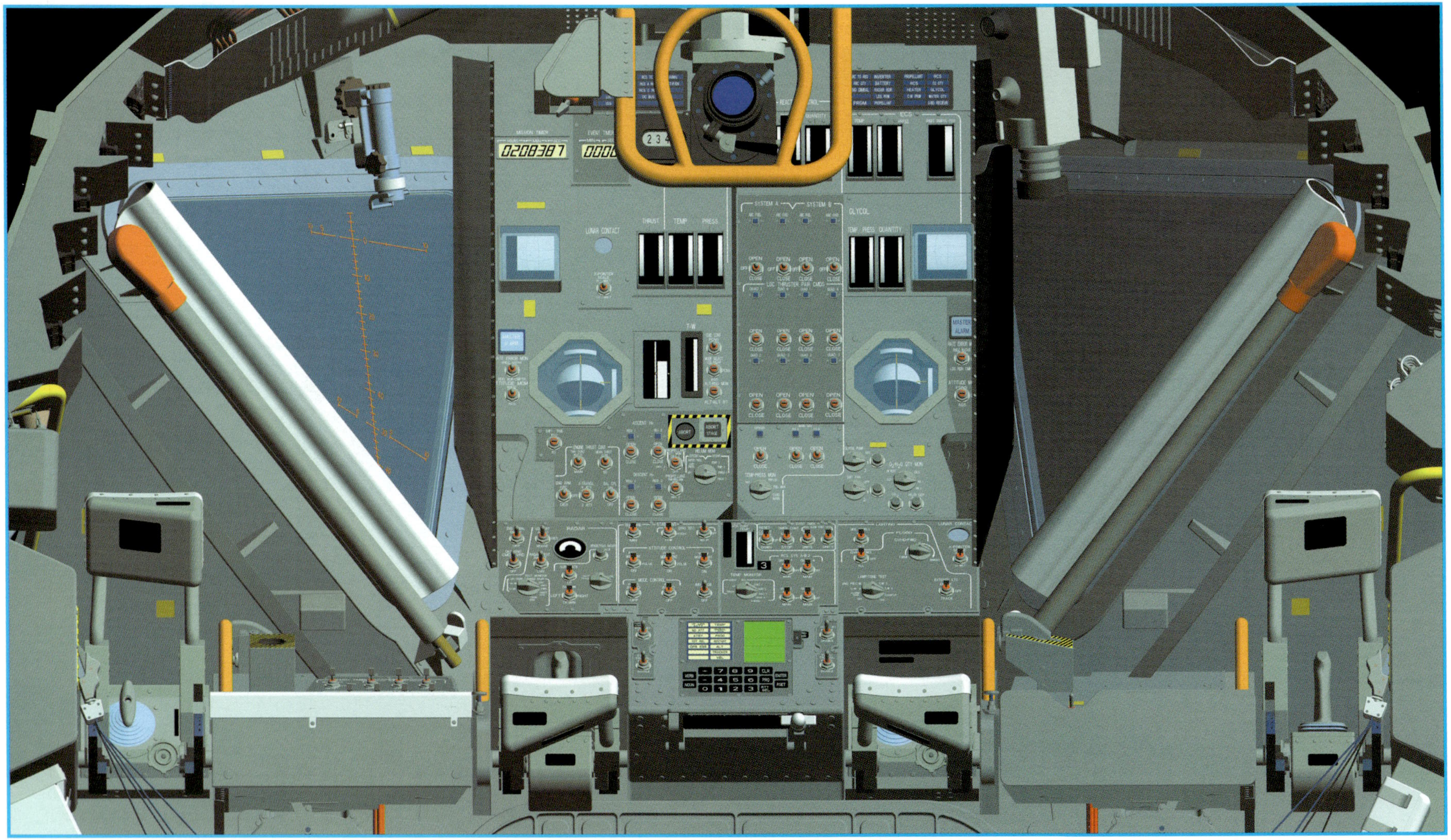

MISSION COMMANDER

FLIGHT STATIONS

LUNAR MODULE PILOT

ASTRONAUT RESTRAINT SYSTEM

The crew restraint system equipment included armrests, handholds, Velcro on the floor to interface with the astronaut's boots, and a restraint system operated by a rope-and-pulley arrangement that secured the astronauts in an upright position under zero gravity conditions. The restraint assembly consisted of cables, restraint clips, and a constant force reel system. The restraint clips attached to D-rings on the astronaut's space suits at waist level. The constant force reel control provided a downward force of approximately 30 pounds. It was locked during landing and docking operations. When the constant force reel was locked, the cables were free to reel in. A ratchet stop prevented paying out of the cables and thus provided restraint during lunar landing.

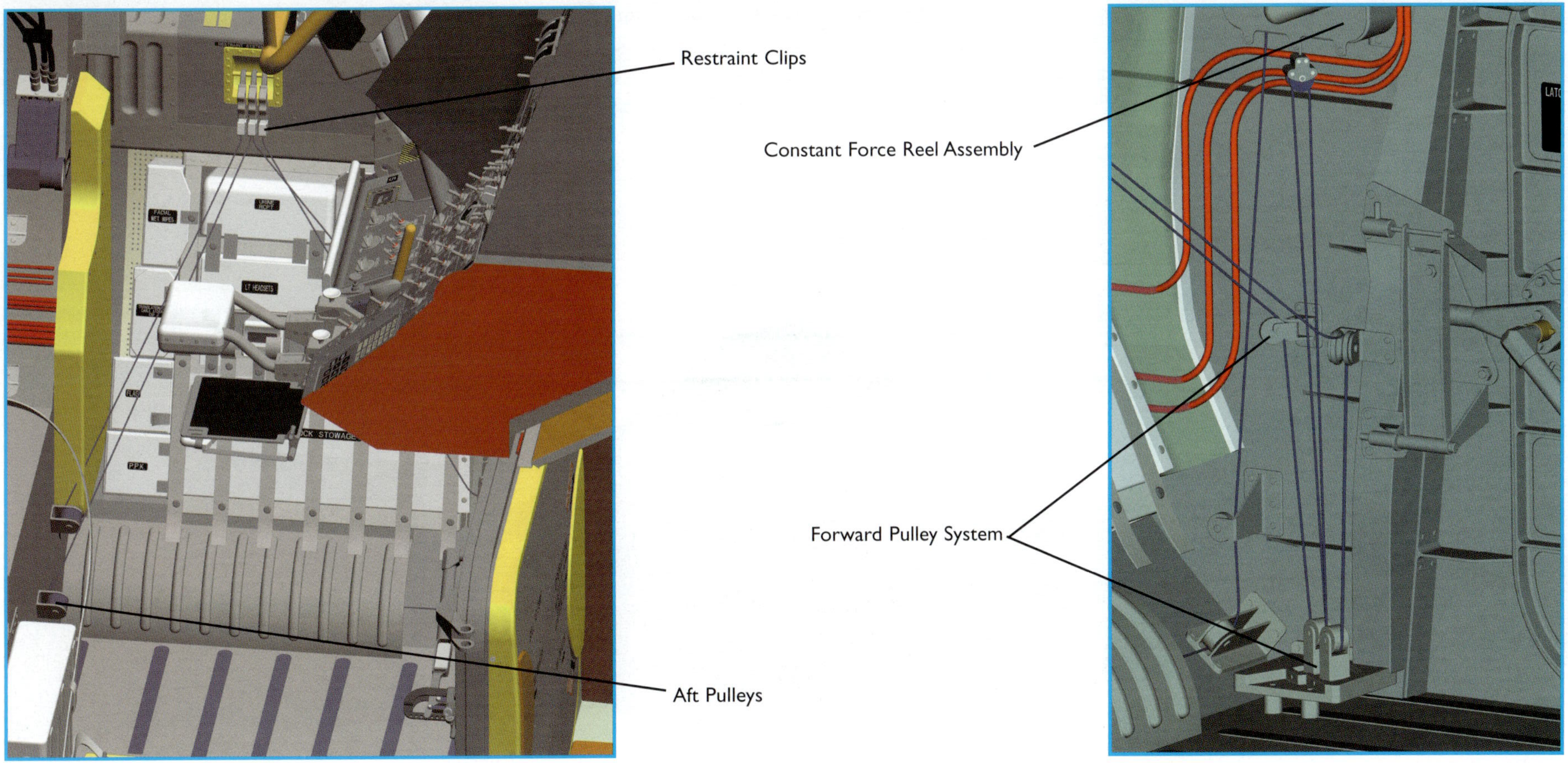

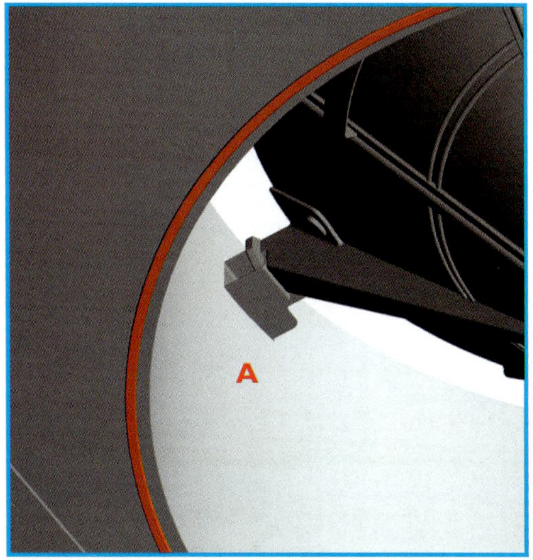

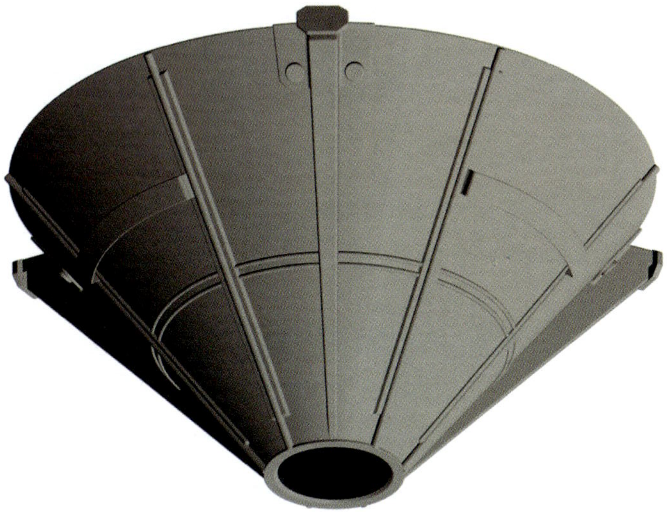

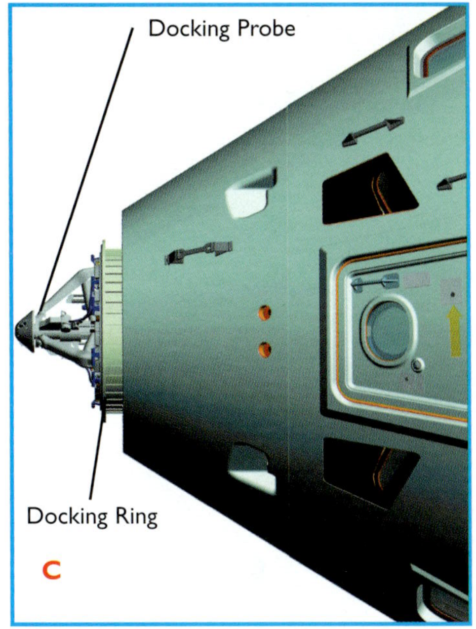

DROGUE ASSEMBLY

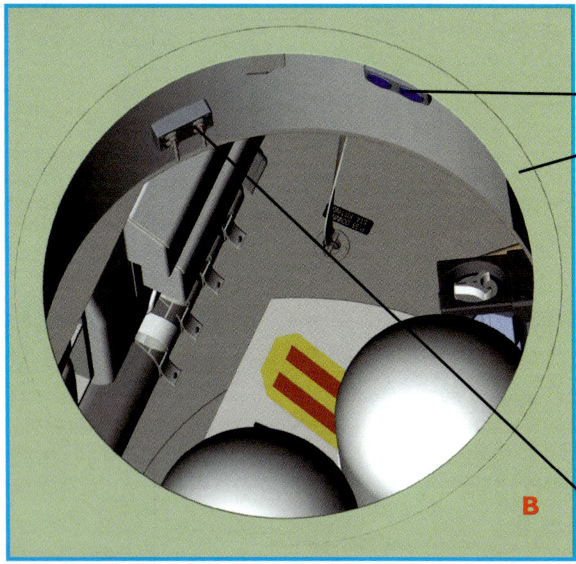

The drogue assembly consisted of a conical structure mounted within the ascent stage docking tunnel. It was secured at three points on the periphery of the tunnel, below the LM's docking ring. The drogue could be removed from either the CM or the LM side of the tunnel.

Docking of the LM and CSM required special equipment and tunnel hardware. Docking equipment included the COAS and docking targets, the drogue assembly, and the probe assembly on the Command Module (C), the CM's forward hatch, and hardware inside the LM's docking tunnel (B). View (A) shows the interface between the drogue assembly and the docking latches inside the LM docking tunnel.

The LM docking tunnel also contained the electrical umbilical connectors, probe mounting ring, dead-face switches, and lights.

COMMAND AND SERVICE MODULES (CSM) AND LUNAR MODULE (LM) DOCKING

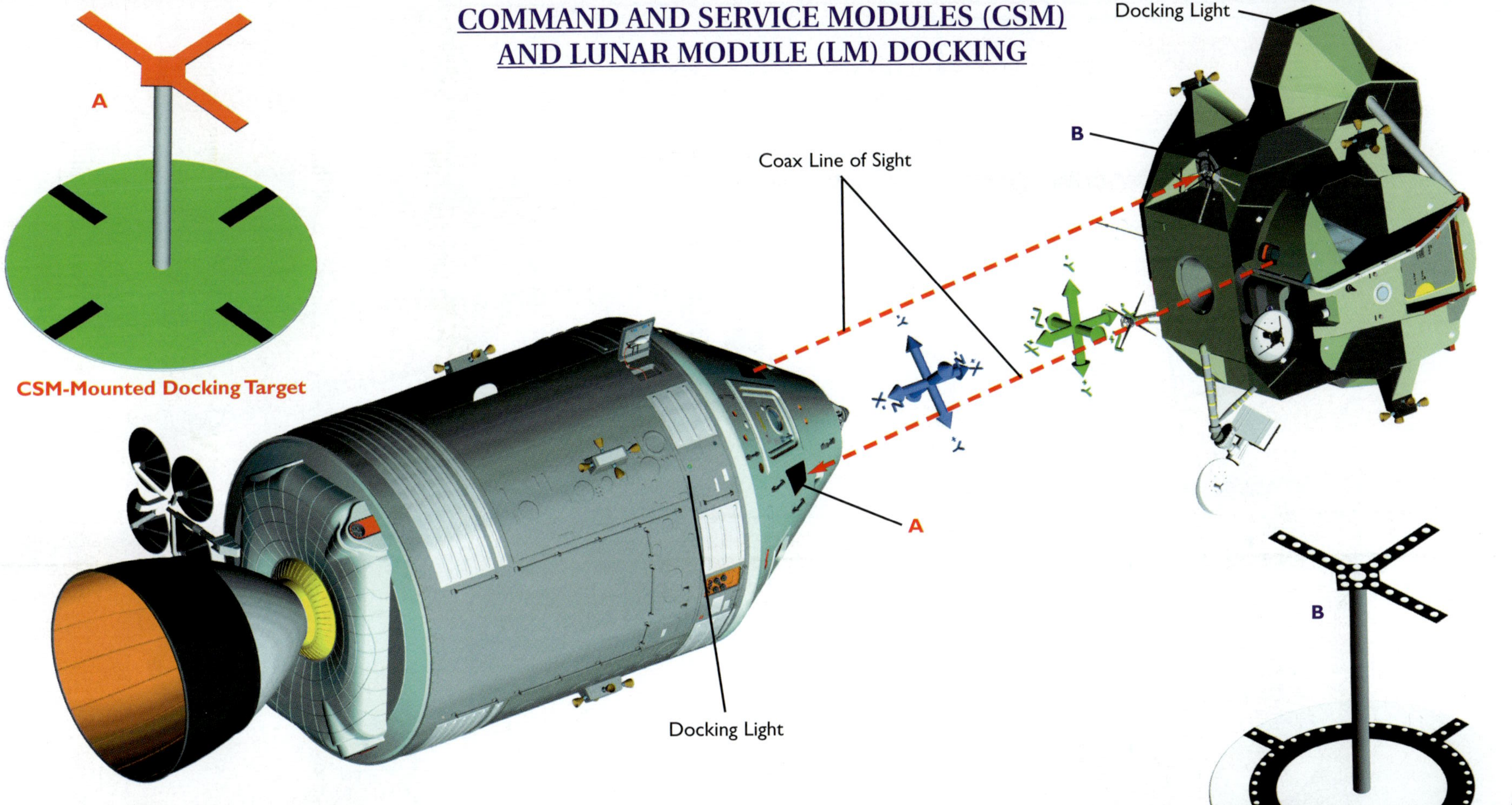

The docking target permitted docking to be accomplished on a 3-dimensional axis basis. The docking probe on the Command Module provided initial CM-LM docking and attenuated impact energy imposed by either vehicle. The COAS provided the commander with gross range cues and closing rates during the docking maneuvers. The closing operation, from 150 feet to the point of contact, required control with a minimum use of fuel and time. The COAS provided the commander with a fixed line-of-sight attitude reference image, which appeared to be the same distance away as the target. The mounting ring, or flange, on the upper part of the LM docking tunnel mated with the 12 docking latches on the Command Module.

– Scott P. Sullivan –

DOCKING TARGET ON LM WAS ALSO CANTED AT 60 DEGREES TO ALIGN WITH CM LINE OF SIGHT

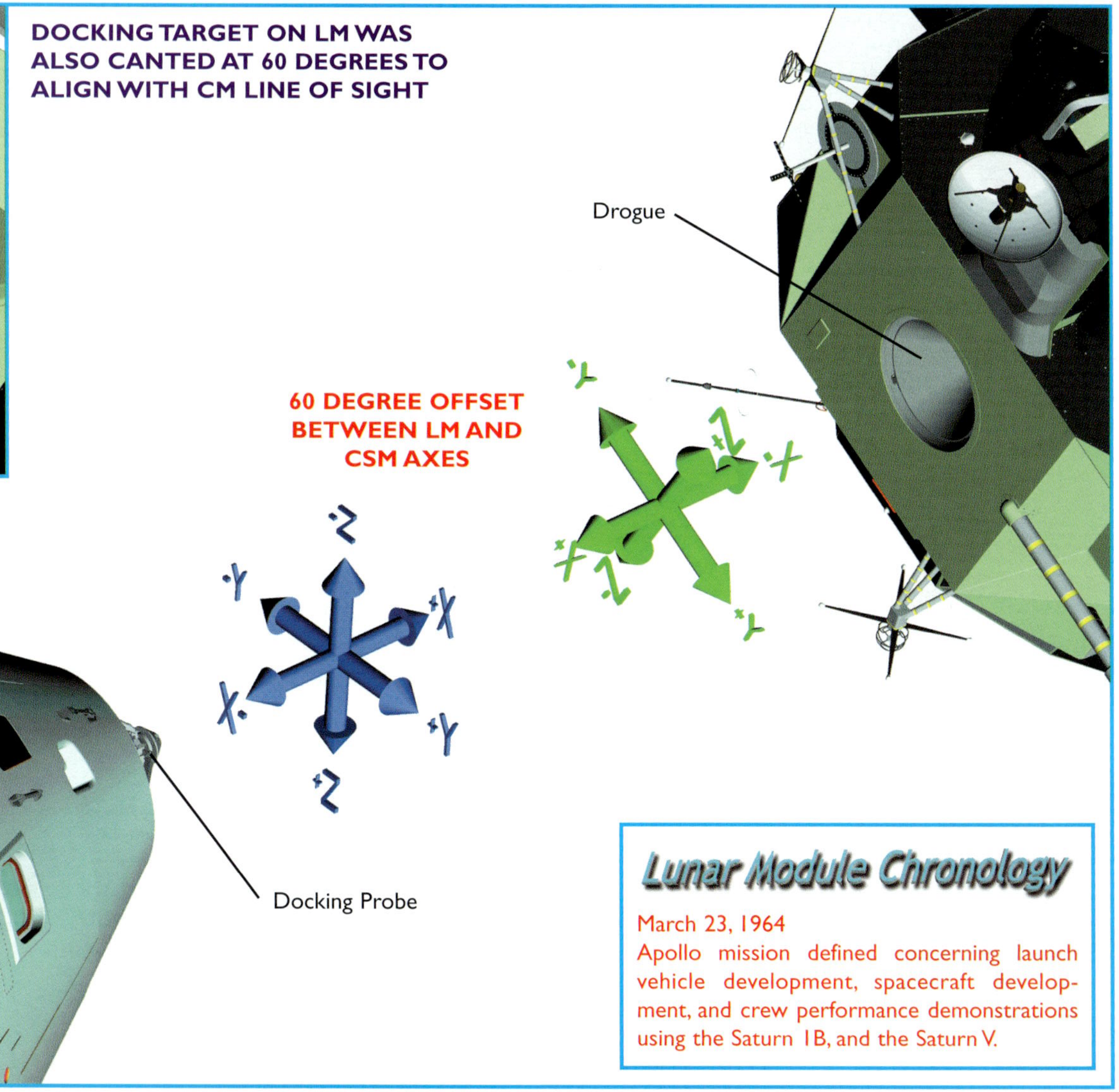

Lunar Module Chronology

March 23, 1964
Apollo mission defined concerning launch vehicle development, spacecraft development, and crew performance demonstrations using the Saturn 1B, and the Saturn V.

LUNAR MODULE ASCENT ENGINE

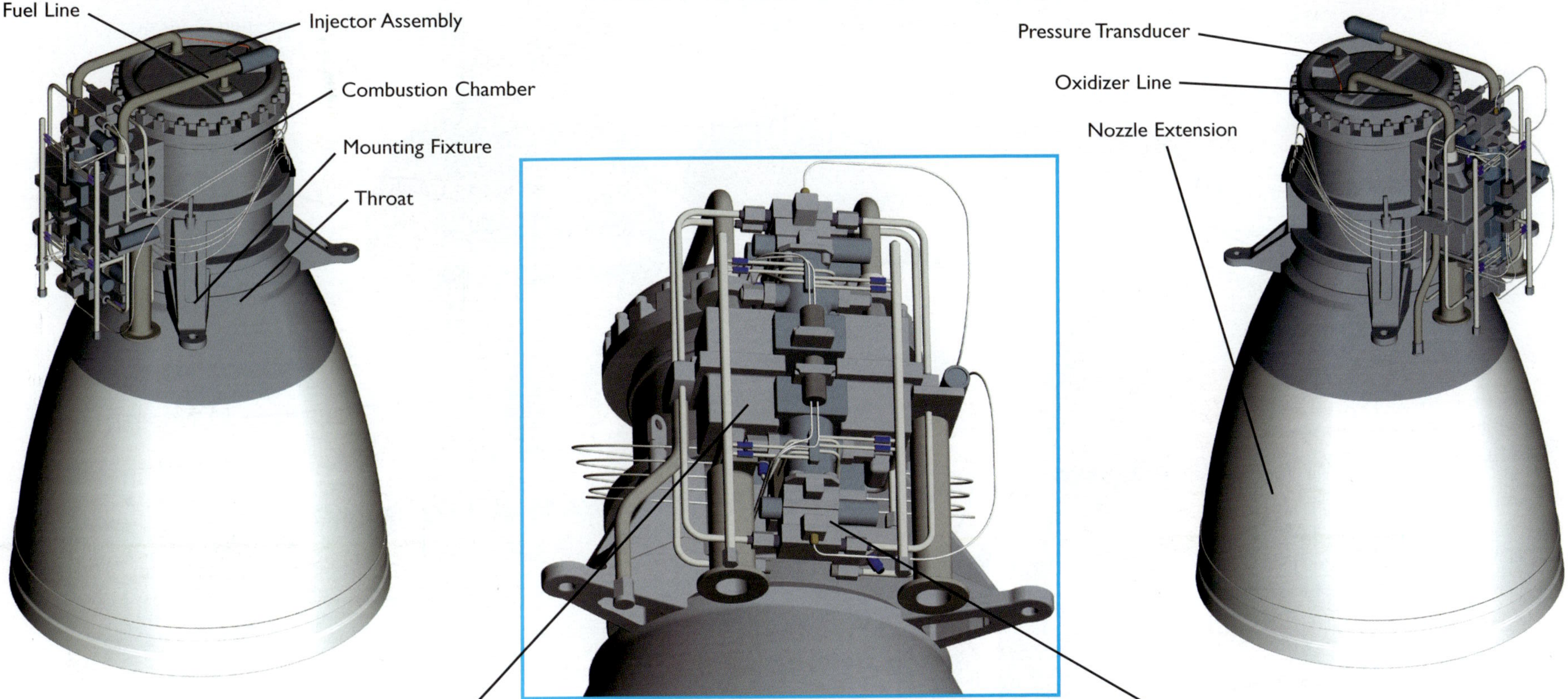

PROPELLANT VALVE PACKAGE ASSEMBLY

The ascent engine was installed in the midsection of the ascent stage; it was tilted so that its centerline was 1.5 degrees from the X-axis, in the direction of the Z-axis. Fuel and oxidizer entering the engine assembly were routed through the filters, propellant shutoff valves, and trim orifices to the injector. The propellants were injected into the combustion chamber, where the hypergolic ignition occurred.

The ascent propulsion section consisted of a constant-thrust, pressure-fed rocket engine, one fuel and one oxidizer tank, two helium tanks, and associated propellant feed and helium pressurization components. The engine developed 3,500 pounds of thrust in a vacuum, and it could be shut down and restarted. Like the descent propulsion section, the ascent propulsion section could be functionally sub-divided into a pressurization section, a propellant feed section, and an engine assembly.

ASCENT STAGE PROPULSION SYSTEM

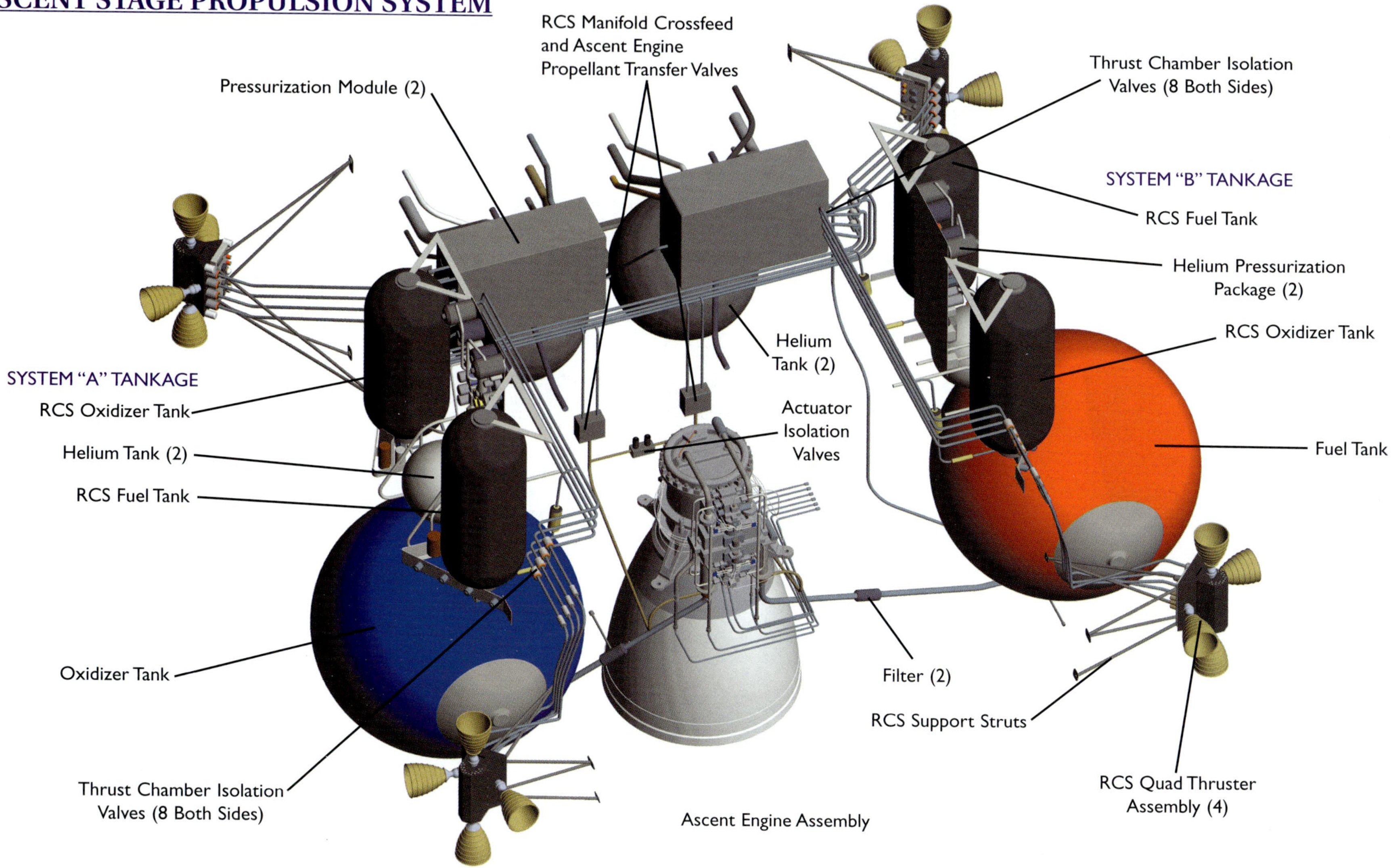

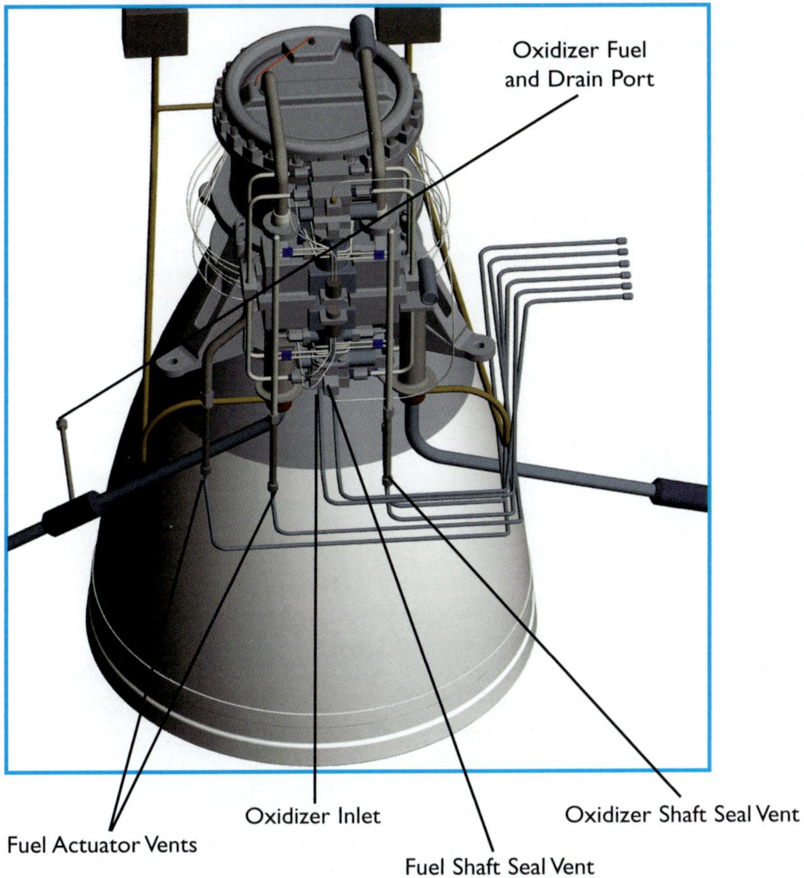

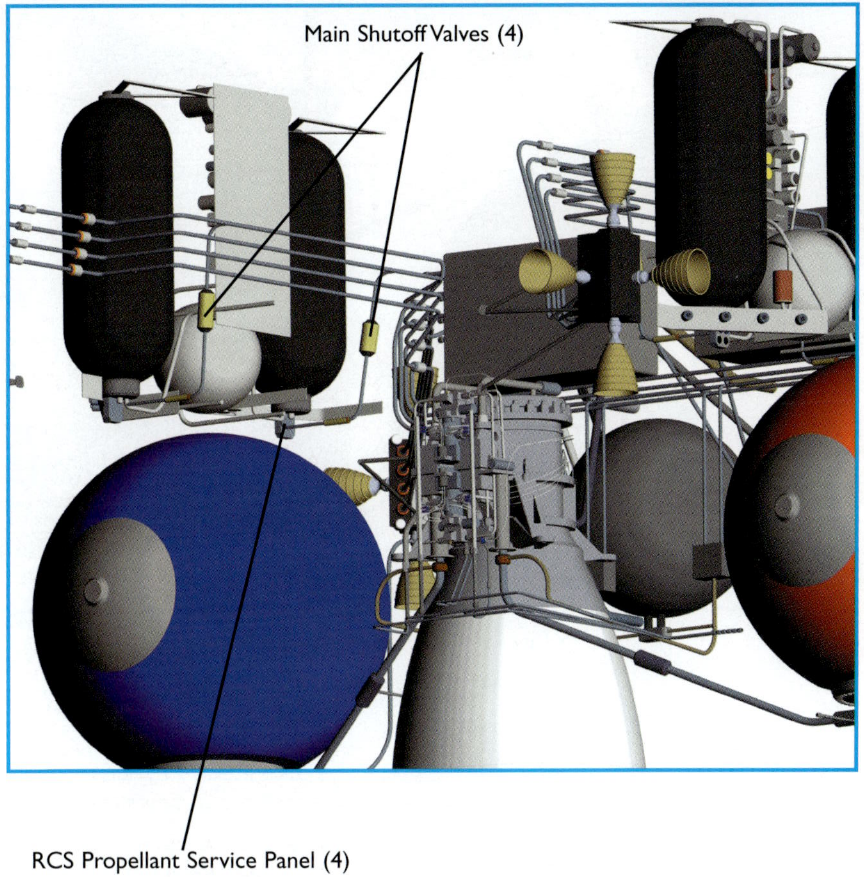

Each helium pressure regulator assembly consisted of two individual pressure regulators connected in series. A rise in outlet pressure decreased demand, thereby reducing the flow into the main stage chamber. An increase in the downstream demand caused a reduction in outlet pressure; this tended to open the pilot valve. The resultant increase in control pressure caused the main stage valve to open, meeting the increased downstream demand.

The propellant supply was contained in two spherical tanks. The tanks were identical in size and construction. One tank contained fuel, the other oxidizer. A helium diffuser at the inlet port of each tank distributed the pressurizing helium uniformly into the tank. A low level sensor in each tank supplied a discrete signal that caused a caution light to go on when the propellant remaining in either tank was sufficient for approximately 10 seconds of burn time (48 pounds of fuel, 69 pounds of oxidizer).

At the propellant feed section / engine assembly interface, the oxidizer and fuel lines led into the valve package assembly. The individual valves that made up the valve package assembly were in a series-parallel arrangement to provide redundant propellant flow paths and shutoff capacity. The valve package assembly consisted of eight propellant shutoff valves and four solenoid-operated pilot valve and actuator assemblies. Valve assemblies consisted of one fuel shutoff valve and one oxidizer shutoff valve.

The injector assembly consisted of the propellant inlet lines, a fuel manifold reservoir chamber, an oxidizer manifold, and an injector orifice plate assembly.

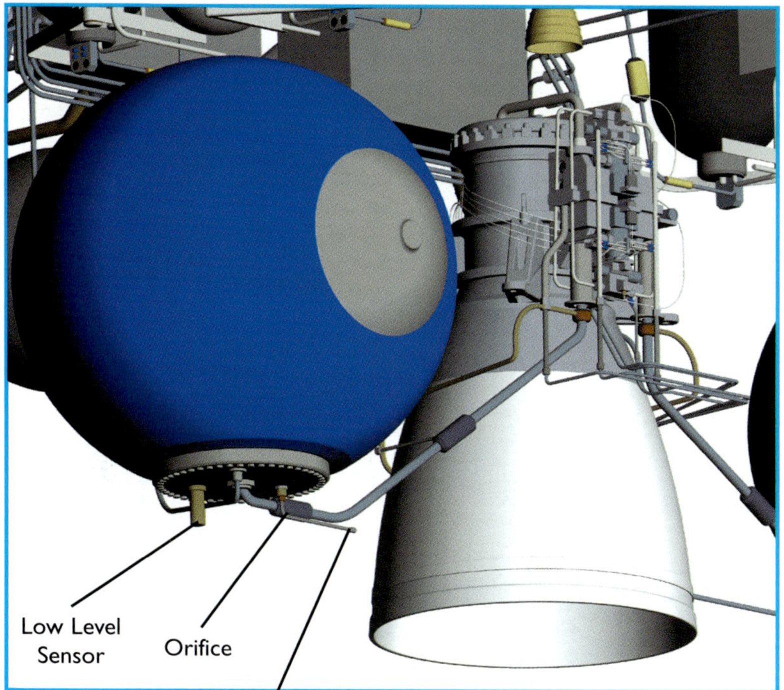

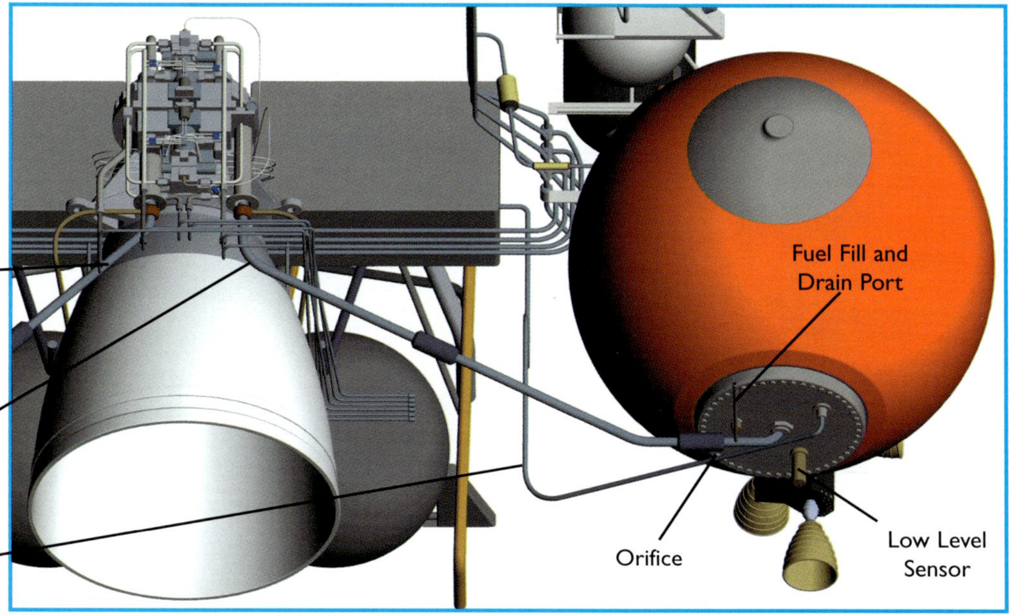

ENVIRONMENTAL CONTROL SYSTEM (ECS)

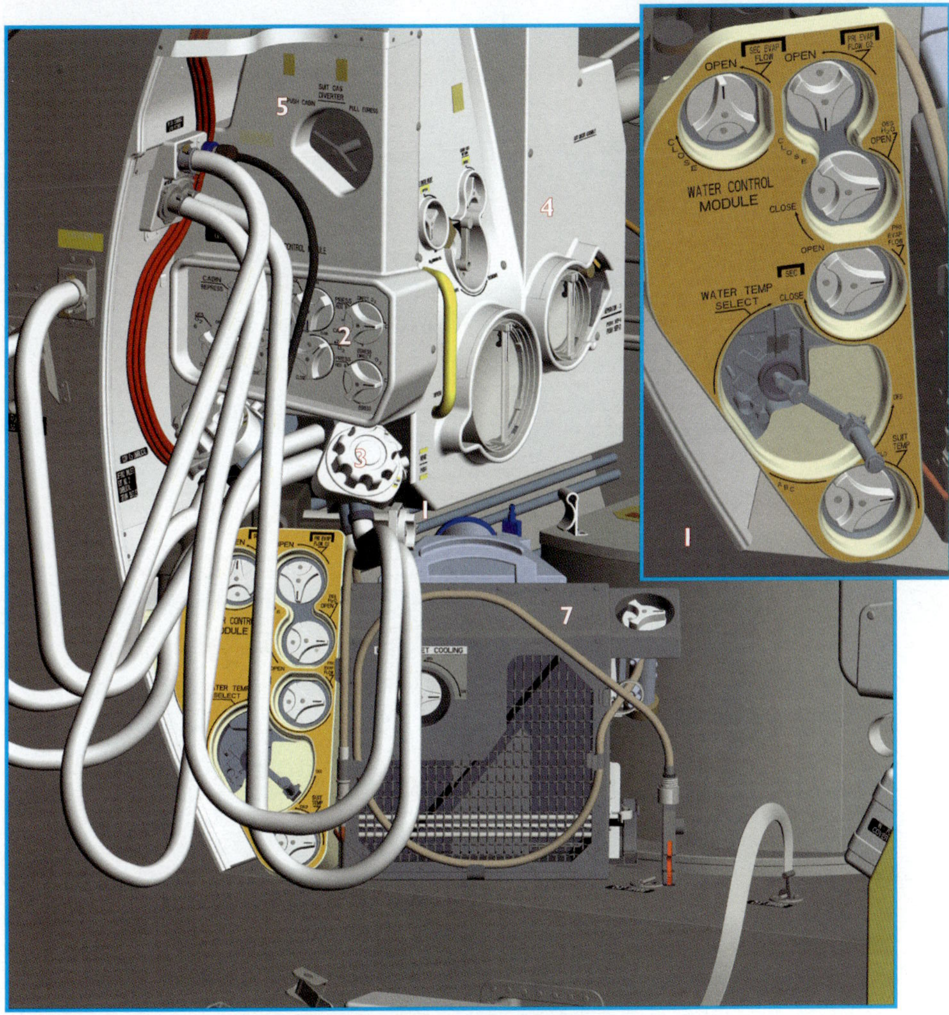

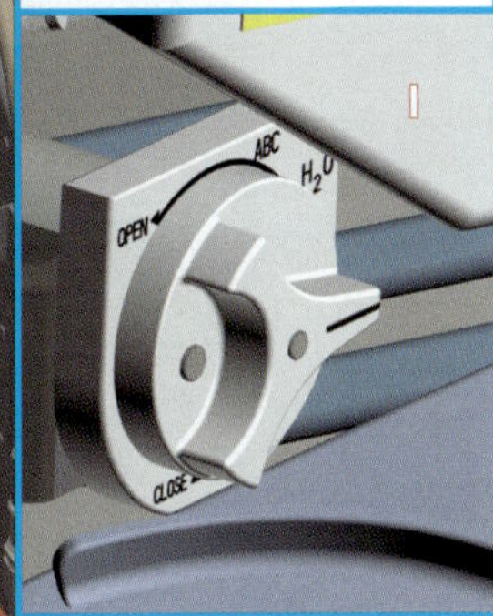

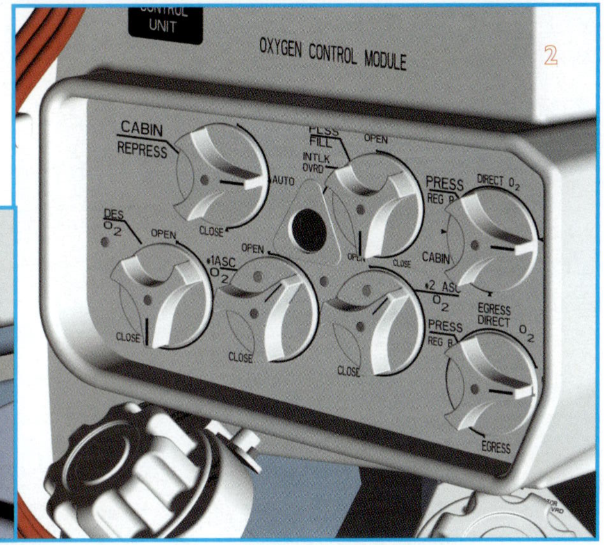

MIDSECTION COMPARTMENT

The environmental control system (ECS) enabled pressurization of the crew cabin and space suits, controlled temperature of electronic equipment, and provided breathable oxygen for the astronauts. It also provided water for drinking, cooling, fire extinguishing, and food preparation, and supplied oxygen and water to the portable life support system (PLSS).

The major portion of the ECS was in the midsection. The peripheral ECS equipment, such as oxygen and water tanks, was located outside the cabin in the ascent and descent stages. The ECS consisted of the following sections: atmosphere revitalization section (ARS), oxygen supply and cabin pressure control section (OSCPCS), water management section (WMS), and heat transport section (HTS).

1 Water Management Controls
2 Oxygen Control Module
3 Suit Flow Controls
4 Suit Circuit and Canister Controls
5 Suit Gas Diverter
6 Cabin Relief and Dump Valve
7 Liquid Garment Cooling

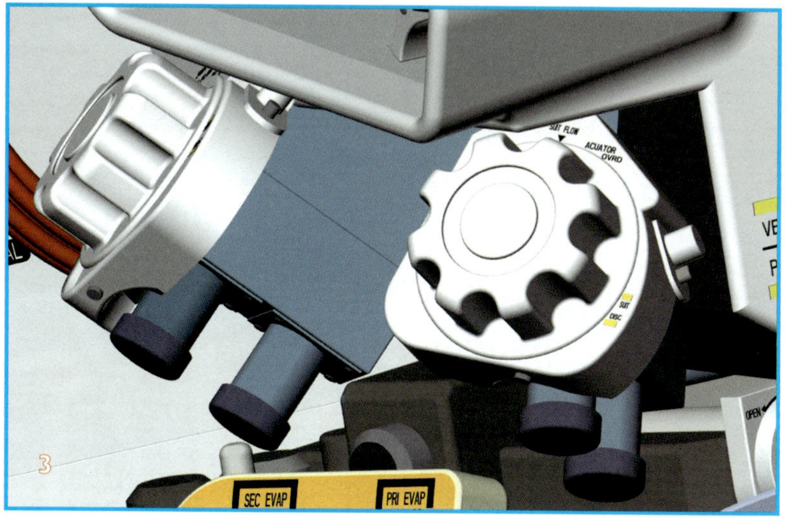
3
SUIT FLOW
ACUATOR OVRD
SEC EVAP
PRI EVAP
OPEN

4
PRIMARY
SECONDARY
PUSH SEP-1
PUSH SEP-2
VENT
PUSH

5
SUIT GAS
DIVERTER
PUSH CABIN
PULL EGRESS

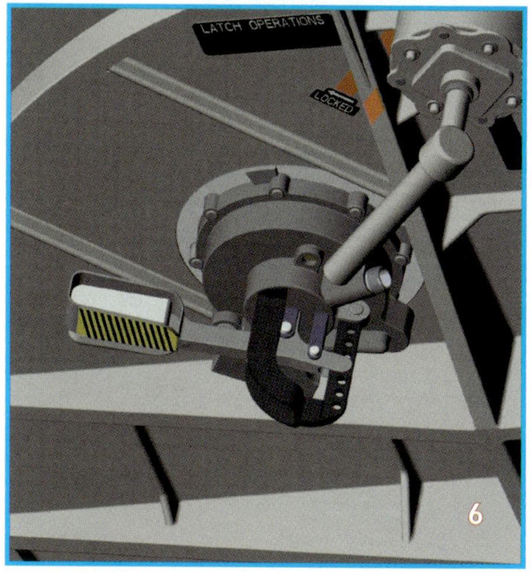
6
LATCH OPERATIONS
LOCKED

7
UID GARMET COOLING

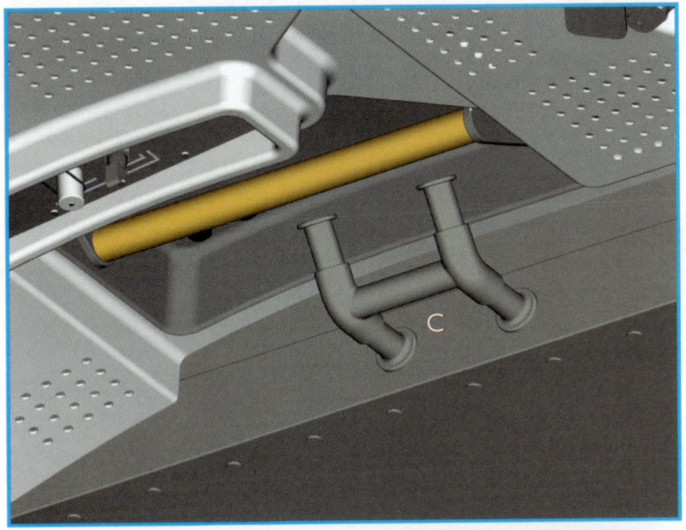

The view to the upper left shows a mounting fixture labeled "C." This fixture was used to attach a cable / pulley system that ran out the open front hatch and down to an astronaut standing on the lunar surface. The cable assembly allowed the astronaut to raise the lunar sample container up and into the ascent stage, after which the other astronaut would detach the container from the cable assembly for stowage in the ascent stage.

The view to the lower left shows the handle marked "D" that was used to raise the EVA antenna at the aft portion of the ascent stage.

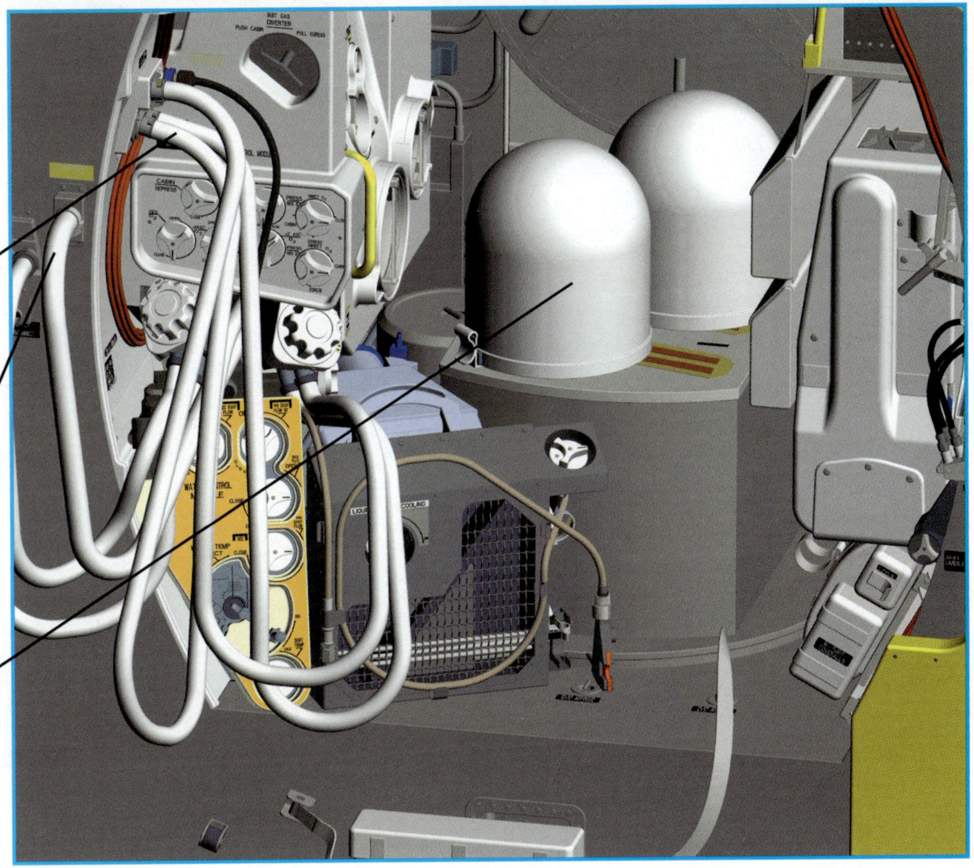

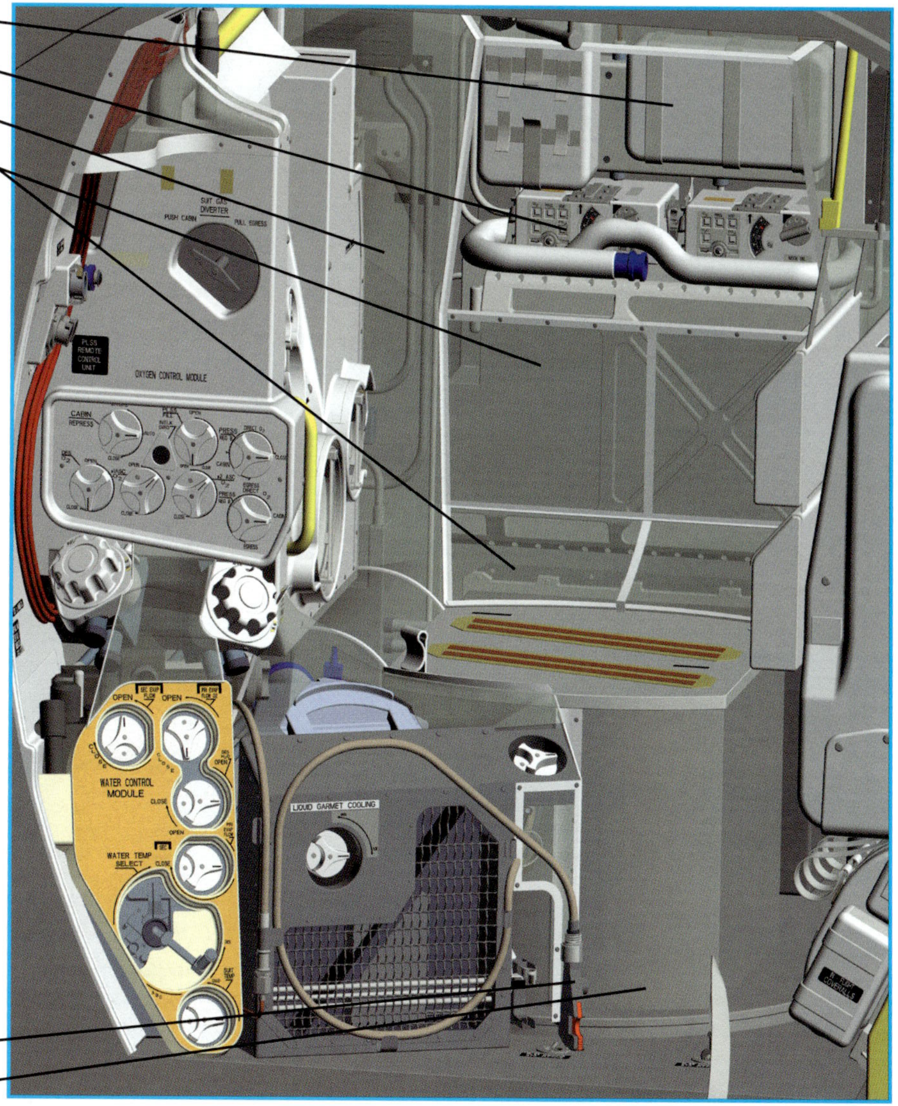

From the flight stations, there was an 18-inch step up into the midsection, which was immediately aft of the crew compartment. Normally, the midsection was not manned and was used for stowage of supplies and equipment and to house the ECS. The midsection was 54 inches deep and approximately 5 feet high. The internal shape was elliptical, with a minor axis of approximately 56 inches. The midsection housed the ascent engine assembly, part of which was ported through the lower deck and covered with a protective housing that acted as an auxiliary stowage position for the astronauts' helmet stowage bags. The top of the engine cover also acted as a platform for the astronauts to observe the lunar landscape while standing in the open docking tunnel prior to EVA. Life support and communications hoses and cables were installed in the right side of the midsection. The equipment used for guidance and navigation was at the rear of the midsection on the aft wall.

Crew personal equipment included a variety of mission-oriented equipment for life support and astronaut safety, and accessories to successfully complete the mission. This equipment ranged from astronaut suits and docking aids to personal items stored throughout the cabin. The portable life support system (PLSS) interfaced with the environmental control system (ECS) for refills of oxygen and water. The pressure garment assembly (PGA) interfaced with the ECS for conditioned oxygen through the oxygen umbilical, and with the bioinstrumentation through the electrical umbilical.

Personal hygiene items consisted of wet and dry cleaning cloths, chemically treated and sealed in plastic covers. Medical equipment consisted of biomedical sensors, personal radiation dosimeters, and emergency medical equipment.

Other, miscellaneous equipment included in-flight data with checklists, emergency tool B and window shades. Tool B (emergency wrench) was a modified Allen-head L-wrench. It was 6.25 inches long and had a 4.25-inch long drive shaft with a 7/16-inch drive. The wrench could apply a torque of 4,175 inch-pounds; it had a ball-lock device to lock the head of the drive shaft. The wrench was used as a contingency for use with the docking probe and drogue, and for opening the Command Module's hatch from the outside.

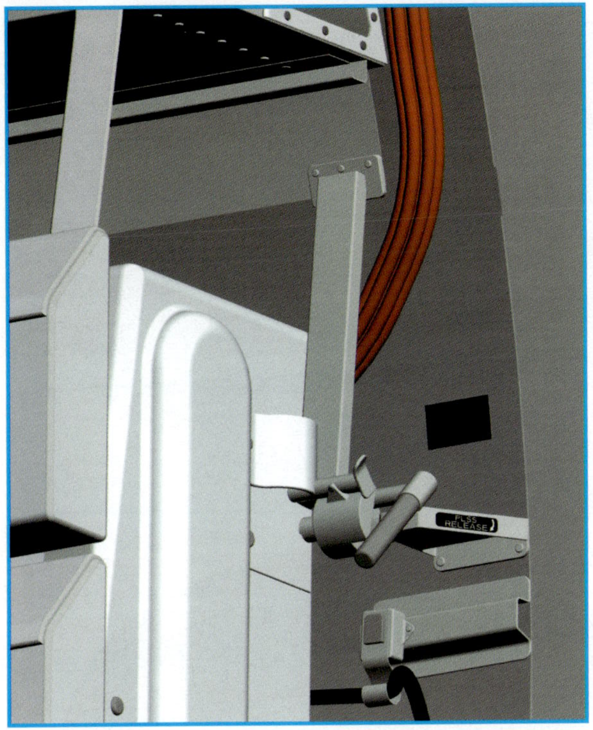

DETAIL VIEW OF PLSS RELEASE HANDLE

DOCKING HATCH

The overhead (docking) hatch was located at the top centerline of the midsection. The hatch was 33 inches in diameter and had an off-center latch that could be opened from either side of the hatch. When the Lunar Module and Command Module were docked, the open hatch permitted transfer of astronauts and equipment. The hatch opened inward by turning the latch handle 90 degrees; a preloaded electrometric silicone compound seal was mounted in the latch frame structure. When the hatch was closed, a lip near the outer circumference of the hatch entered the seal, providing a pressure tight contact. Normal cabin pressure was enough to force the hatch to seal, and to open the hatch the cabin would need to be depressurized by opening a cabin relief pressure valve on the hatch, which could be opened from either side.

The docking tunnel immediately above the hatch provided a structural interface between the LM and CM structure. The tunnel was 32 inches in diameter and 16 inches long. A ring (or flange) at the top of the tunnel allowed the 12 locking latches from the Command Module to "grab hold" of the LM and pull it into it's own docking ring.

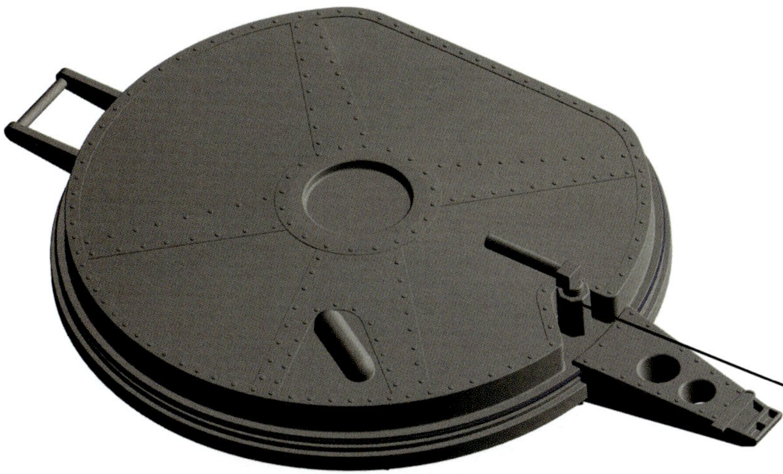

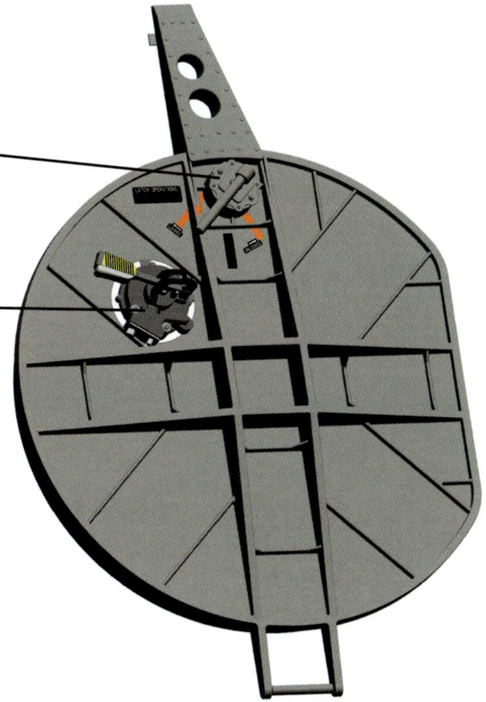

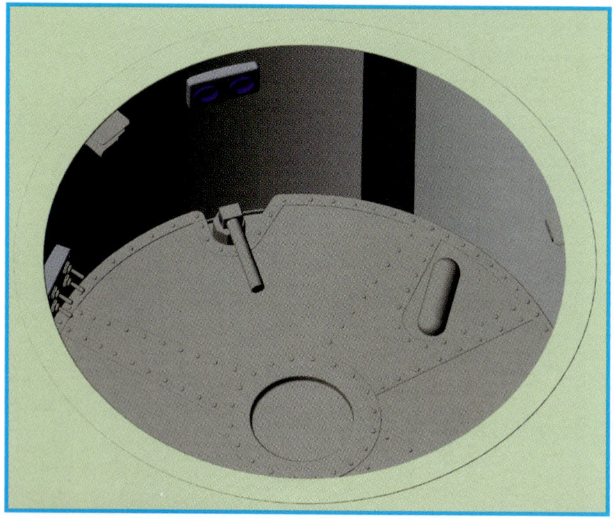

VIEW FROM OUTSIDE LM LOOKING INTO DOCKING TUNNEL

THE ASCENT STAGE FROM THE GROUND UP

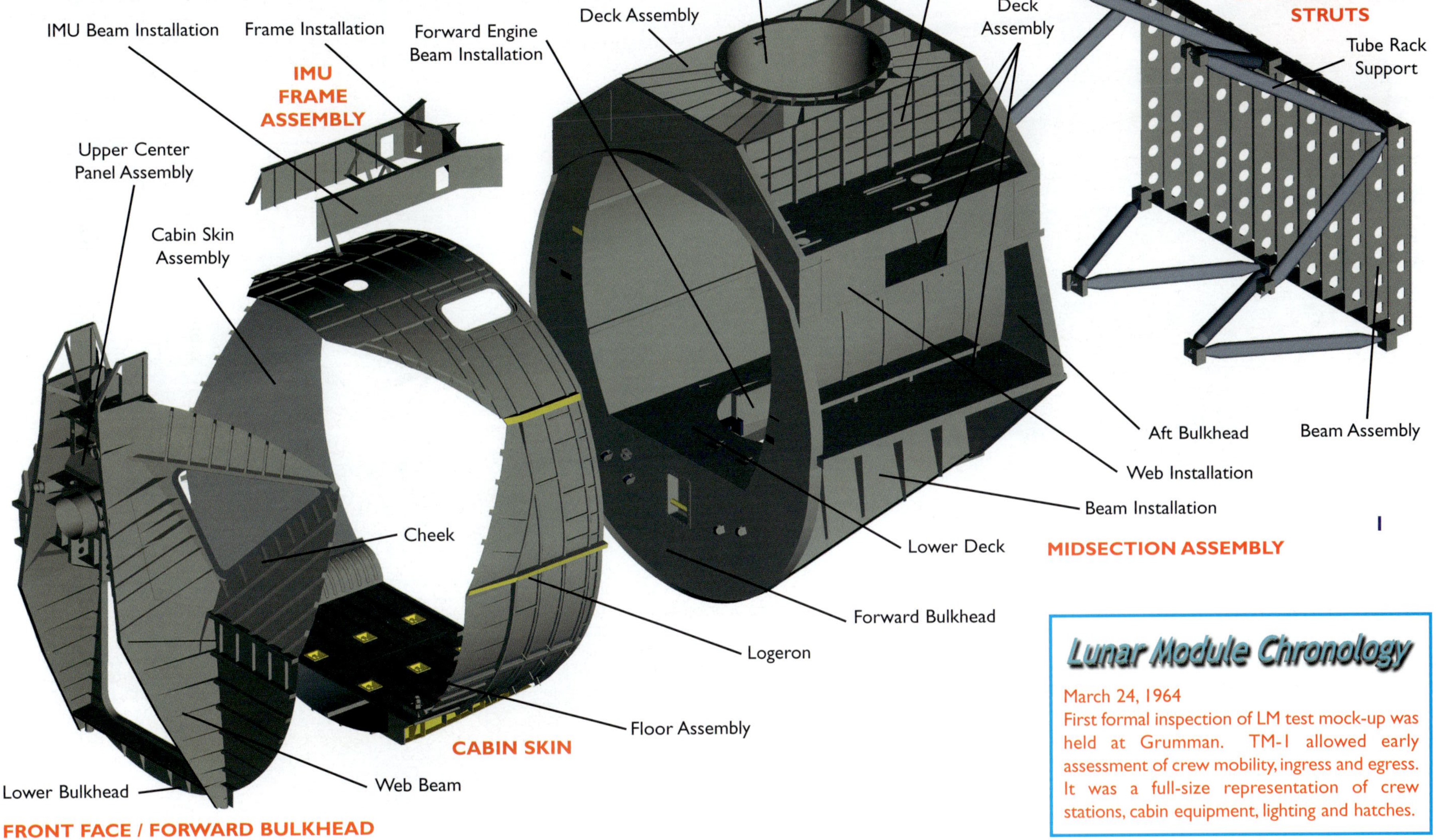

Lunar Module Chronology

March 24, 1964
First formal inspection of LM test mock-up was held at Grumman. TM-1 allowed early assessment of crew mobility, ingress and egress. It was a full-size representation of crew stations, cabin equipment, lighting and hatches.

Lunar Module Chronology

April 17, 1964
Grumman conducts LM drop test to determine the LM crew's ability to land the spacecraft from a standing position.

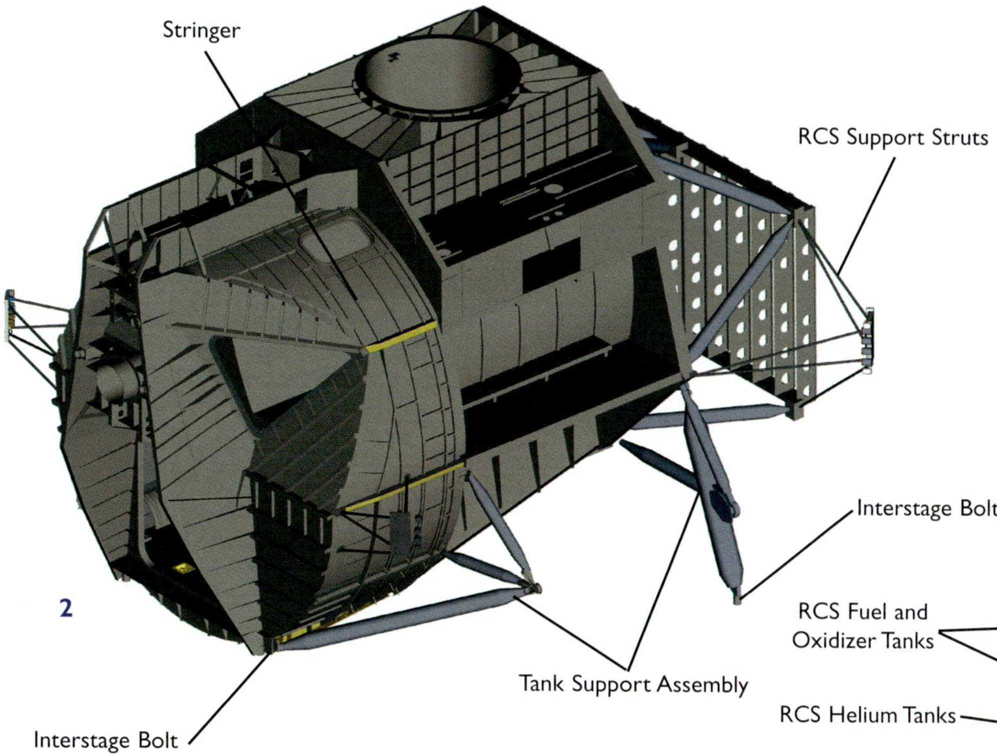

2

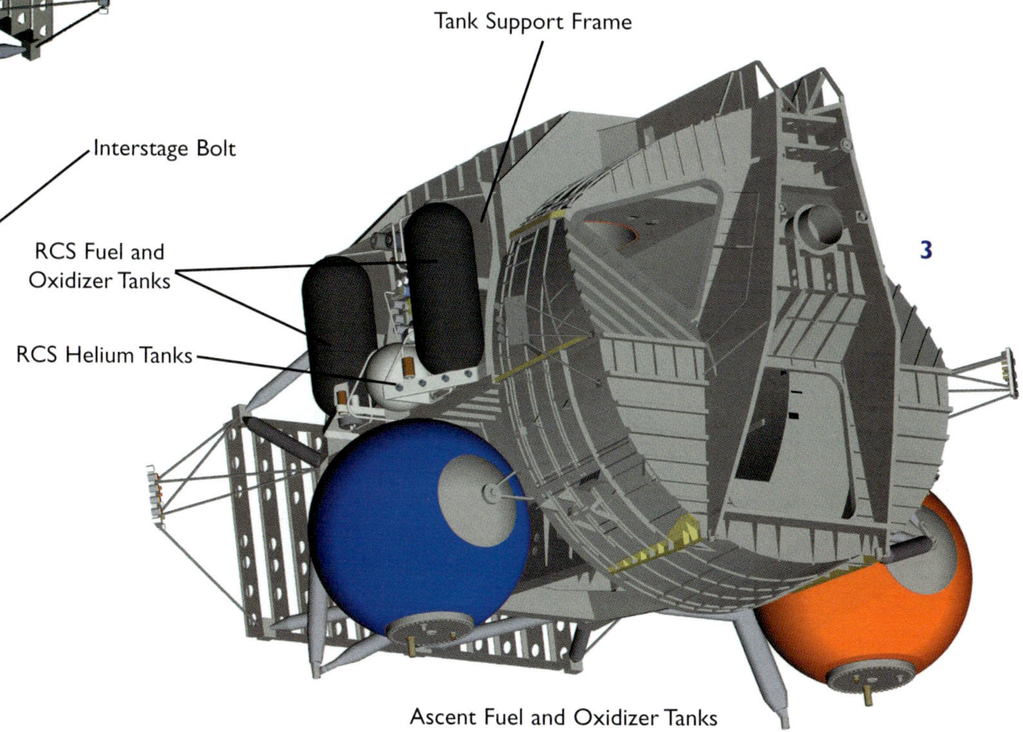

3

Lunar Module Chronology

April 2, 1964
Space Technology Laboratories begins using its new San Juan Capistrano, California test facility to static fire the firm's LM descent engine.

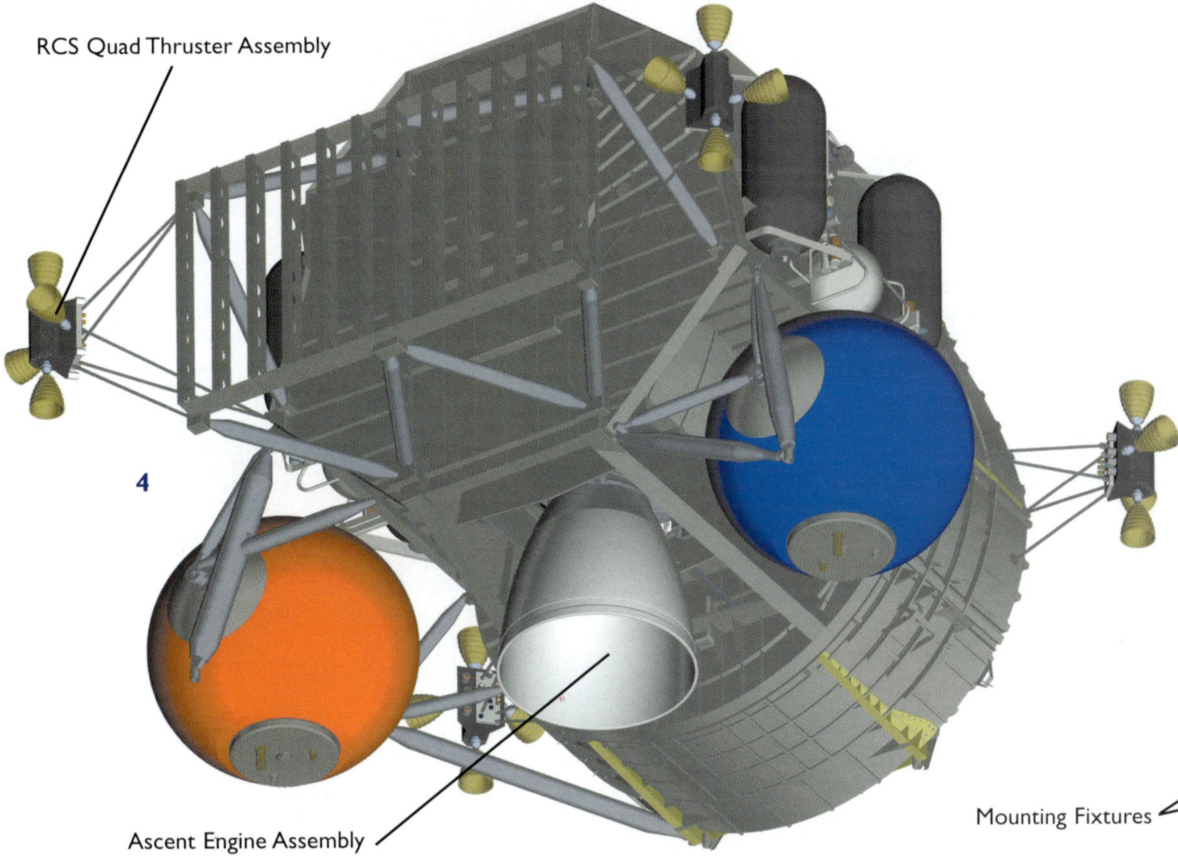

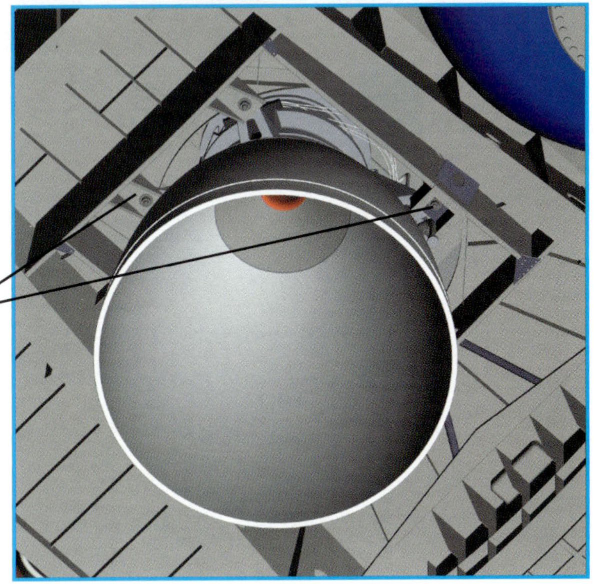

DETAIL VIEW OF ASCENT ENGINE WELL

Lunar Module Chronology

April 23, 1964
First firing of a prototype thrust chamber for the LM descent engine by Rocketdyne.

Lunar Module Chronology

April 24, 1964
Deletion of the LM's front docking capability and addition of an overhead (docking) window to allow for docking at the spacecraft's upper hatch.

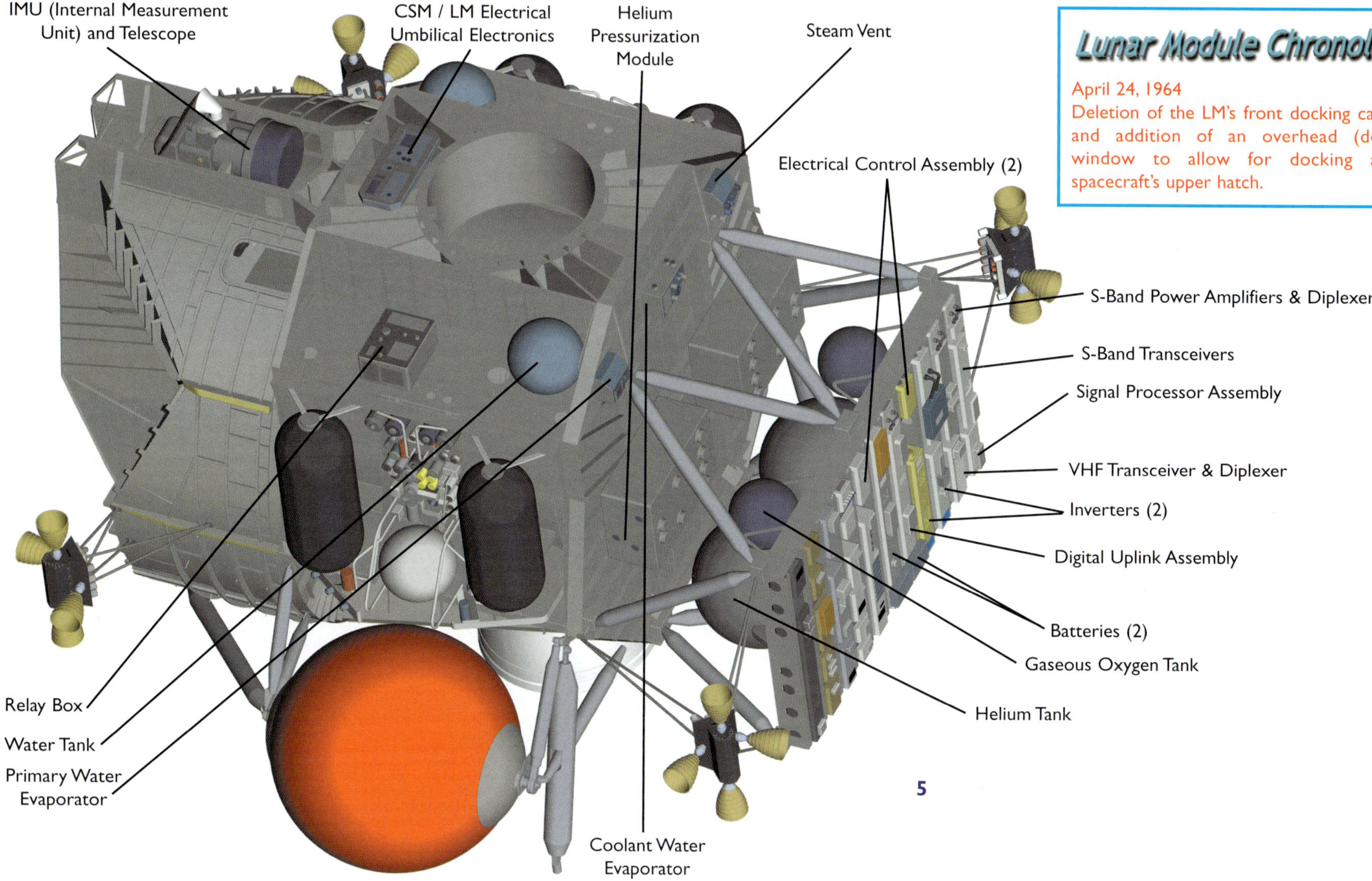

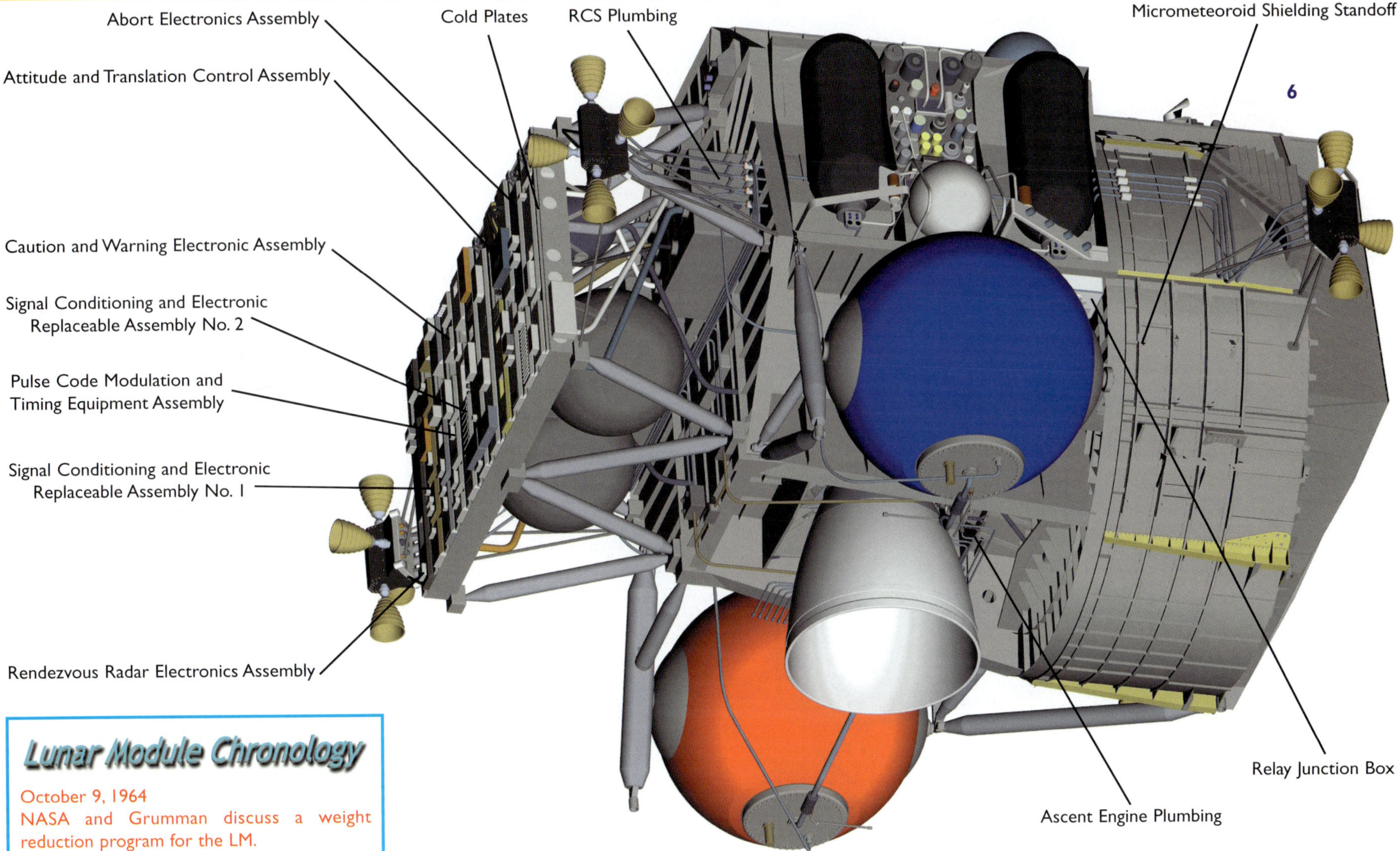

Lunar Module Chronology

October 9, 1964
NASA and Grumman discuss a weight reduction program for the LM.

Lunar Module Chronology

October 28, 1964
Test firing of the first flight-weight 15-cell stack of the LMs fuel cell assembly.

Lunar Module Chronology

November 6, 1964
NASA anticipated five major milestones for the forthcoming year:

- Review of the entire LM
- Start of production
- Delivery of LTA-2
- Start of vibration testing
- Sea level and altitude qualification

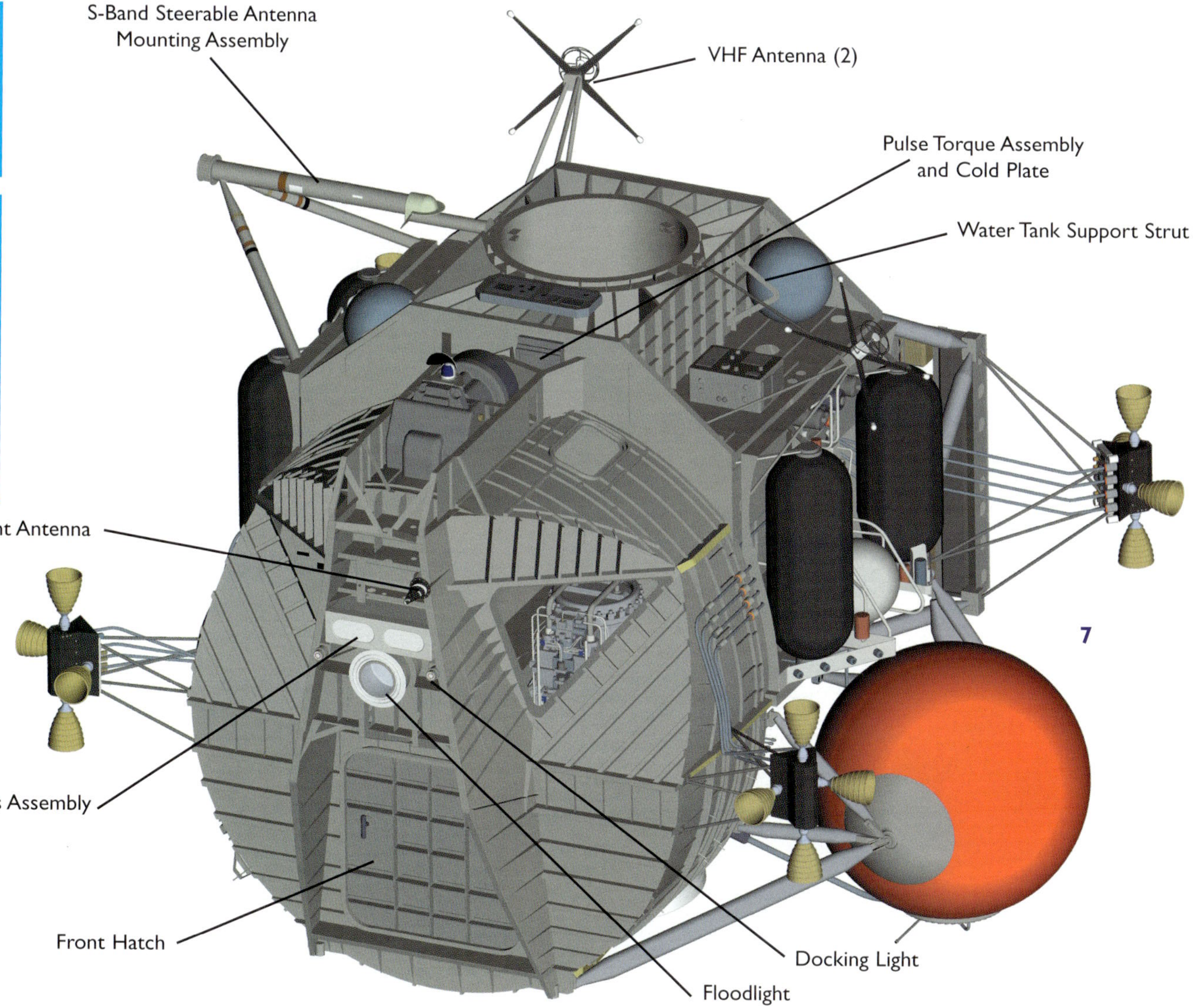

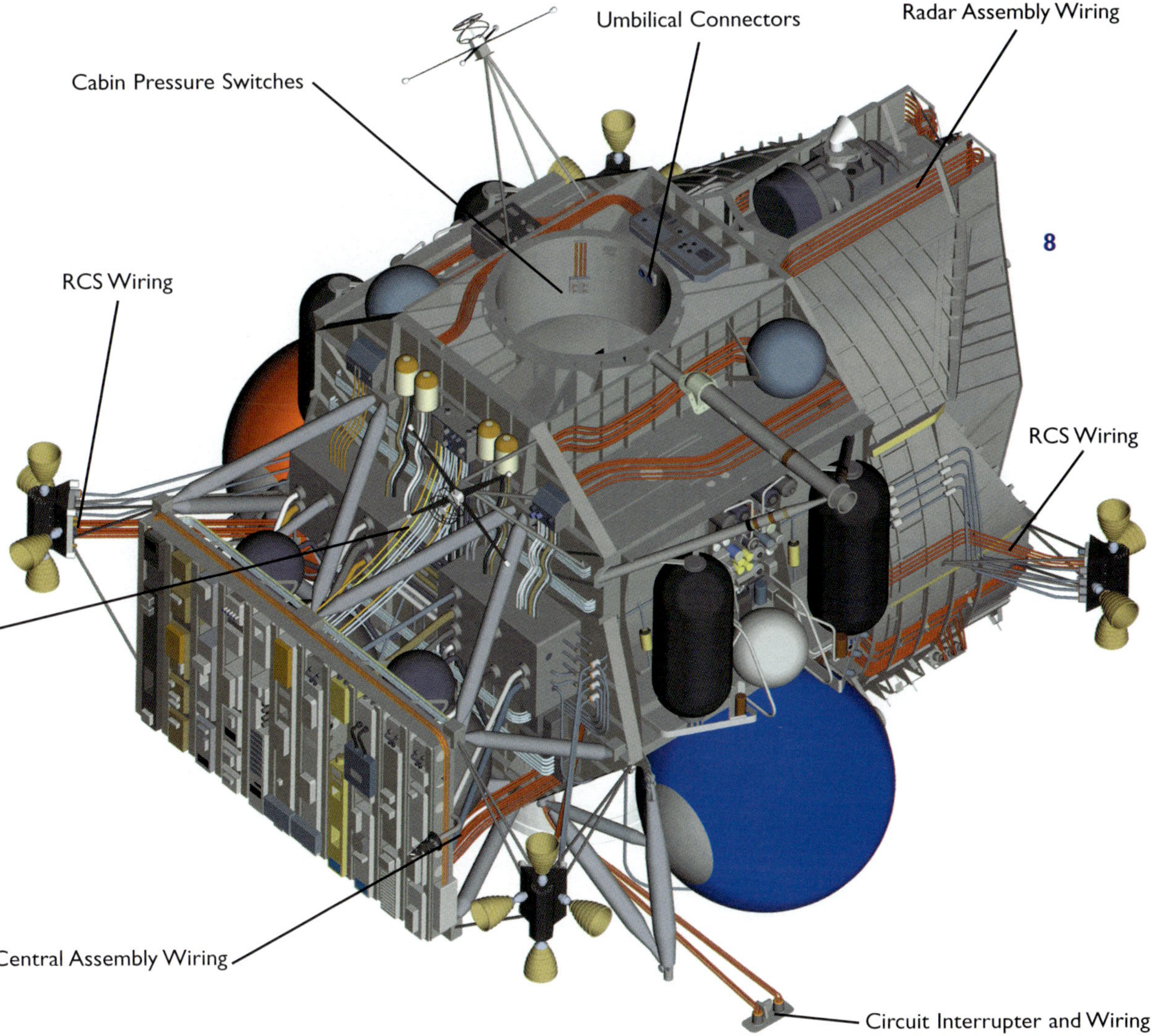

Lunar Module Chronology

November 12, 1964
Program to resize the LM emphasizes the fuel tanks be resized for separation and lunar liftoff.

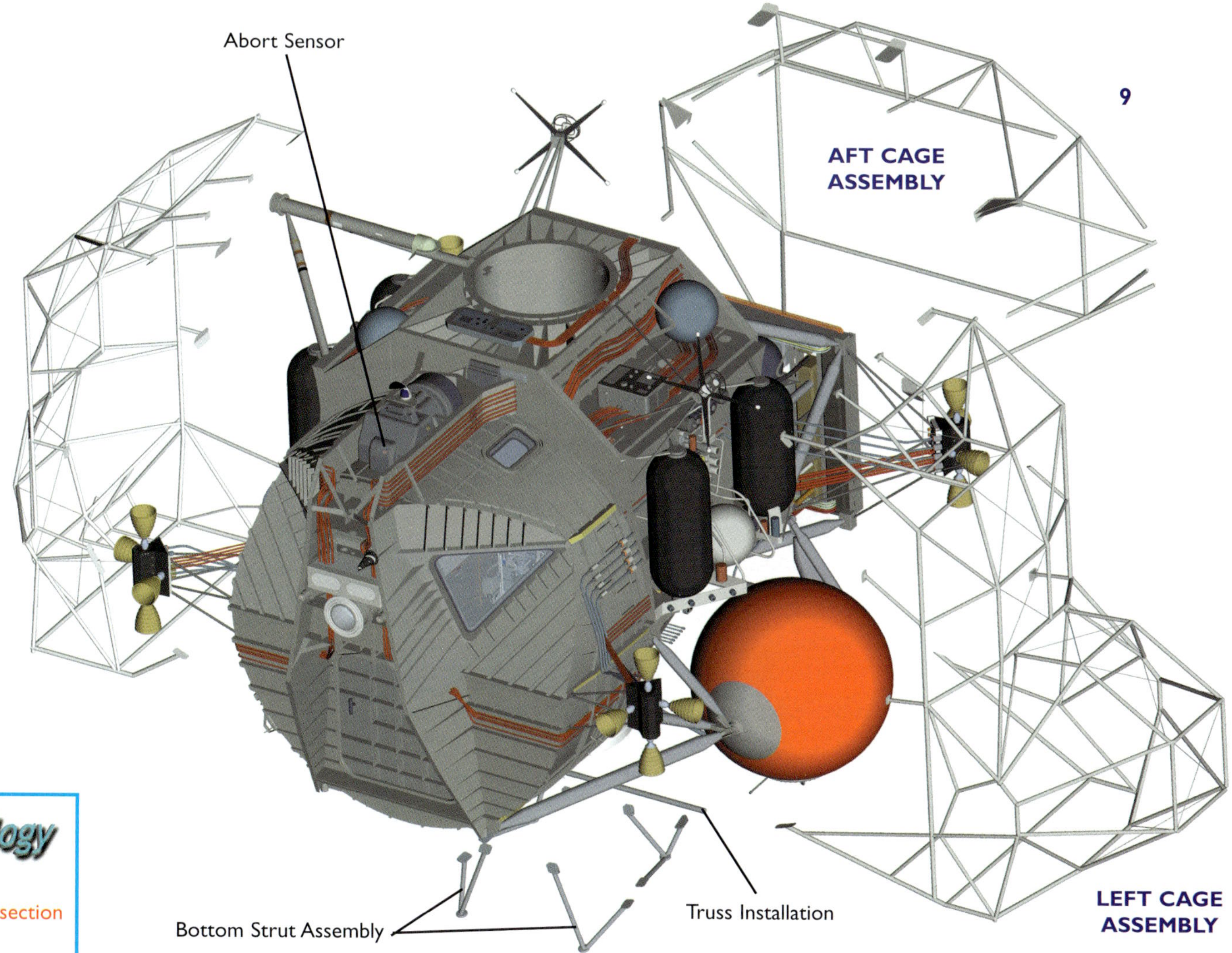

Lunar Module Chronology

December 11, 1964
Grumman given LM abort guidance section parameters.

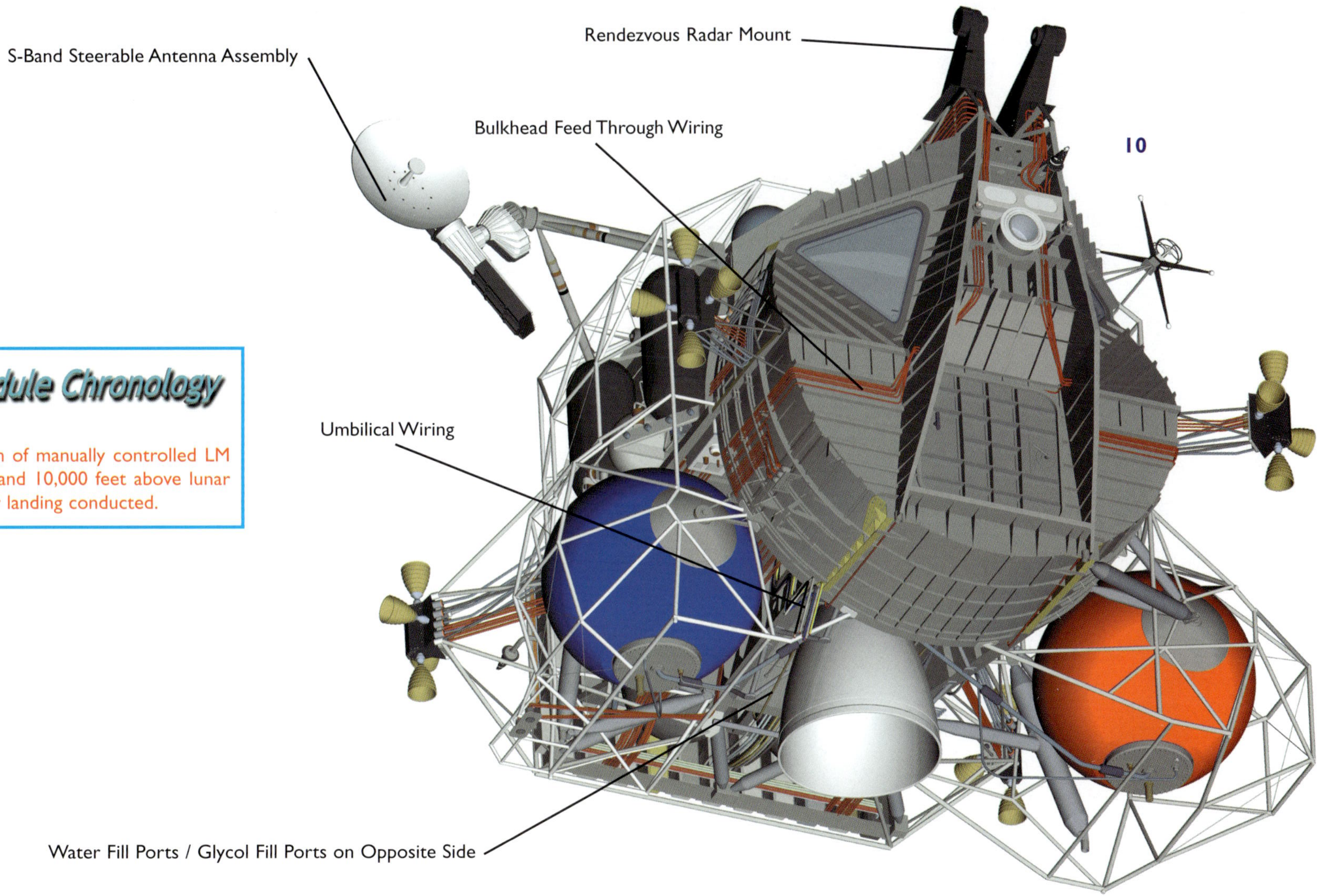

Lunar Module Chronology

January 5, 1965
Piloted simulation of manually controlled LM between 15,000 and 10,000 feet above lunar surface, and lunar landing conducted.

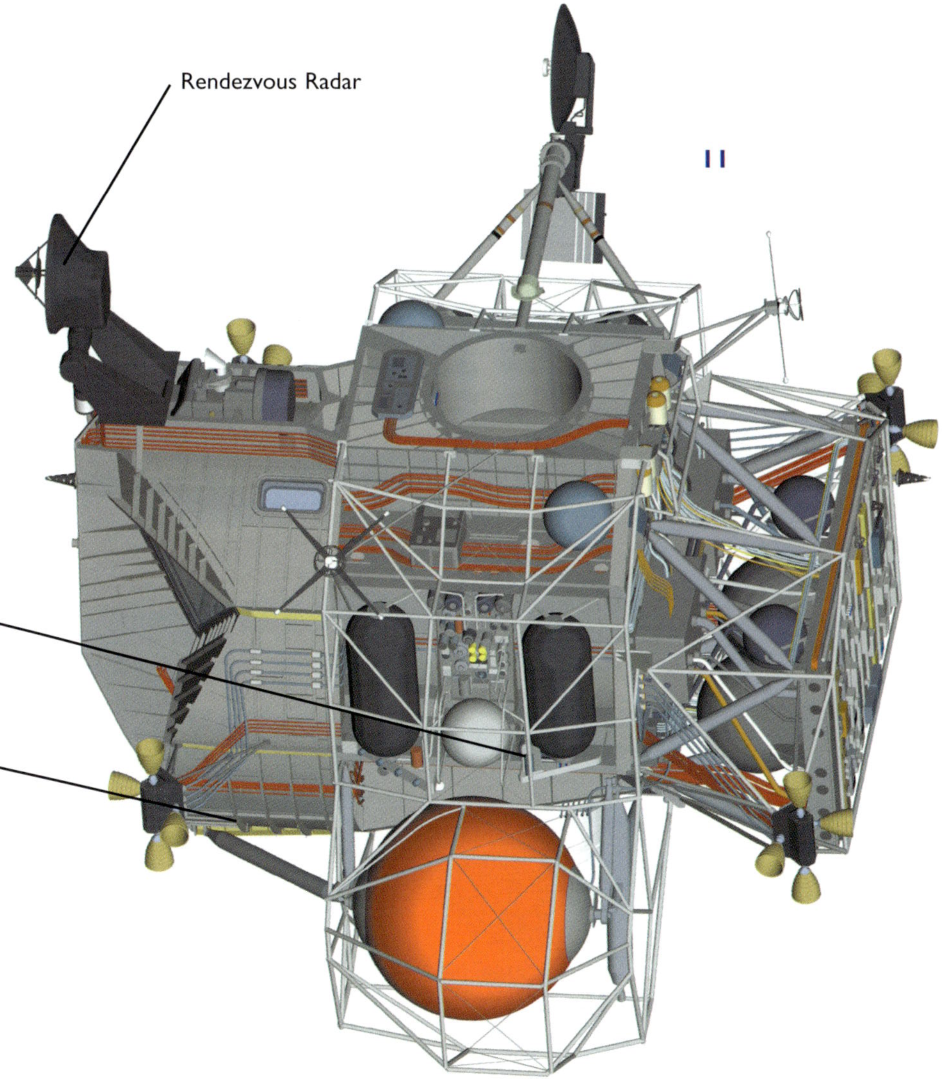

Lunar Module Chronology

January 21, 1965
Parallel development of the LM descent engine halted. Space Technologies Laboratories is named to be the sole contractor.

Lunar Module Chronology

January 31, 1965
Grumman modifies the forward hatch from a round to a square shape to facilitate easier ingress and egress.

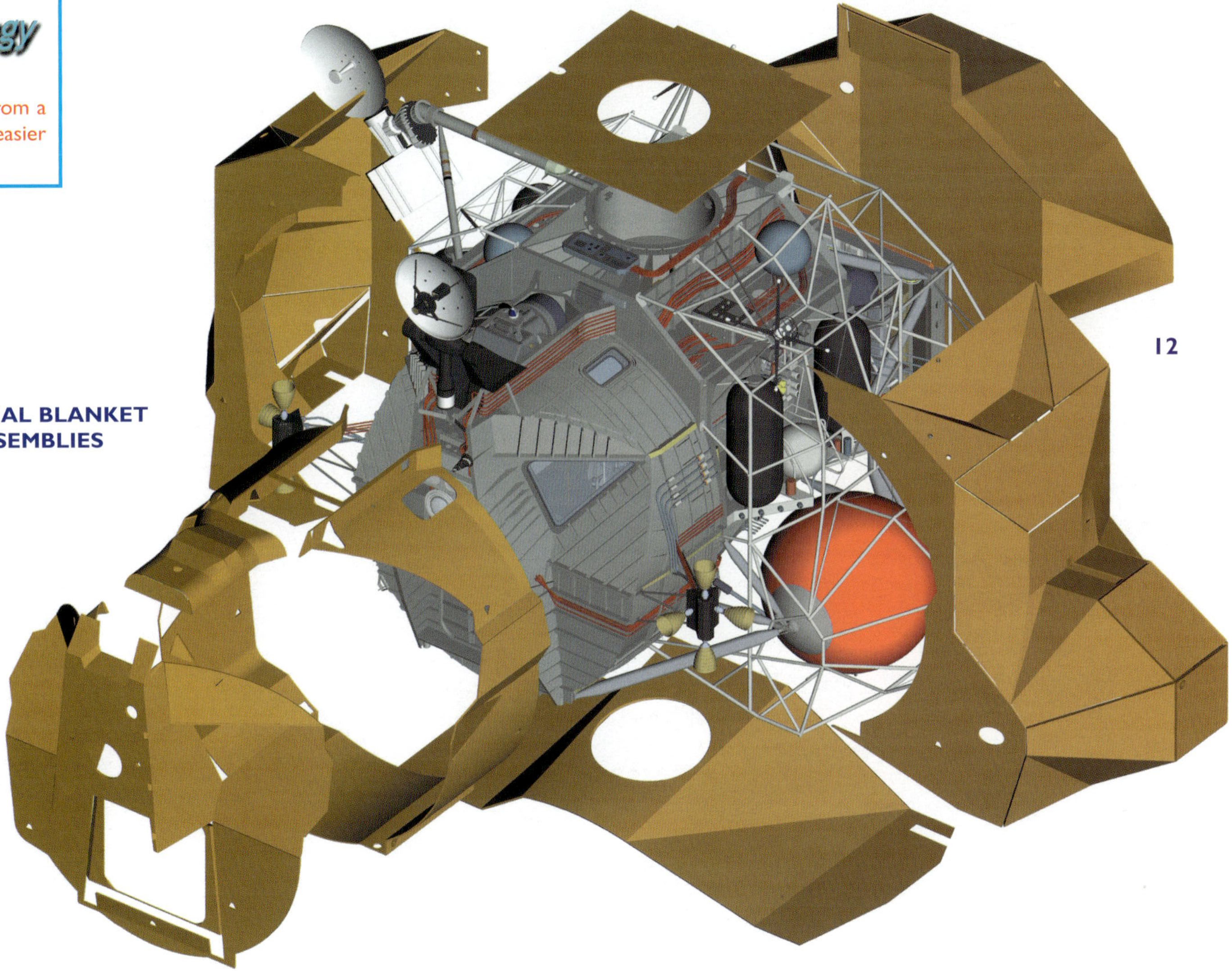

THERMAL BLANKET ASSEMBLIES

MICROMETEOROID SHIELDING ASSEMBLIES (OUTER SKIN)

13

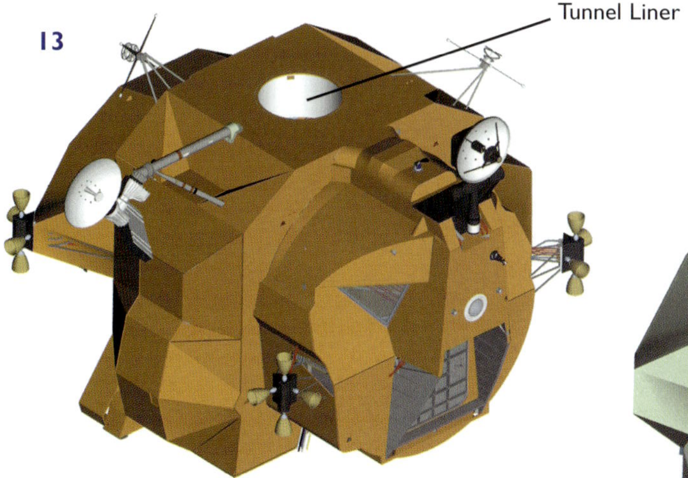

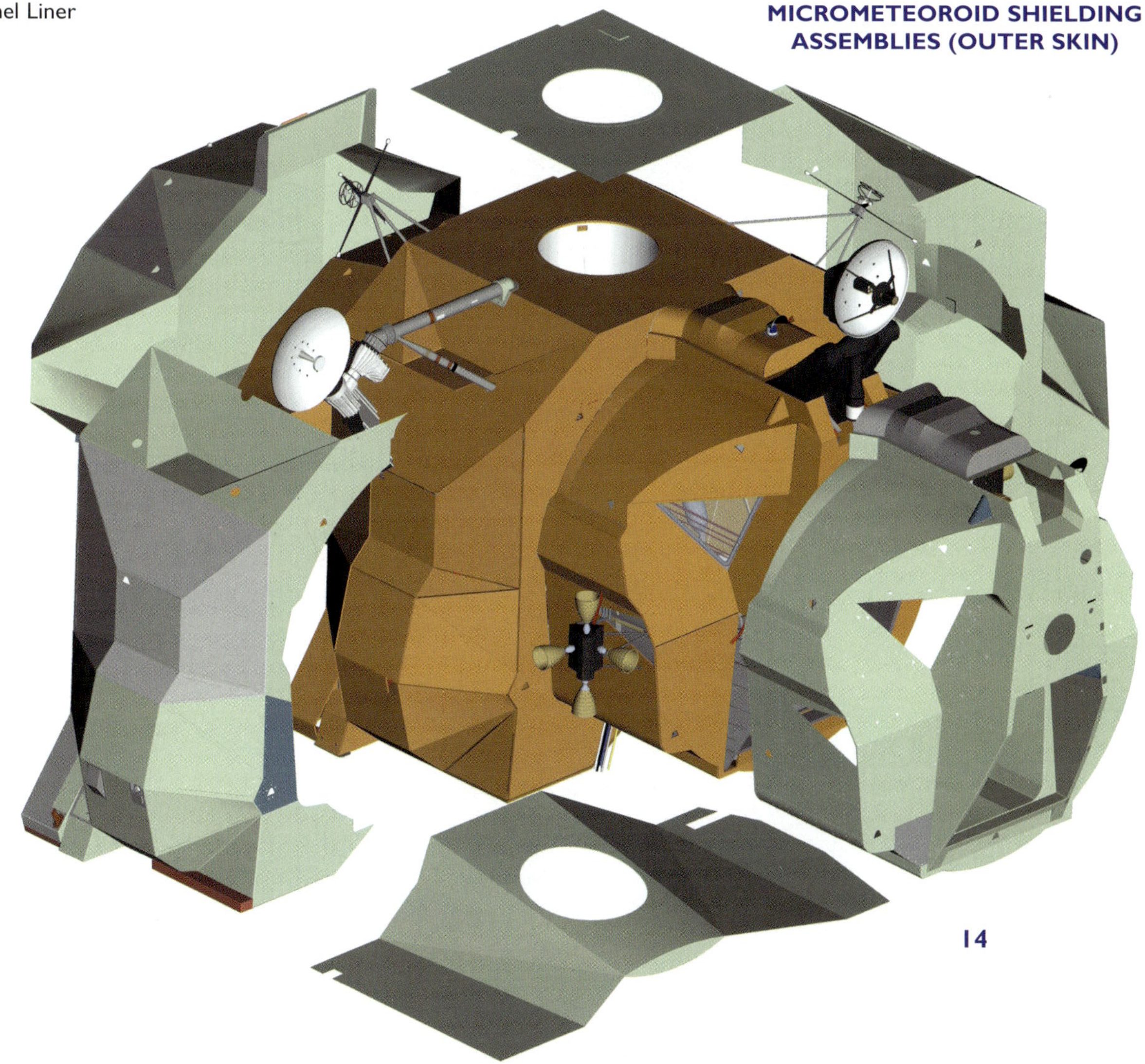

14

Lunar Module Chronology

February 11, 1965
Evaluation of three footpad probes on the LMs landing gear conducted to ensure proper shut-off of descent engine on lunar landing.

The cabin skin assembly was fabricated from chemically milled skin panels that were welded and mechanically fastened. The front face of the ascent stage was also fabricated from chemically milled skin panels that were welded and mechanically fastened. Sealing mechanical joints trimming the forward face contour, and adding formed longerons and stringers completed the operations for this assembly.

The midsection consisted of two machined bulkheads, an upper deck tunnel weldment, a lower engine deck weldment, and chemically milled skins. The front face of the cabin skin subassembly was mechanically joined with the midsection and was sealed to form the cabin pressure shell of the ascent stage.

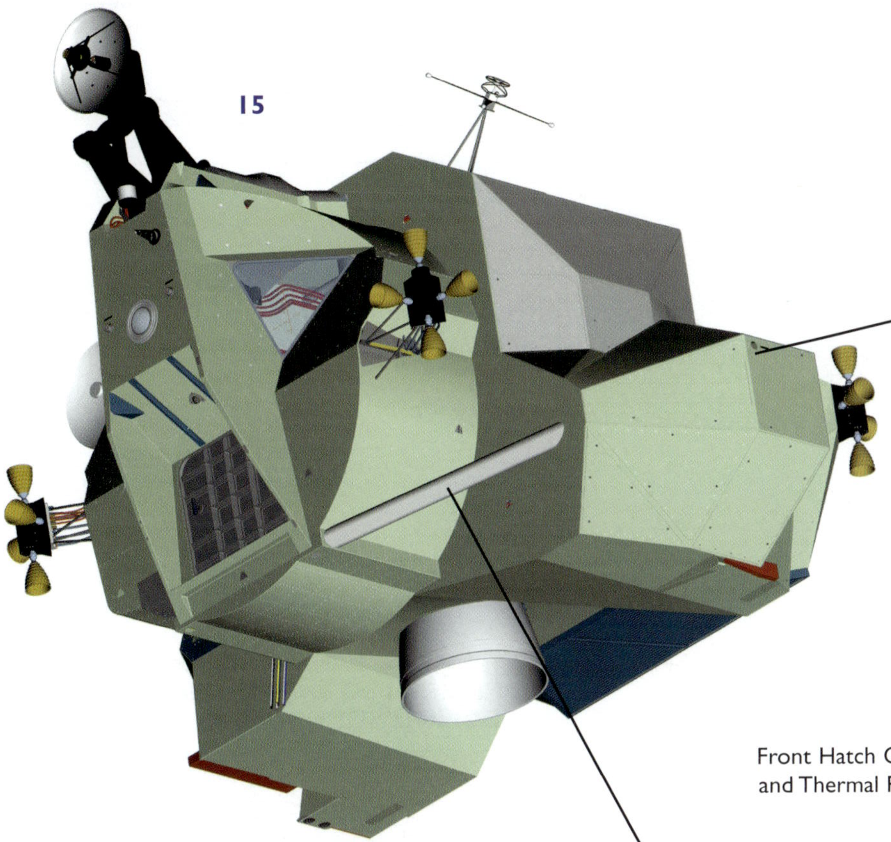

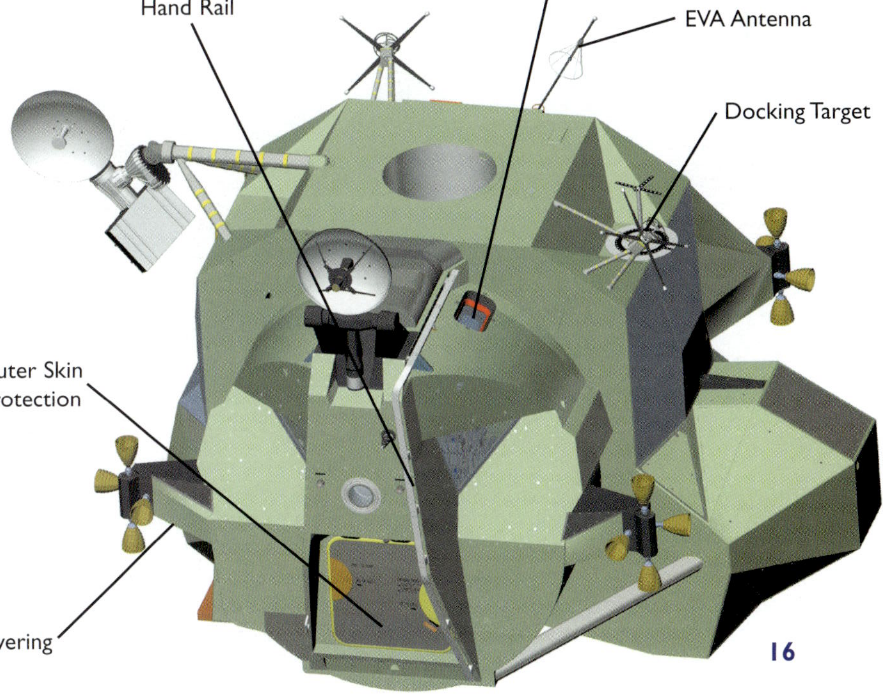

Lunar Module Chronology

February 18, 1965
Apollo Lunar Module Test Article 2 shipped to Marshall Space Flight Center to undergo a series of Saturn booster vibration tests.

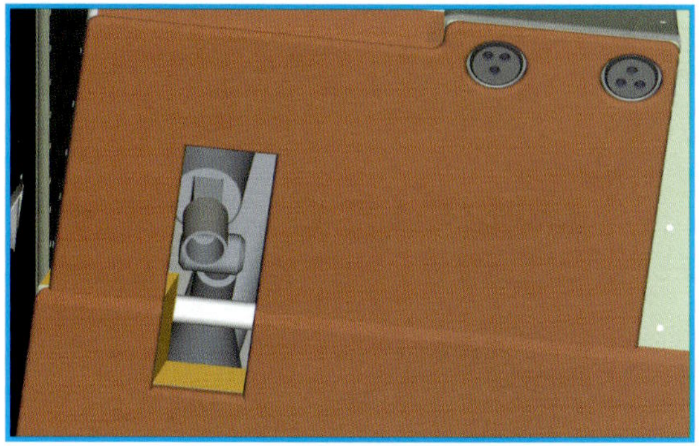

DETAIL VIEW OF CIRCUIT INTERRUPT CONNECTORS AND ONE OF THE FOUR HARDPOINT CONNECTIONS

The aft equipment bay was an unpressurized area, formed by the aft midsection bulkhead and the equipment rack, which was cantilevered approximately 33 inches aft of the bulkhead. The equipment rack assembly had integral cold rails that transferred heat from electronic and electrical equipment (components of the GN&CS, EPS, and communications subsystems) mounted on the rack. The cold rails were mounted vertically in the rack structural frame. Water-glycol flowed through the cold rails. Two gaseous oxygen tanks and two helium tanks were secured to the truss members between the midsection aft bulkhead and the equipment rack. ECS and main propulsion subsystem components that did not require a pressurized environment or access by the astronauts were mounted on the outboard side of the aft bulkhead.

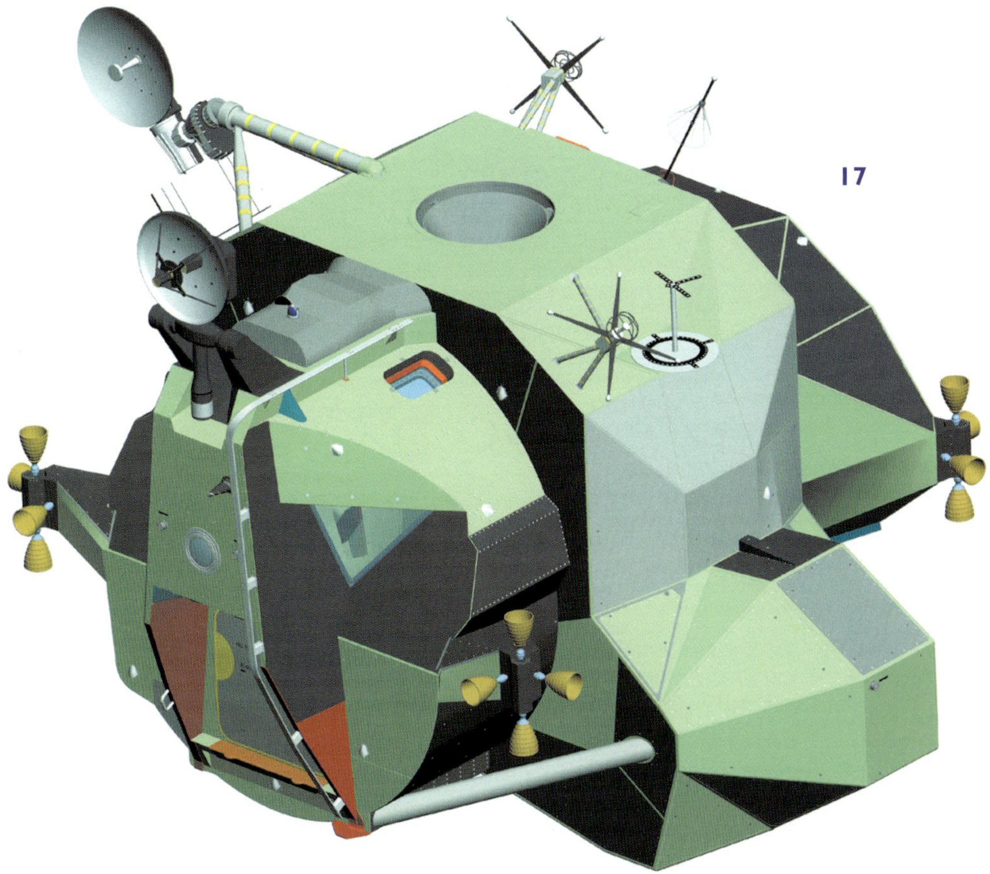

17

FINAL EXTERNAL THERMAL PROTECTION, PYRO PAINT, AND DROGUE ASSEMBLY

Lunar Module Chronology

March 2, 1965
LM fuel cells replaced by batteries.

LEFT SIDE CUT-AWAY

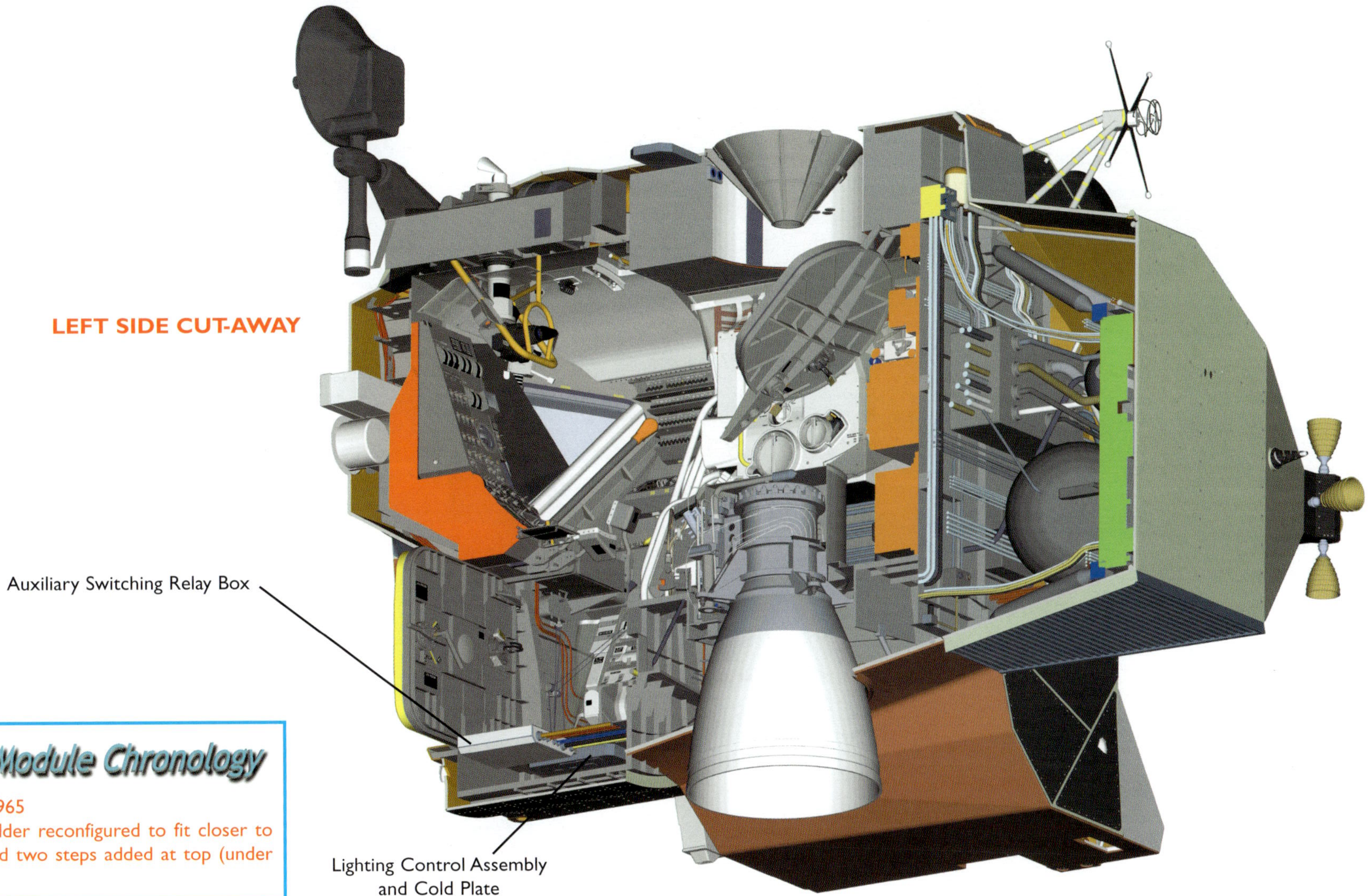

Lunar Module Chronology

March 25, 1965
Boarding ladder reconfigured to fit closer to the strut and two steps added at top (under porch).

Lunar Module Chronology

March 31, 1965
Major problems with the descent engine in attaining high performance and good erosion characteristics over the entire throttling range.

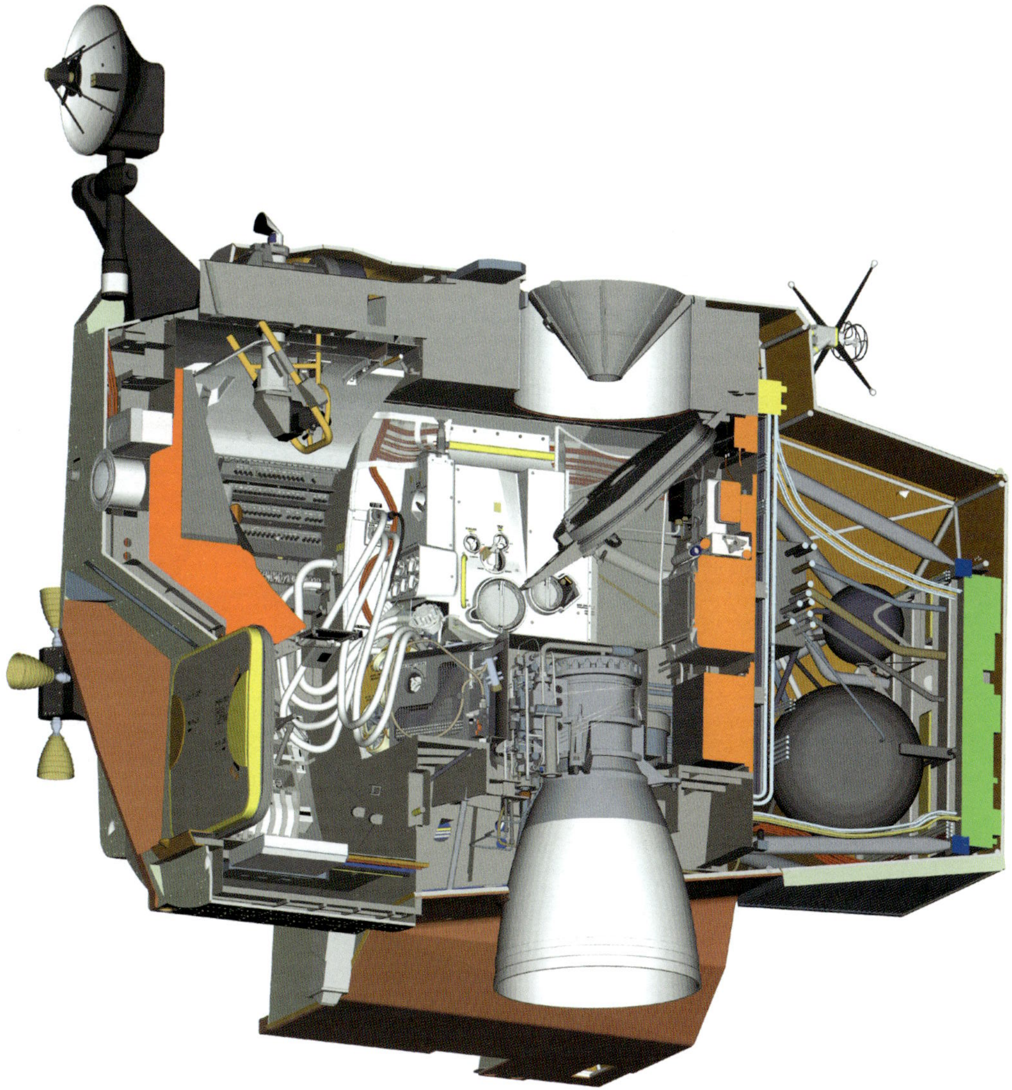

LEFT SIDE CUT-AWAY

Lunar Module Chronology

April 1, 1965
Grumman presents its recommendations for an all-battery powered system for the LM to MSC.

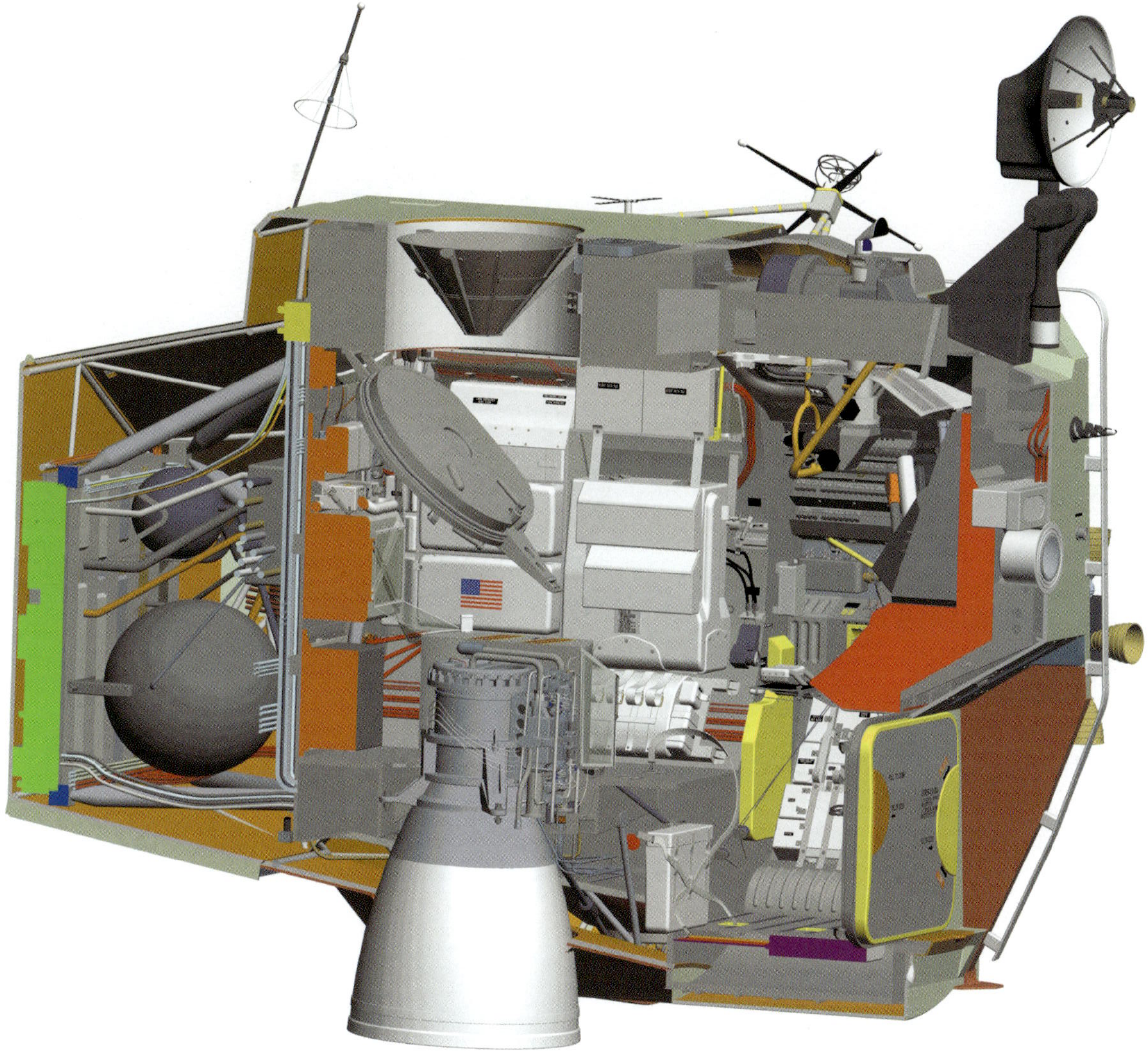

RIGHT SIDE CUT-AWAY

Lunar Module Chronology

April 15, 1965
First firing of the LM ascent engine test rig was successfully conducted at White Sands Missile Range.

RIGHT SIDE CUT-AWAY

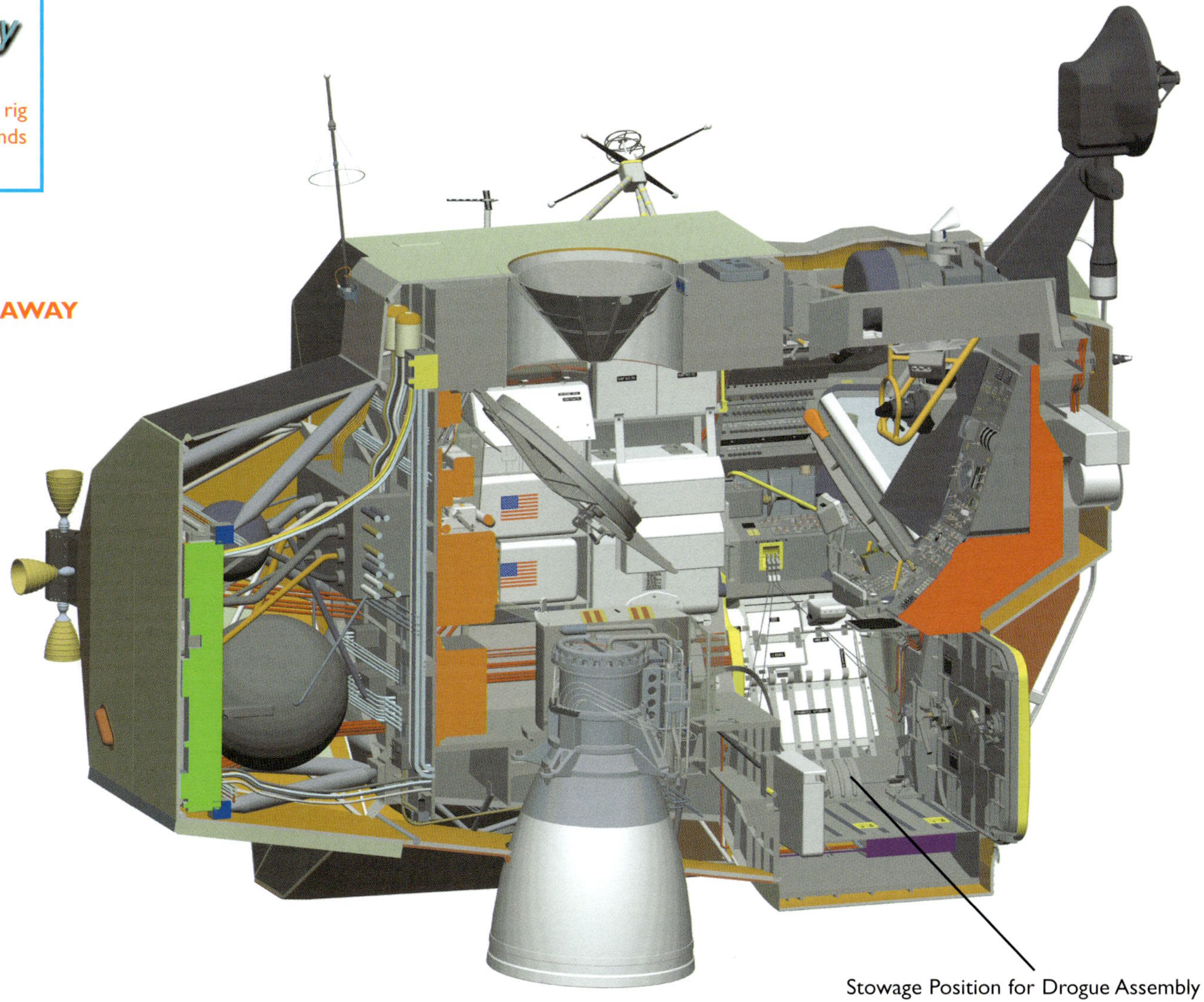

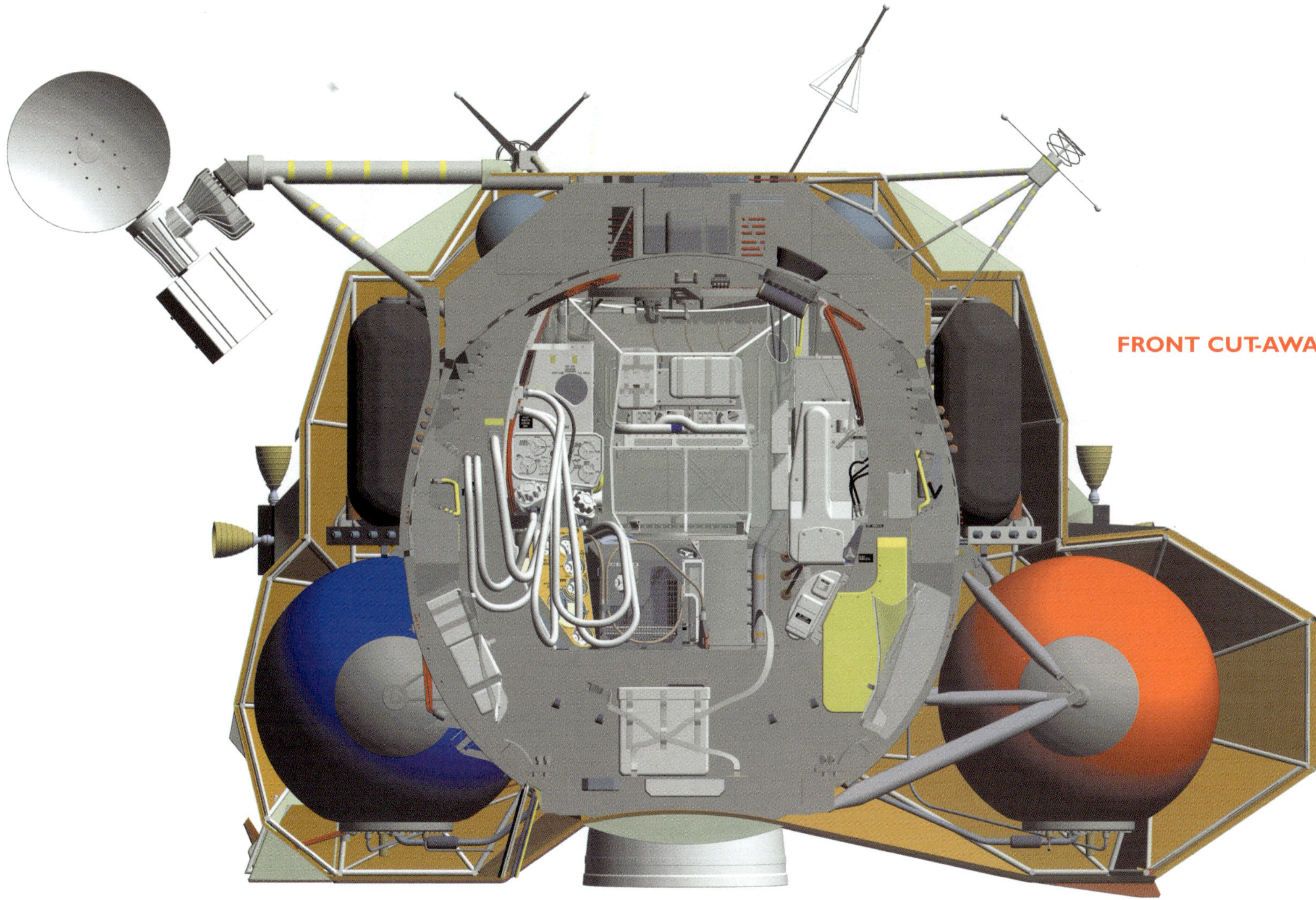

FRONT CUT-AWAY

FRONT CUT-AWAY

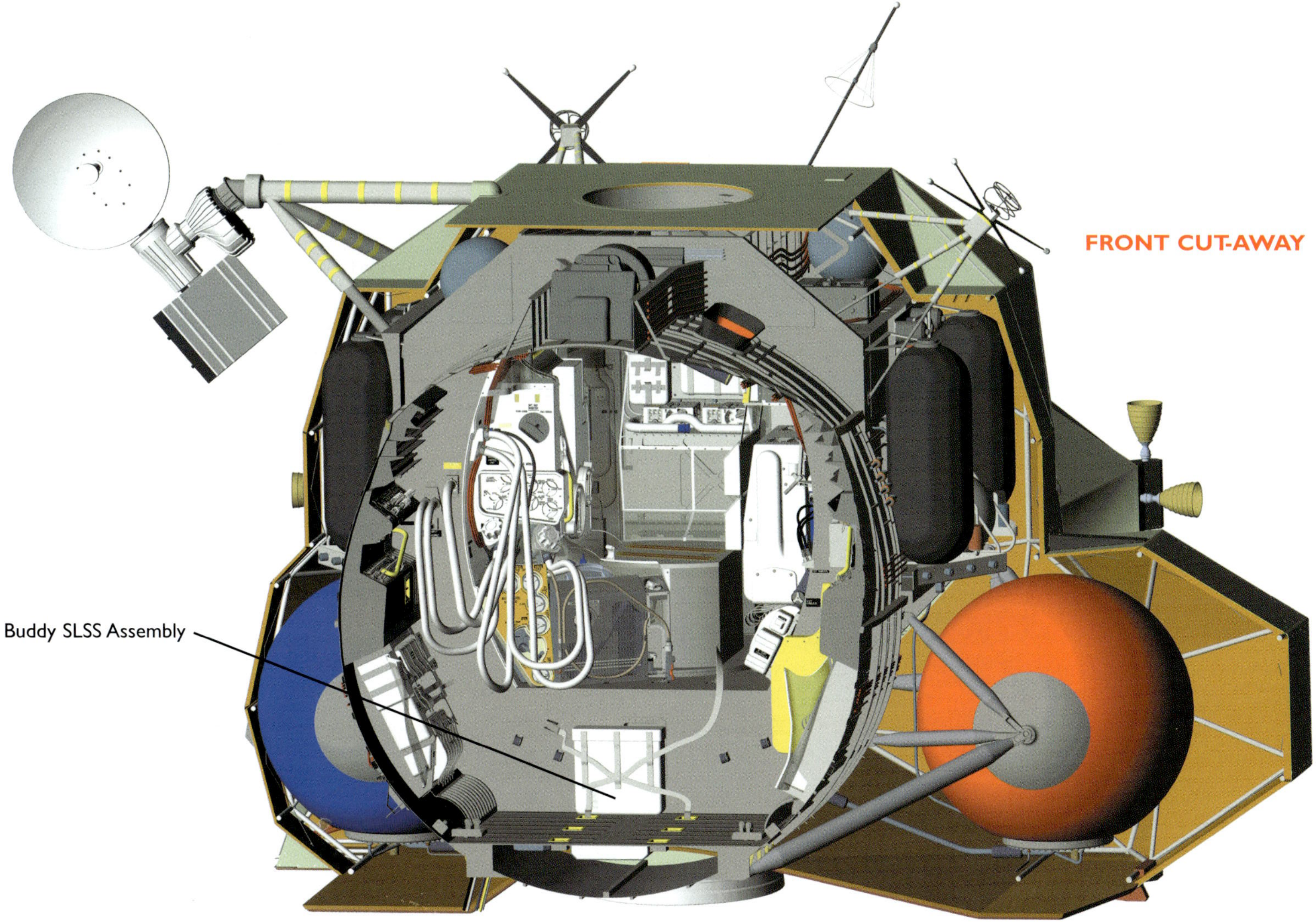

REAR CUT-AWAY

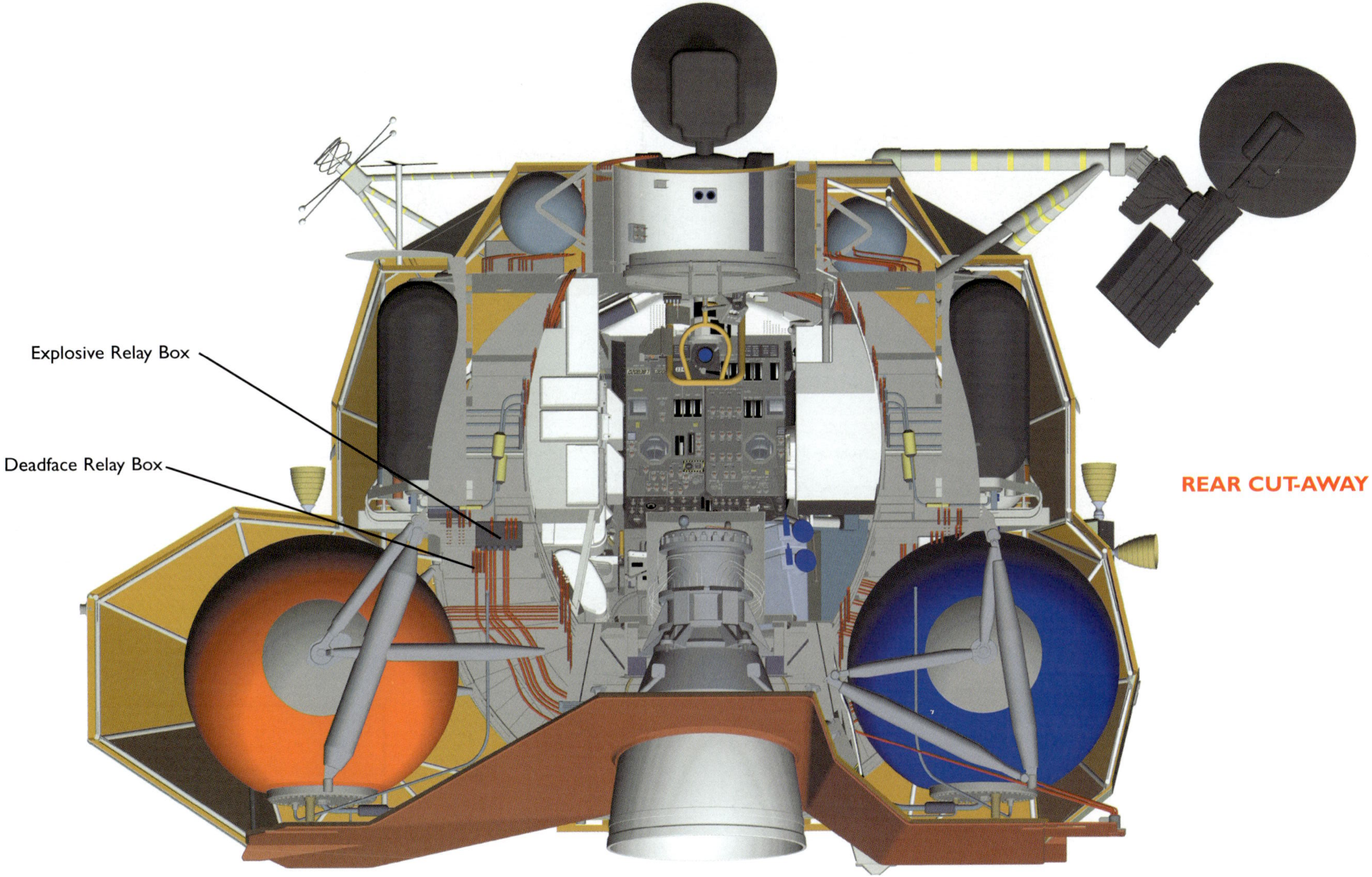

Lunar Module Chronology

May 26, 1965
LM control weight revised. Ascent stage to be 4,835 lbs., and descent stage to be 4,775 lbs. Total weight fueled to be 32,000 lbs.

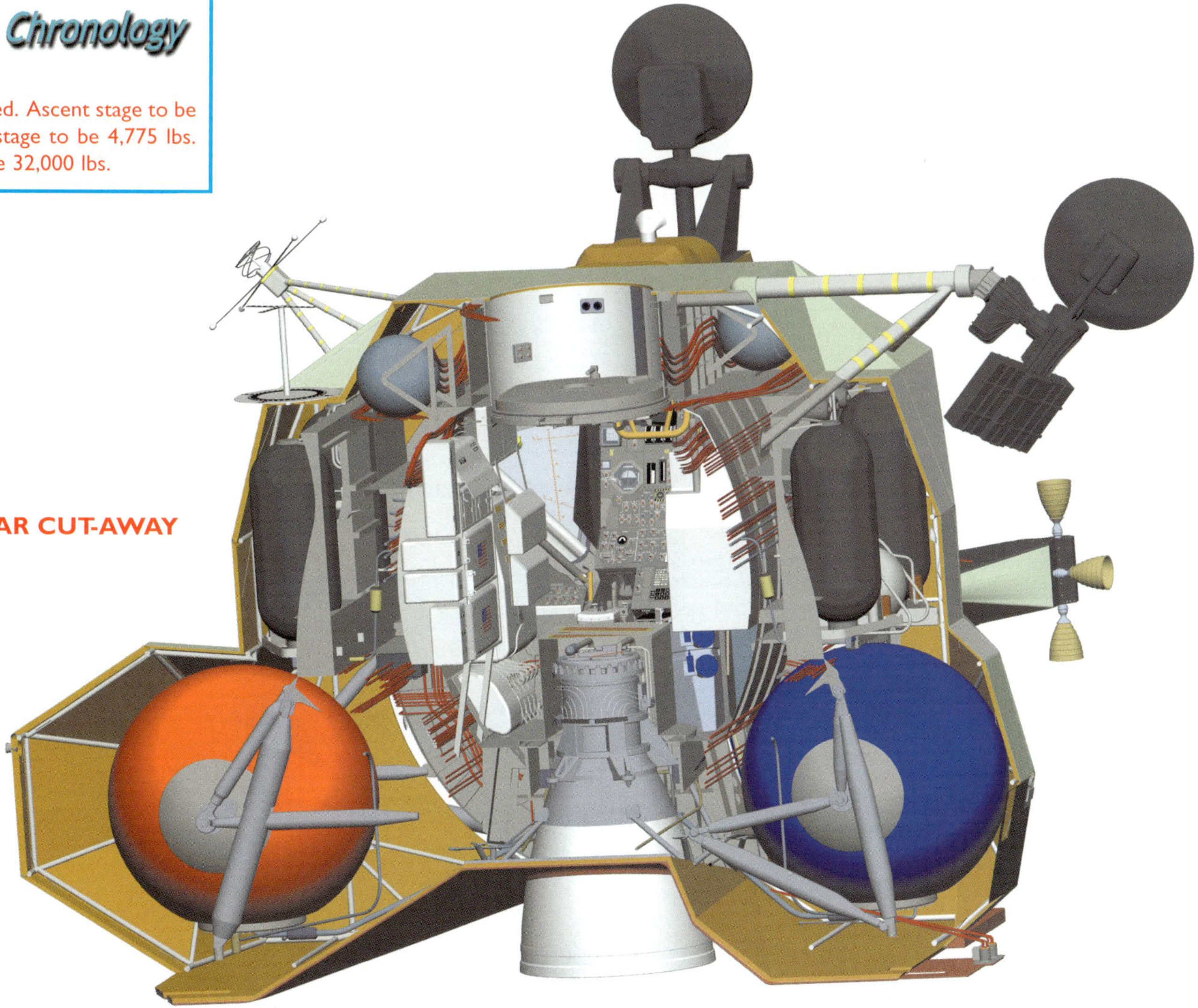

REAR CUT-AWAY

Lunar Module Chronology

July 16, 1965
Lunar Module to be used as back-up for the service propulsion engine on the CSM.

TOP CUT-AWAY

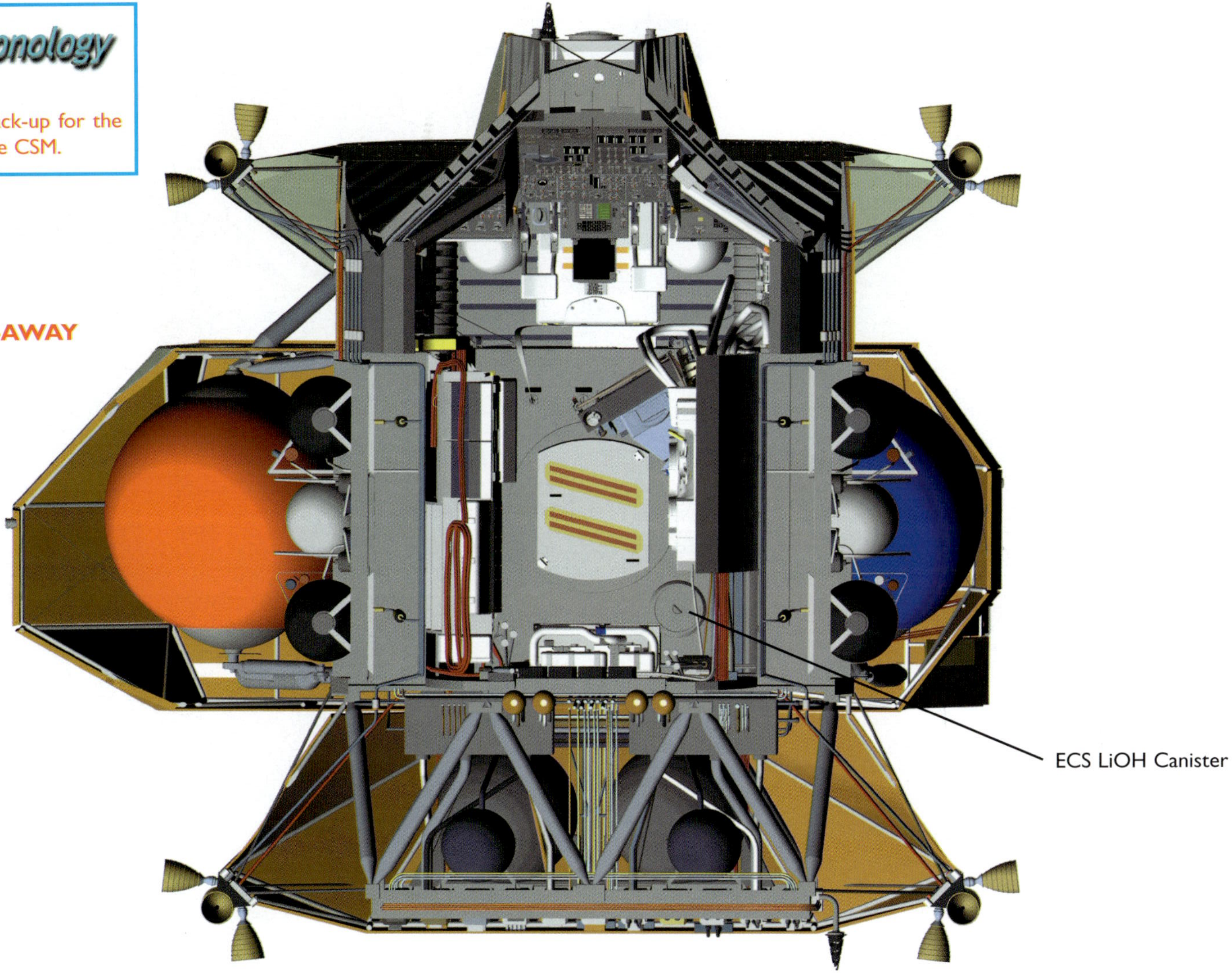

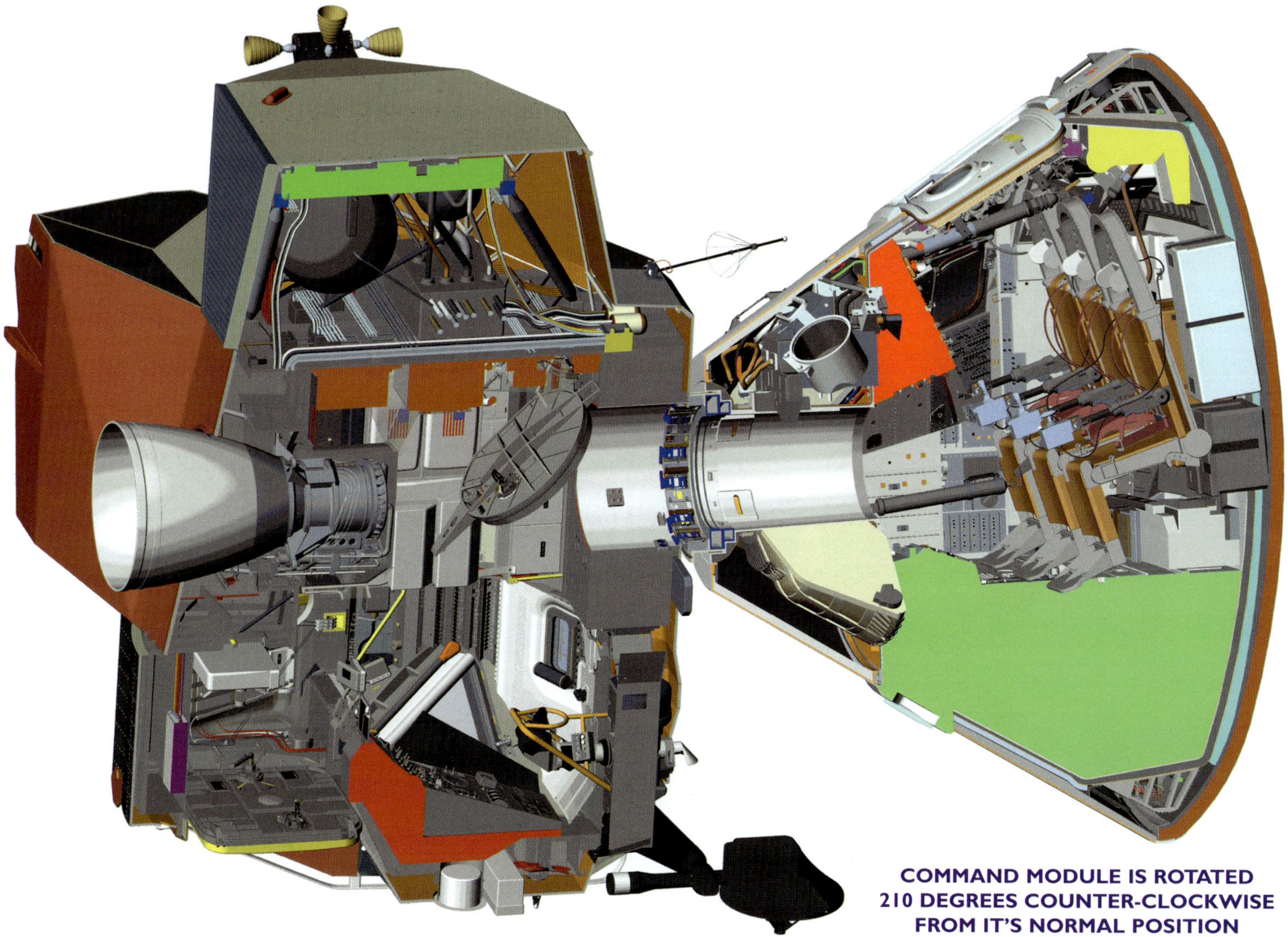

COMMAND MODULE IS ROTATED
210 DEGREES COUNTER-CLOCKWISE
FROM IT'S NORMAL POSITION

APOLLO LUNAR MODULE THERMAL PROTECTION

The entire Lunar Module was enveloped by a thermal and micrometeoroid shield. Due to the fact that weight was a paramount concern when designing the LM (there was a 400-to-1 fuel to payload ratio – which meant that for every one pound of payload required to make the lunar journey, 400 pounds of fuel was required), engineers decided to use a completely different thermal shielding from what was used on the Command Module. The LM would not experience the extreme temperatures that the Command Module would during Earth entry, so design requirements were less demanding. The thermal covering was not a metal foil, but a plastic film, thinly coated with aluminum, which would reflect heat and insulate the spacecraft. The LM's thermal coverings varied to some extend from one LM to another, but the basic configuration remained the same – the structural skin of the spacecraft, a thermal blanket, and either another layer of aluminized Kapton, a sheet of aluminum, or inconel that was the outer skin and micrometeoroid shield. The thermal coverings shown here are a typical example.

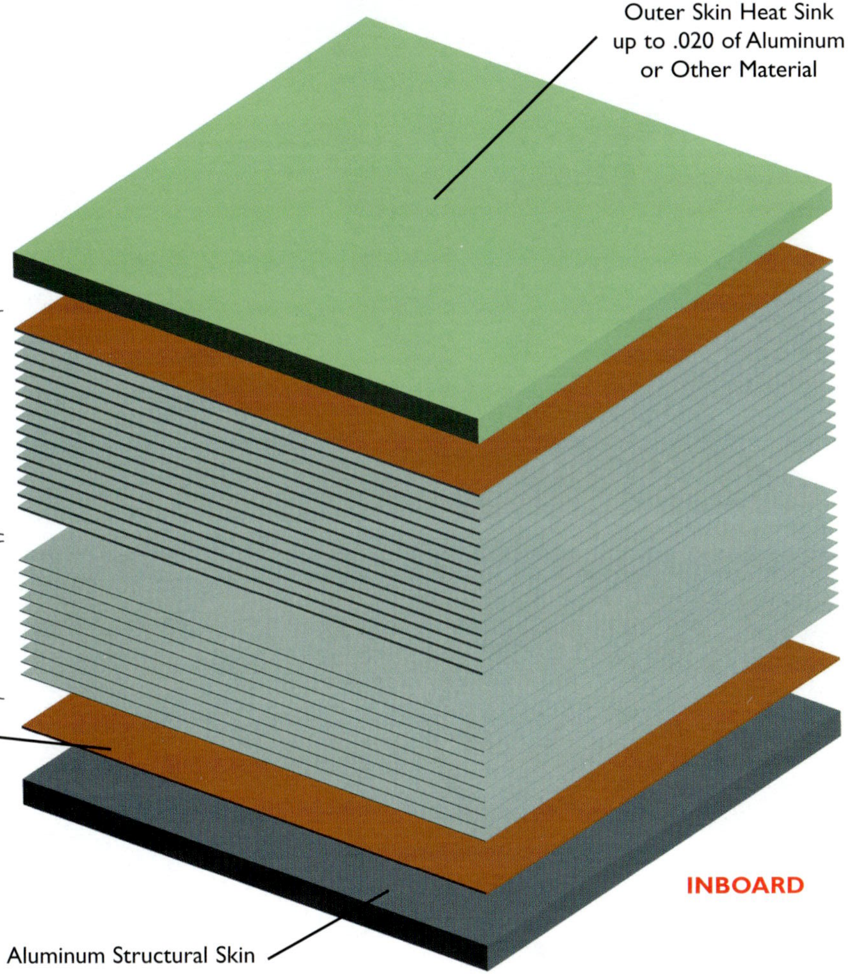

TYPICAL THERMAL SHIELD AND BLANKET

A typical thermal blanket was formed in various shapes and sizes to conform to the faceted shape of the LM and consisted of (outboard to inboard) 15 layers of aluminized Kapton, 10 layers of aluminized Mylar, and another single layer of aluminized Kapton. A few ascent stage areas had different thermal requirements, and the number of layers in their blankets varied slightly. The areas around the RCS thrusters normally saw more heat, so the blanket in those areas consisted of (outboard to inboard) a inconel skin .0015 thick, and one or more layers of inconel mesh alternated with .0005 inches of nickel foil. The number of inconel mesh and nickel foil layers, and the thickness of the inconel skin, would vary depending on the thermal requirements in that area.

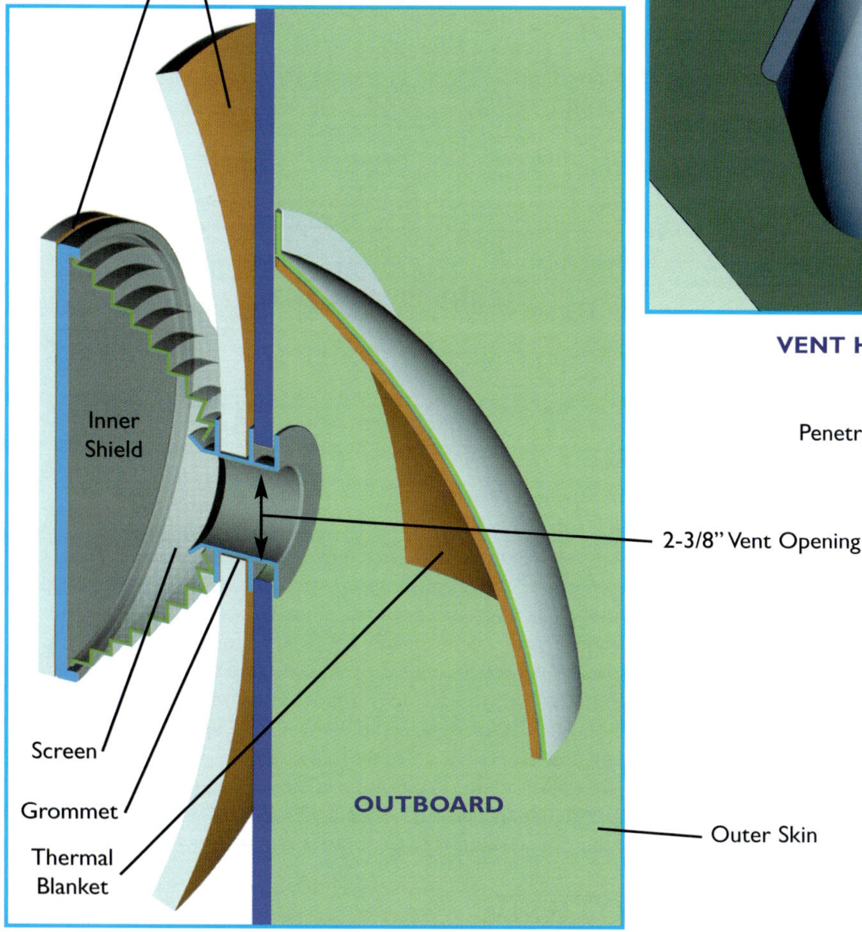

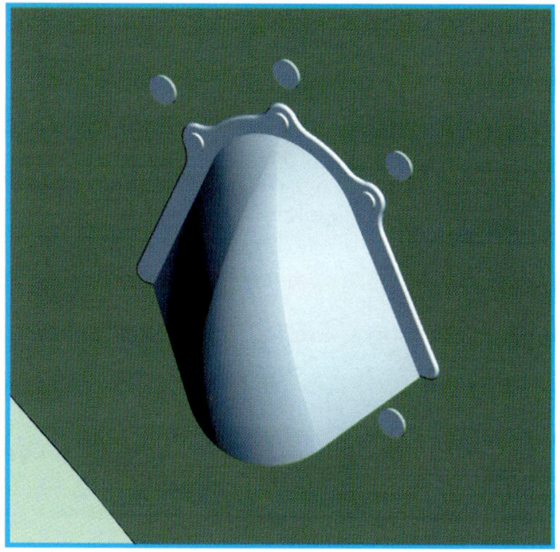

VENT HOOD (Typ)

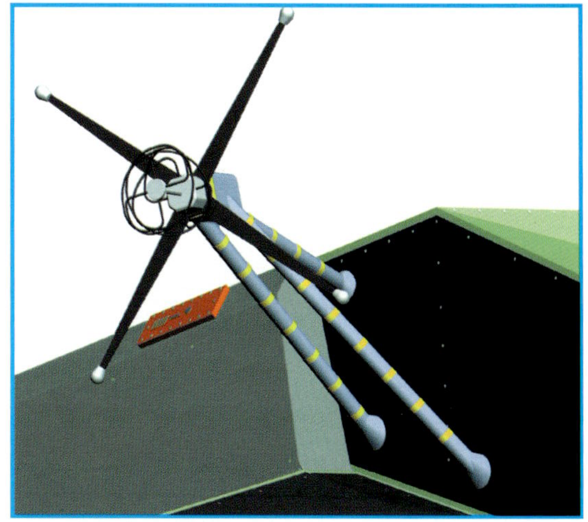

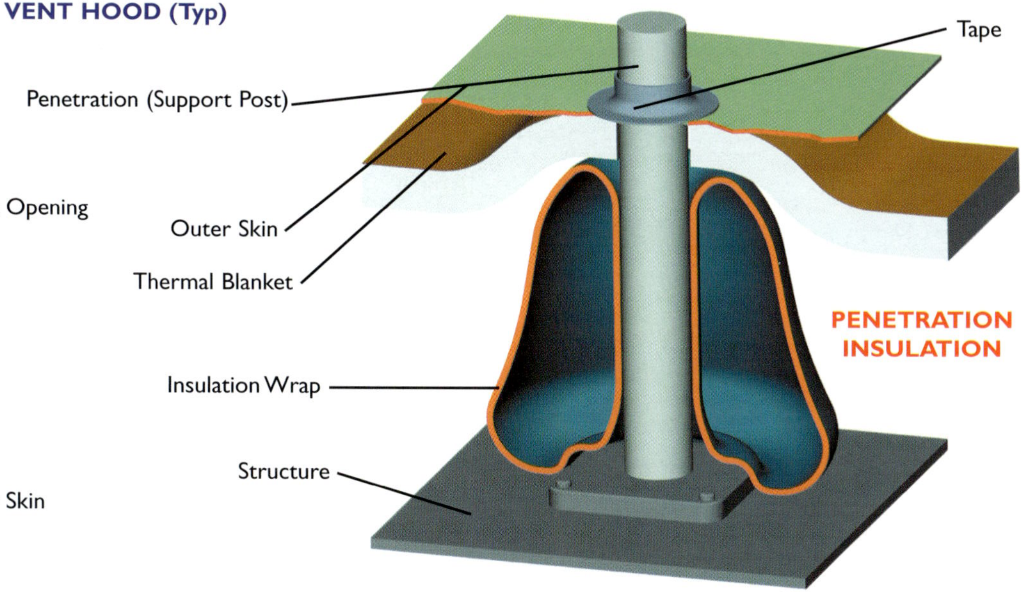

PENETRATION INSULATION

INBOARD

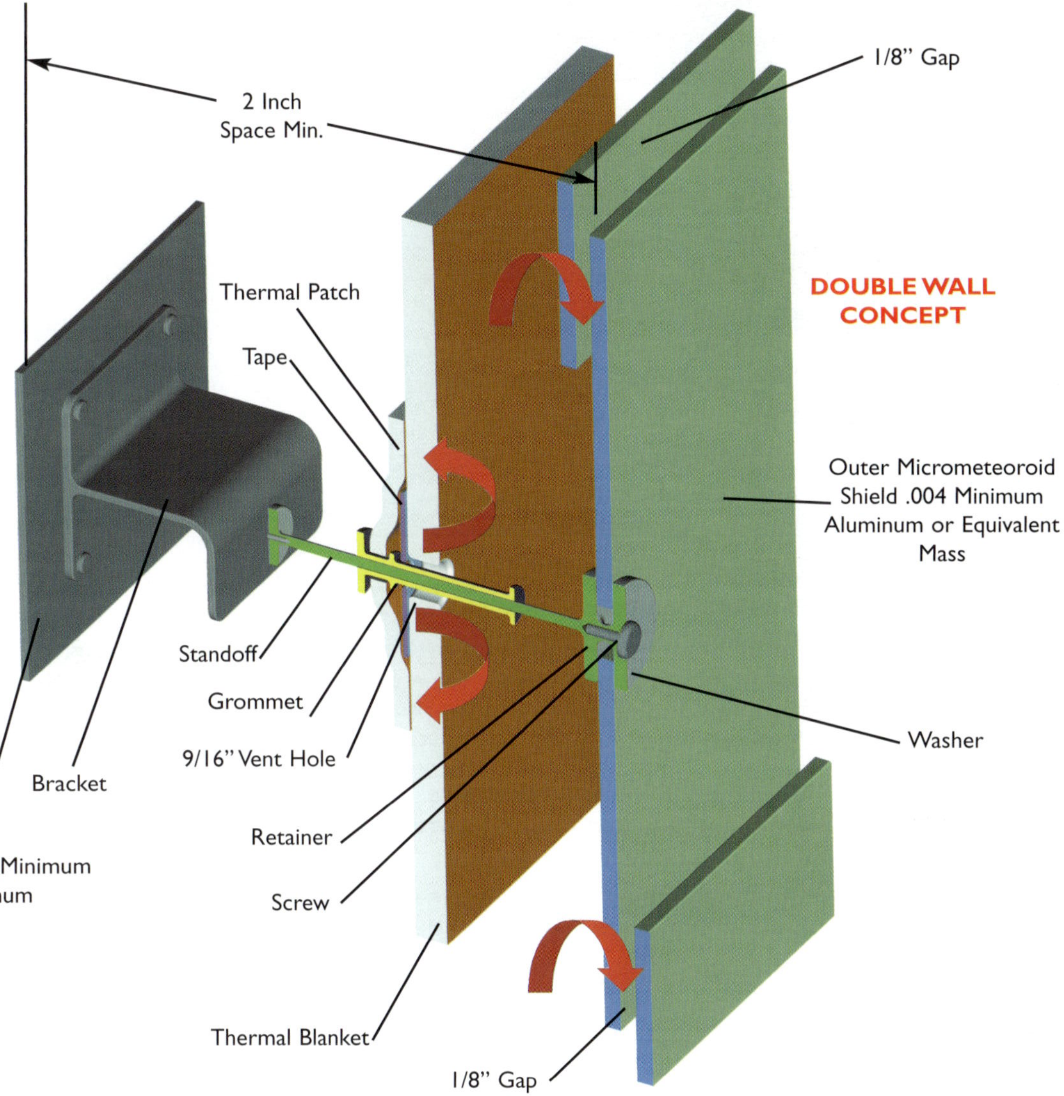

THE SINGLE WALL CONCEPT CONSISTED OF 1/8" ALUMINUM OR EQUIVALENT MASS

STAND-OFF INSTALLATION, BLANKET VENTING AND MICROMETEOROID PROTECTION

The combined thermal blanket and micrometeoroid shield was mounted on low-thermal-conductive supports (standoffs), which kept it at least 2 inches from the main structure. Where subsystems were mounted external to the ascent stage structure, the standoffs were mounted in aluminum frames surrounding the components. Where the blankets met, the mating edges were sealed with Mylar tape. Vent holes were provided in the blanket to prevent heat from building up in the voids between the structural skin and the thermal blanket. The thermal blankets insulated the LM against temperatures of up to 350 degrees F. Access panels were designed for areas where components on the ascent stage needed to be readily accessible during prelaunch activities.

The plume deflectors required a higher level of thermal protection because of the heat from the RCS thrusters. The thermal coverings on the deflectors got thicker toward the bottom of the deflector shield, and consisted of (outboard to inboard) a single layer of nickel foil 0.5 millimeters thick, several layers of inconel mesh, and then a single layer of inconel 1.25 millimeters thick.

The descent stage used the same thermal blanket, but its micrometeoroid shield consisted of either a layer of aluminized Kapton or a single layer of inconel painted with a black nickel-steel alloy. The black sheets absorbed heat and radiated it back into space. The shield was mounted on low thermal conductive supports, which kept it at least ½ inch away from the primary structure. A titanium blast shield secured to the upper side of the descent engine compartment deflected ascent engine exhaust during the lunar launch. The bottom of the descent stage and engine compartment were subjected to extremes in temperature due to radiation from the descent engine. A base shield composed of titanium, with a blanket of nickel foil and Fiberfil outside, and a blanket of aluminized Kapton inside, protected the bottom of the descent stage from engine heat. A flange-like ring of columbium backed with fibrous insulation was attached directly to the engine nozzle extension and joined to the base heat shield by an annular bellows of aluminized Kapton. This bellows arrangement allowed for engine gimbaling, but prevented the heat from the engine leaking into the engine compartment.

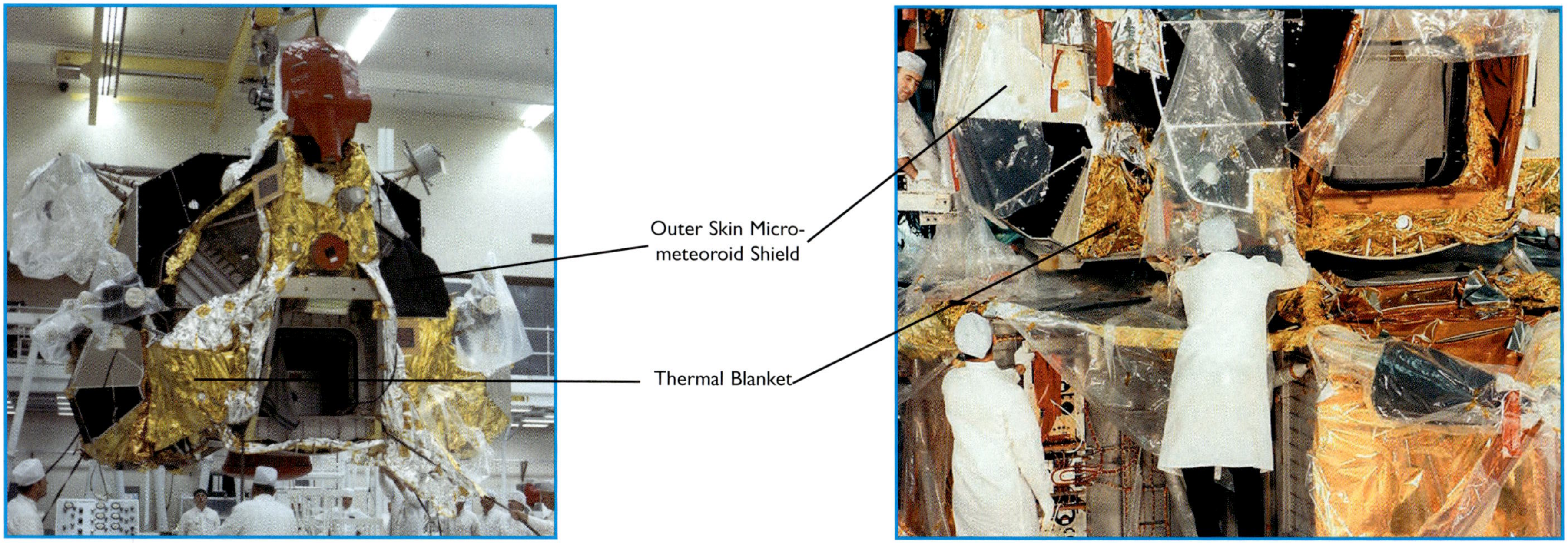

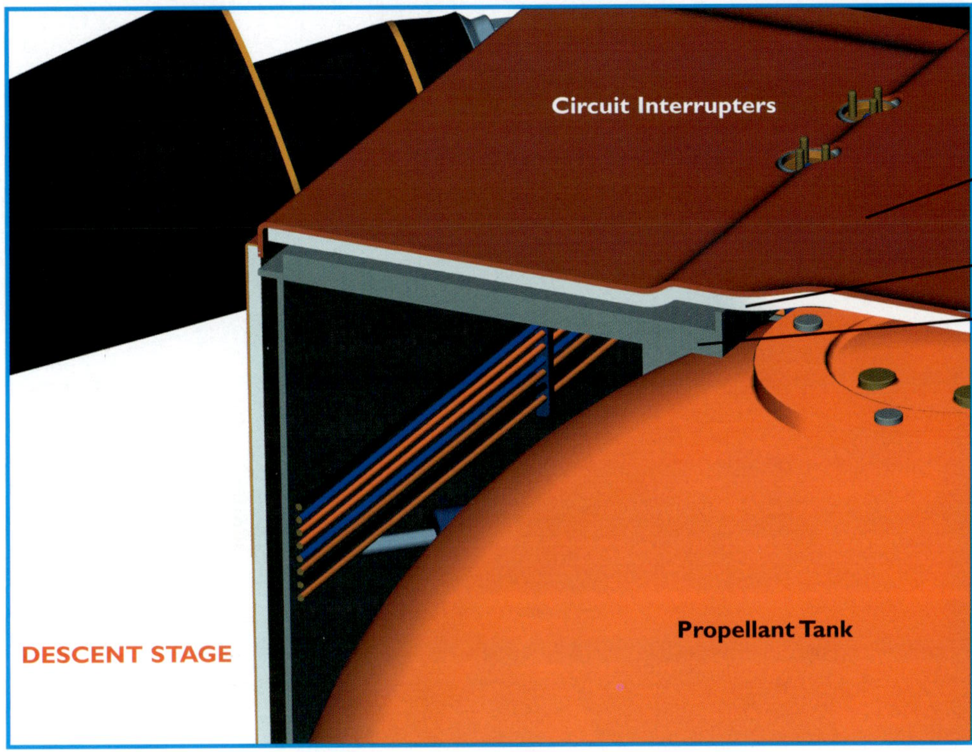

REAR VIEW OF ASCENT STAGE SHOWING THE UNDERLYING THERMAL BLANKET AND THE MICROMETEOROID SHIELDING BEING ASSEMBLED OVER IT

Propellant Tank
Columbium 10 mils
Aluminized Kapton Blanket ½ mil, 25 Layers
Titanium Backface Structure
Bellows Aluminized Kapton ½ mil, 25 Layers
Engine Heat Shield Min-K 1 inch
Nickel-Fiberfil Insulation (2 mil / 1/2 mil) (24 mils)
20 Layers
18 Layers
DESCENT STAGE BASE HEAT SHIELD

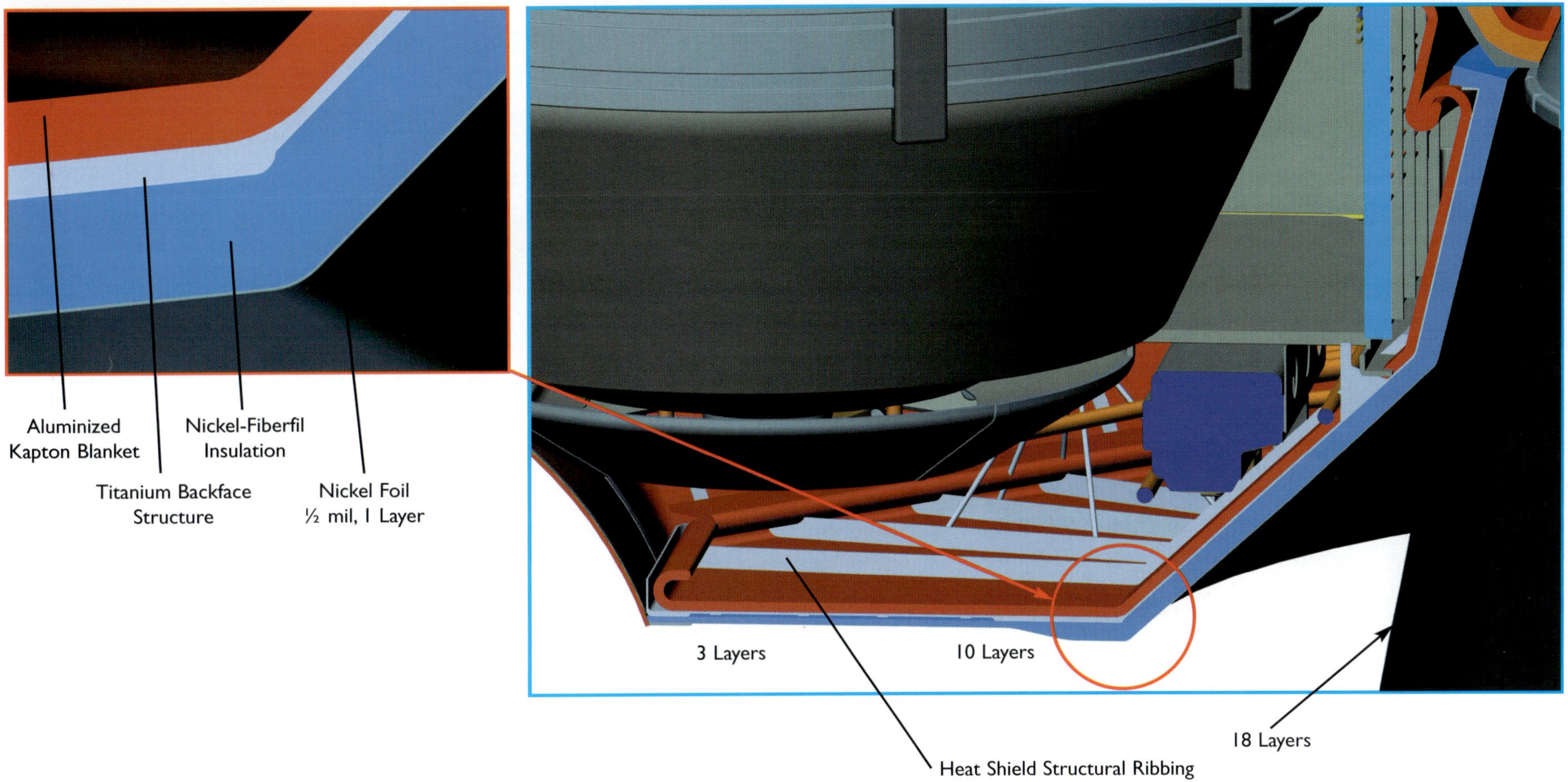

DESCENT ENGINE HEAT SHIELD DETAIL VIEWS

THE APOLLO LUNAR ROVING VEHICLE (LRV) OVERVIEW

The Lunar Roving Vehicle (LRV) was an electric vehicle designed to operate on the low-gravity vacuum of the lunar surface and to be capable of traversing the lunar surface, allowing the astronauts to extend the range of their extra vehicular activities. There were three LRVs built and taken to the Moon, plus seven test and training units. On *Apollo 15* the LRV was driven a total of 27.8 km in 3 hours and 2 minutes. On *Apollo 16* the LRV was driven a total of 26.7 km in 3 hours and 26 minutes. On *Apollo 17* the LRV was driven a total of 35.9km in 4 hours and 26 minutes.

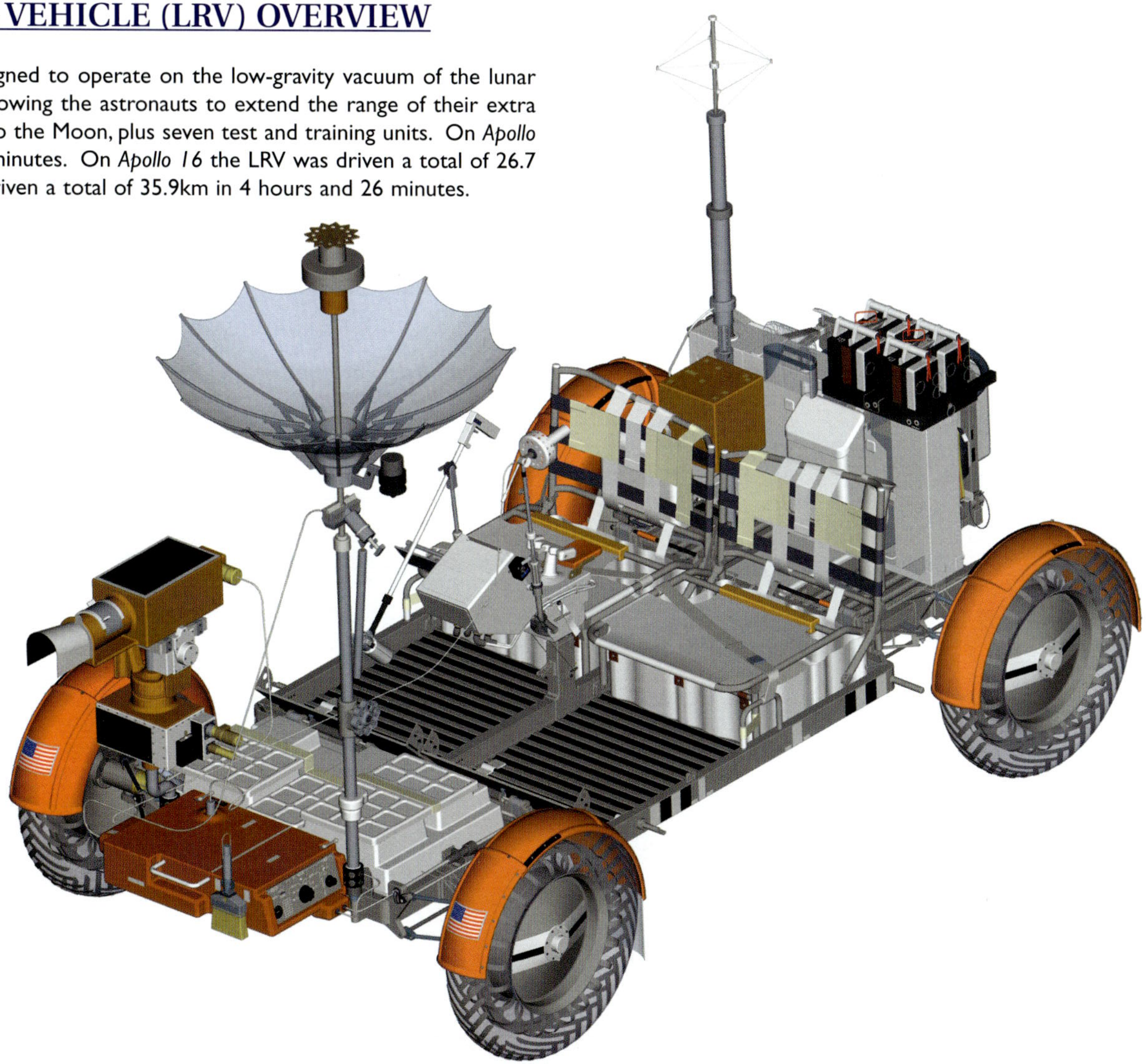

WEIGHT UNLOADED: 460 POUNDS

WEIGHT FULLY LOADED: UP TO 1,140 POUNDS

MAJOR SUBSYSTEMS:

(A) Navigation
(B) Drive Control
(C) Suspension System
(D) Traction Drive
(E) Steering System
(F) Thermal Control
(G) Payload Interface
(H) Chassis
(I) Crew Station
(J) Power System

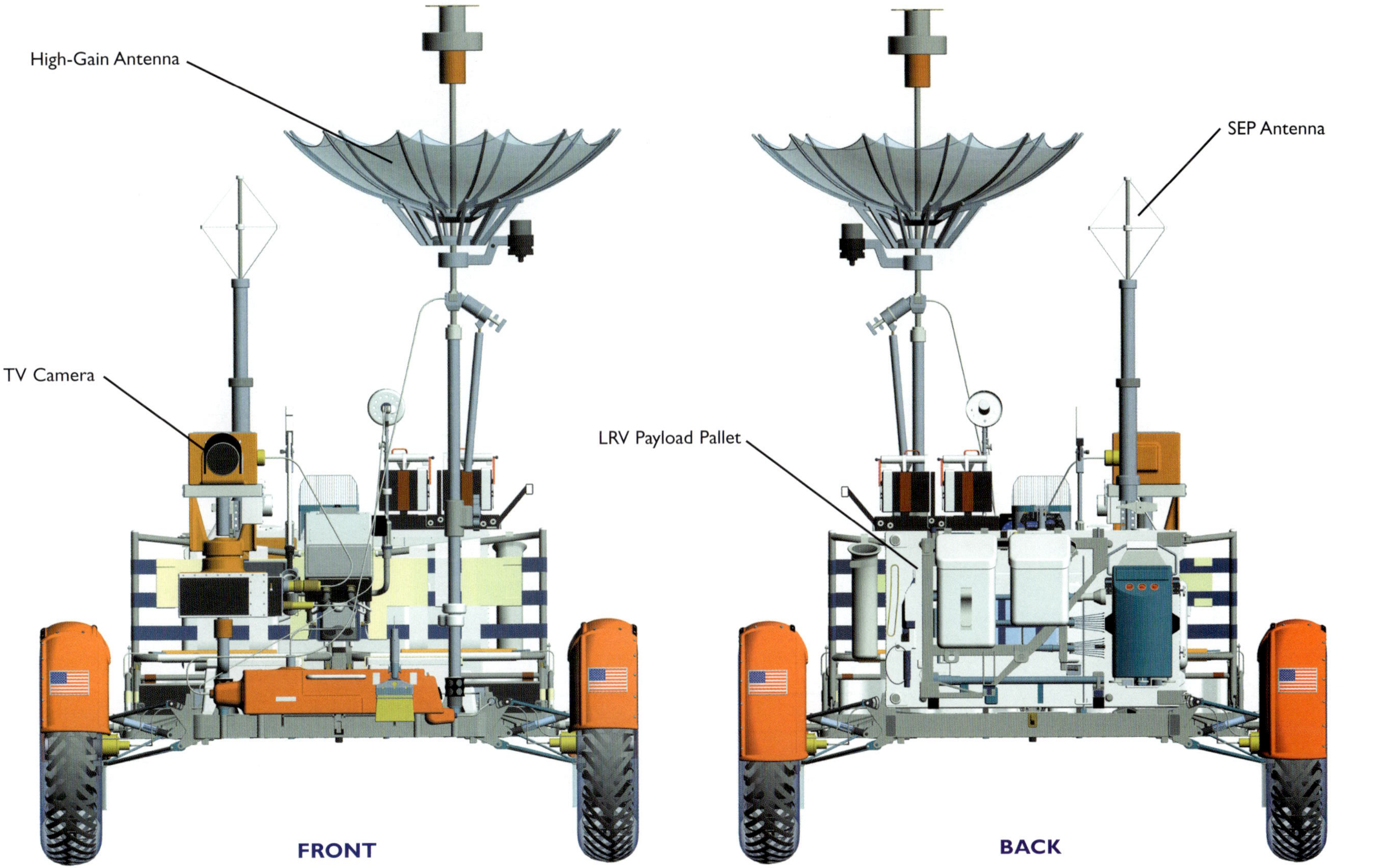
High-Gain Antenna
TV Camera
LRV Payload Pallet
SEP Antenna
FRONT
BACK

Lunar Module Chronology

August 18, 1965
Grumman put into effect "Operation Scrape," an effort to lighten the spacecraft by 125 pounds.

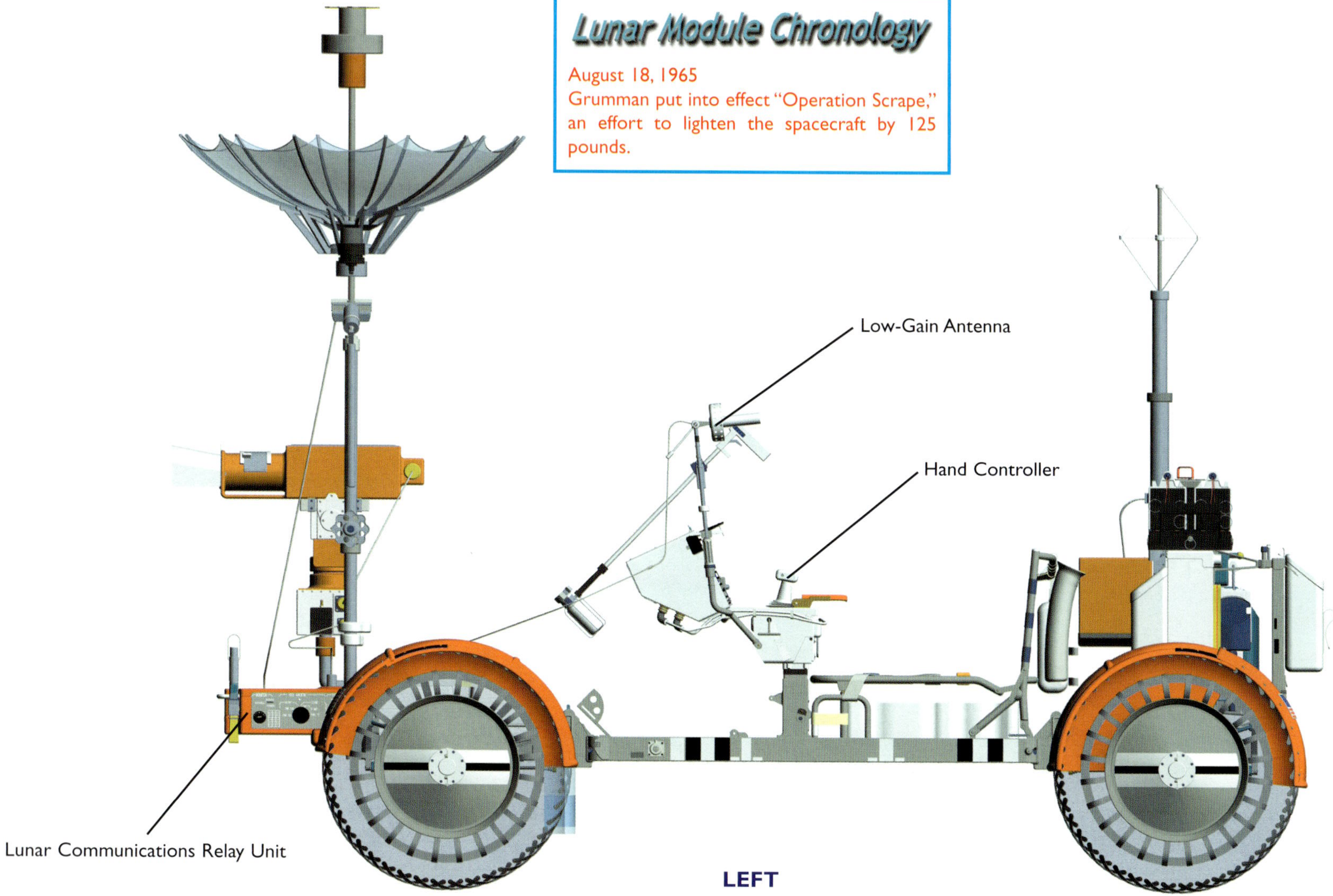

Lunar Module Chronology

August 19, 1965
Two Apollo Lunar Module Test Articles (numbers 10 and 2) to be used on missions SA-501, and A-502 respectively were delivered.

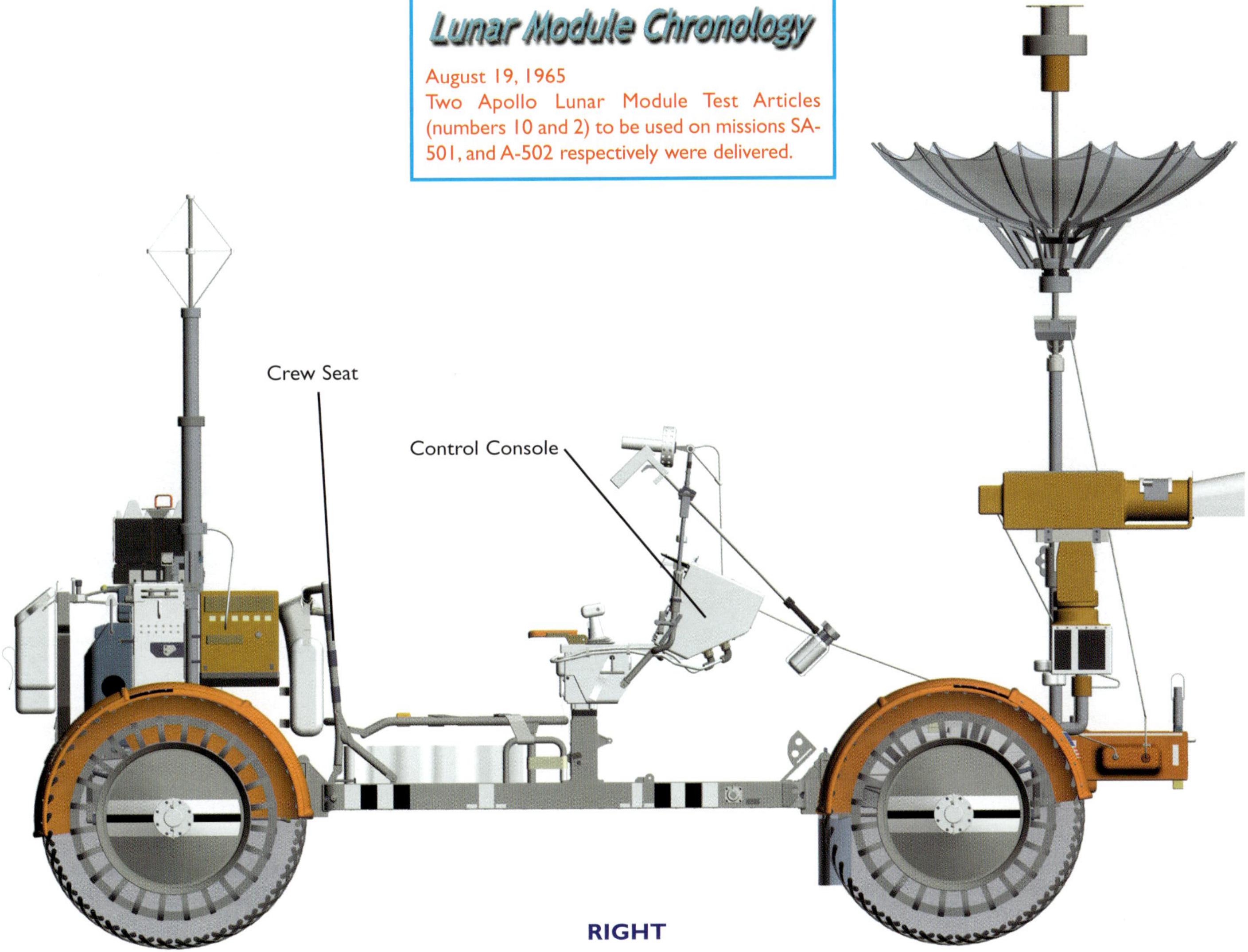

RIGHT

The LRV was 122 inches long with a wheel base of 72 inches. The maximum height was 44.8 inches and ground clearance was 14 inches when the LRV was fully loaded. The frame was made of aluminum alloy 2219 tubing welded assemblies and consisted of a three-part chassis which was hinged on both the forward and rear sections to allow the entire assembly to fold in on itself to be stowed in Quad 1 of the descent stage. The LRV was designed to operate for 78 hours during the lunar day. It could make several exploration sorties up to a cumulative distance of 40 miles (65 km). Because of the limitation of the astronauts' PLSS backpacks, however, the vehicles range was restricted to a radius of about 6 miles (9.5 km) from the Lunar Module. This provided a walk-back capability to the LM should the LRV become immobile at the maximum radius from the LM.

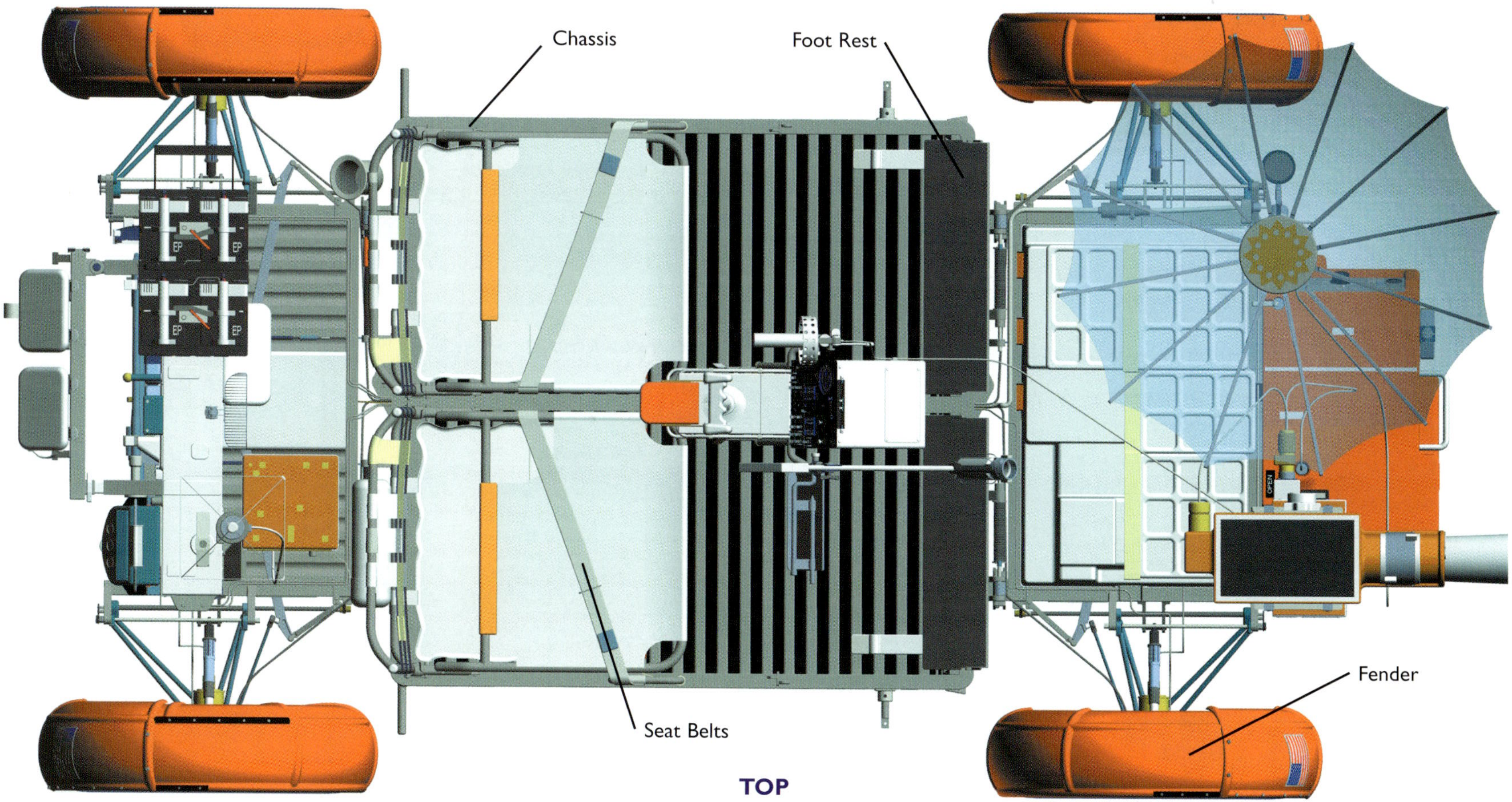

TOP

THE ORIGINAL COST OF THE LRV PROGRAM WAS 38 MILLION DOLLARS

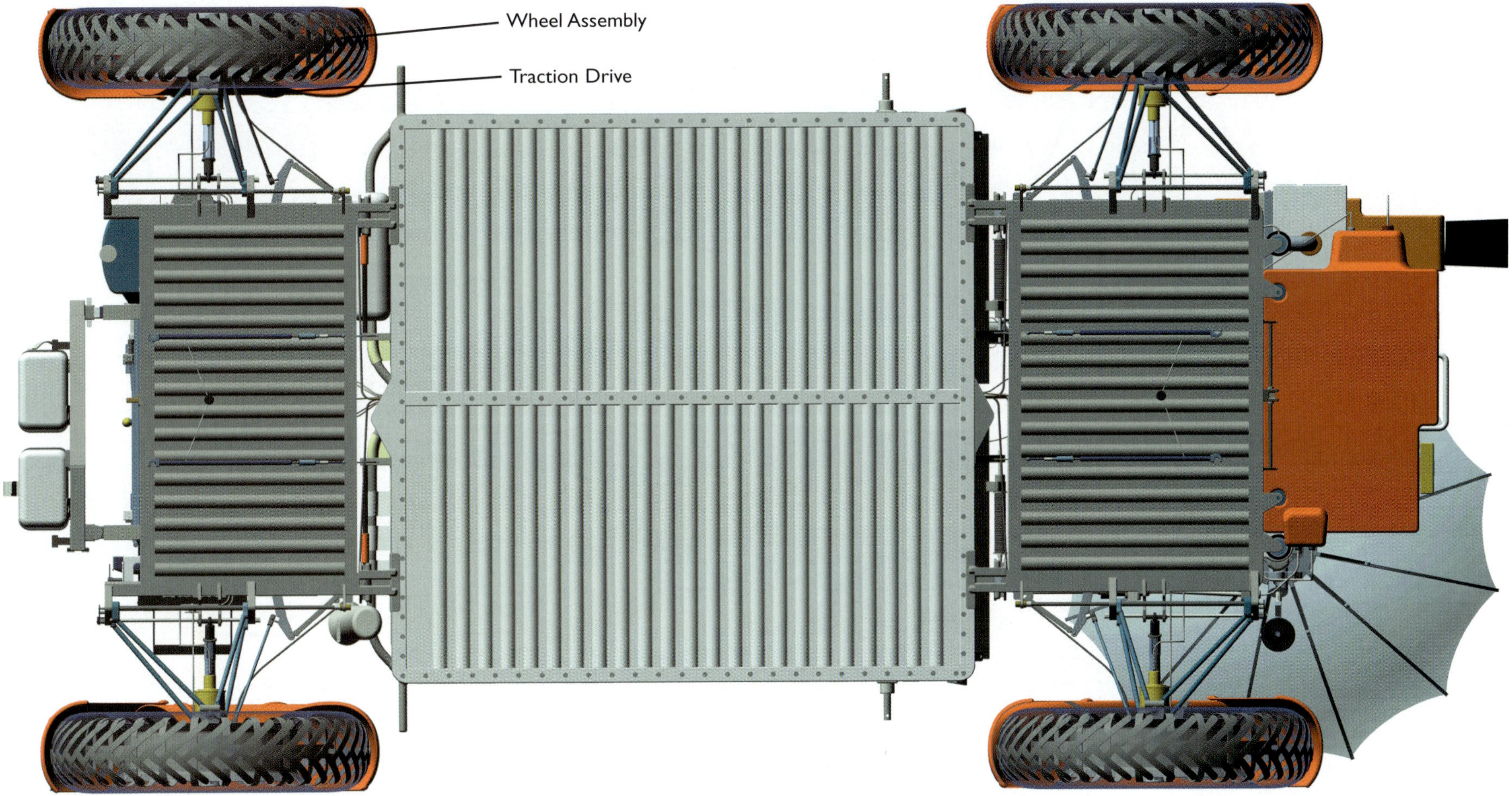

BOTTOM

LRV STOWED CONFIGURATION

The LRV would remain folded and locked in Quad I of the LM descent stage from the time of Earth launch until the astronauts deployed it from its protective berth on the surface of the Moon.

FOLDED HEIGHT: 36.3 INCHES
FOLDED WIDTH: 66.00 INCHES
FOLDED DEPTH: 59.50 INCHES

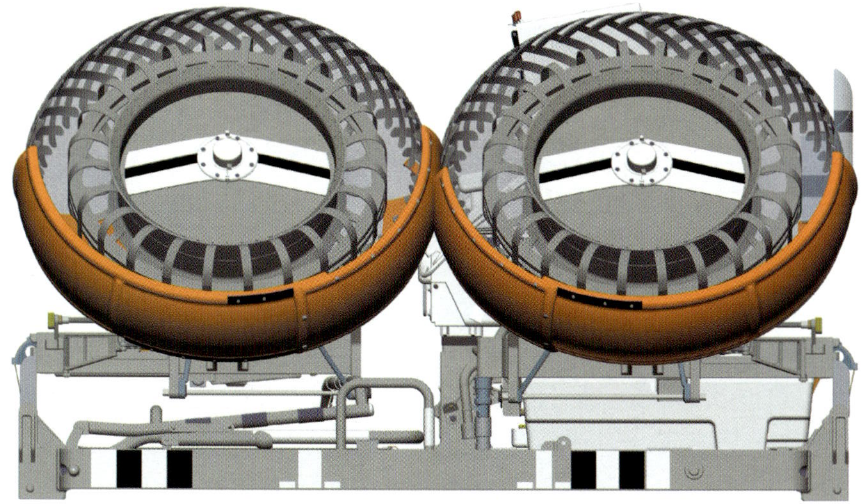

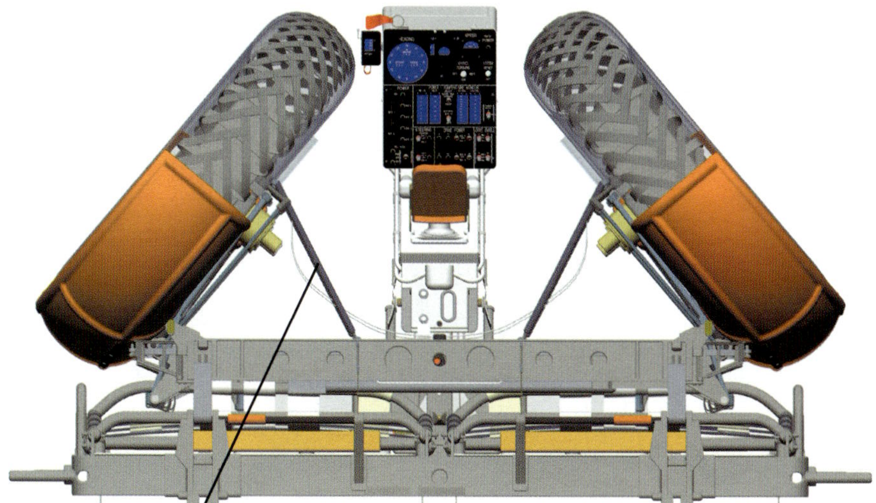

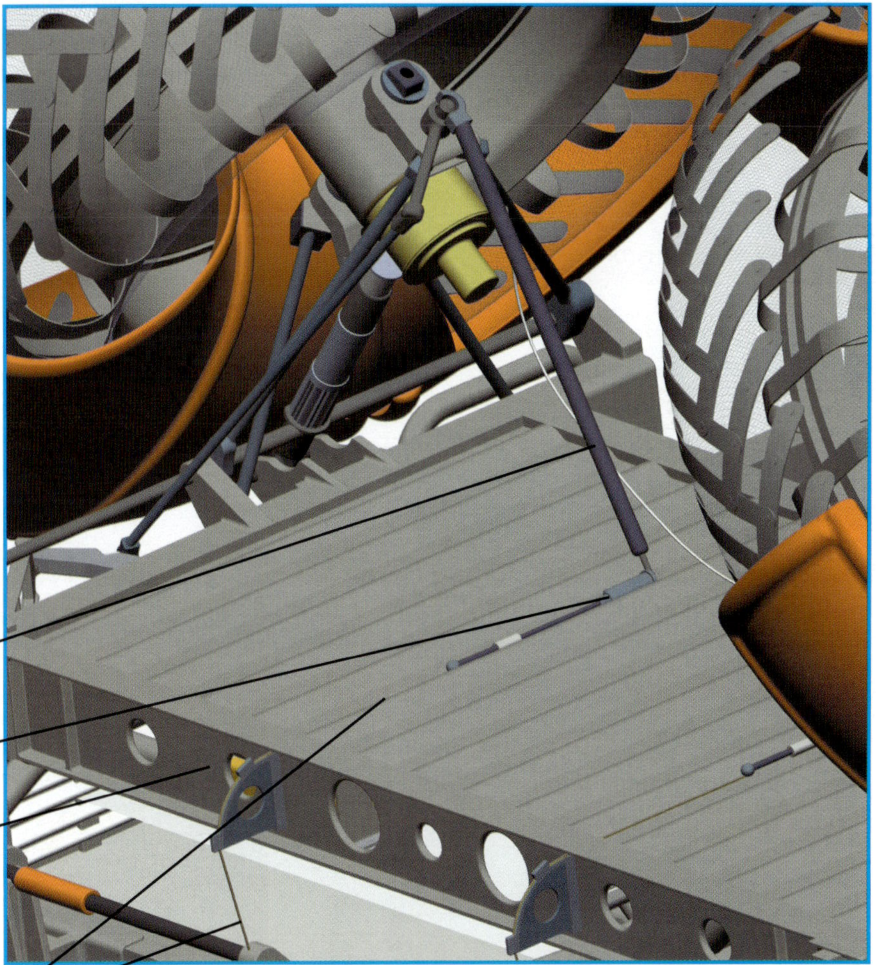

THE WHEEL STRUT RELEASE CABLE WRAPPED AROUND FITTINGS GETTING TIGHT AND PULLING THE WHEEL LOCK STRUT RELEASE PIN AS THE FORWARD CHASSIS ROTATED DURING UNFOLDING, ALLOWING THE WHEELS TO UNFOLD

LRV DEPLOYMENT

DEPLOYMENT ACTUATOR (D-HANDLE) ASSEMBLY

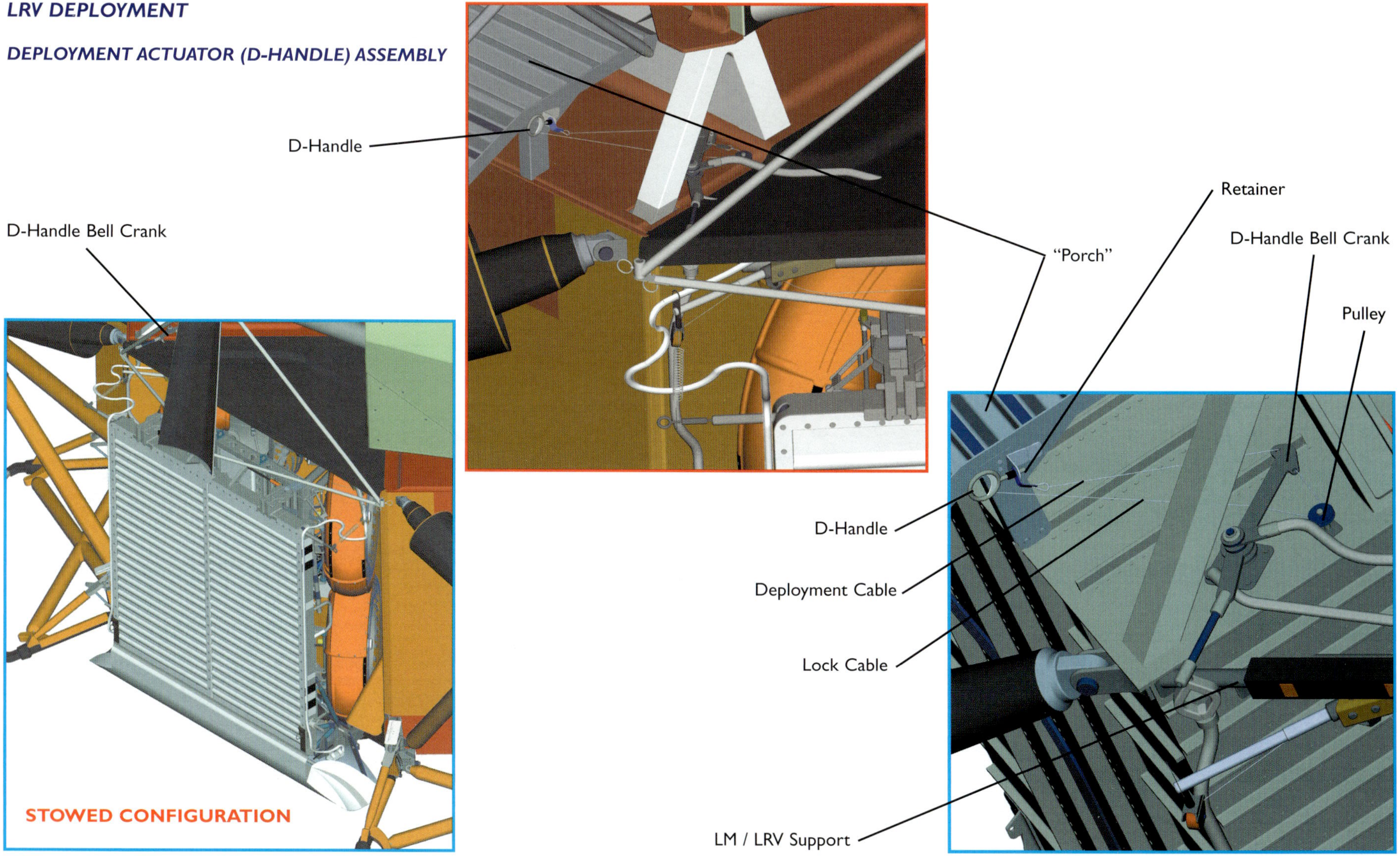

SPACE SUPPORT EQUIPMENT
(LM / LRV SUPPORT STRUCTURE)

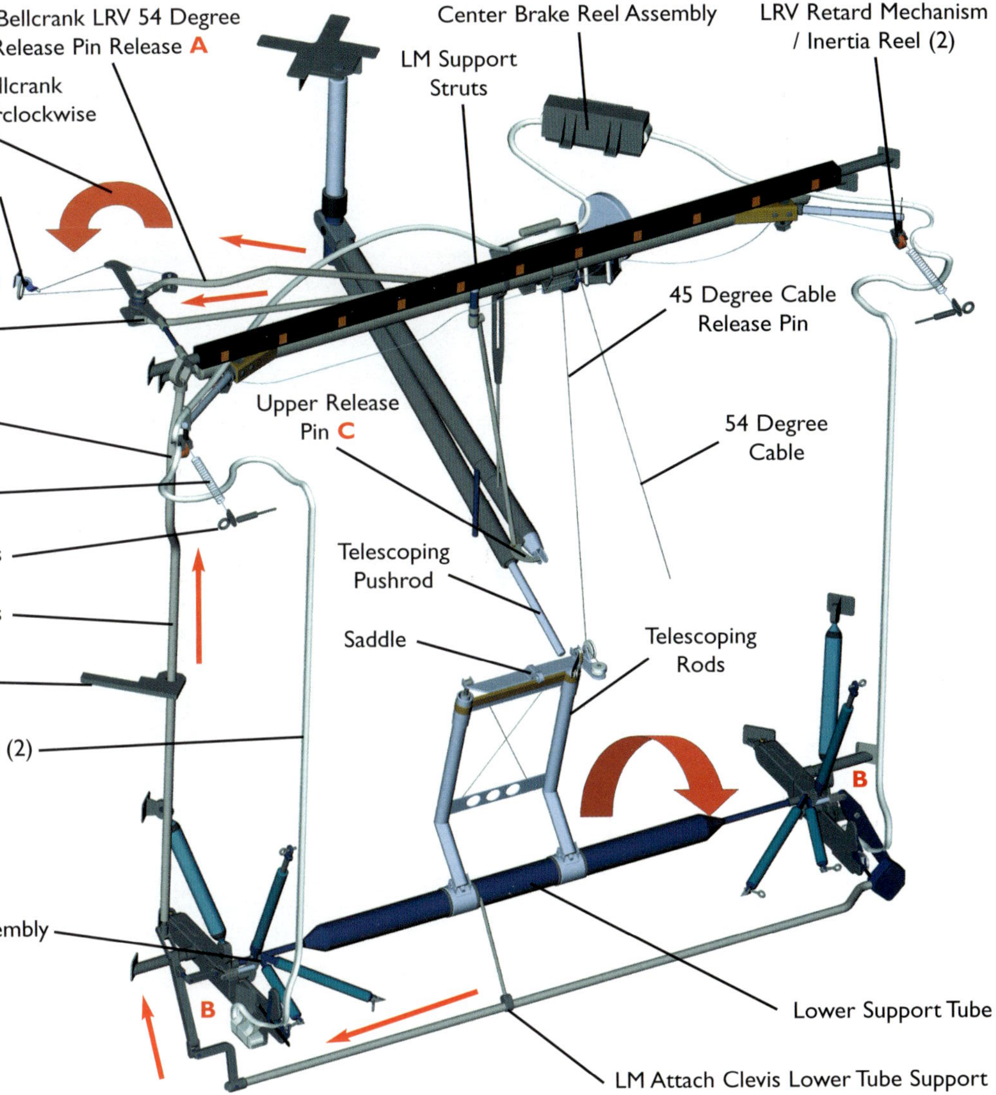

> **Lunar Module Chronology**
>
> December 15, 1965
> Grumman was invited by NASA to provide estimates for four additional Lunar Modules subsequent to *Apollo 11* (LM 5).

VIEW FROM INSIDE OF QUAD I

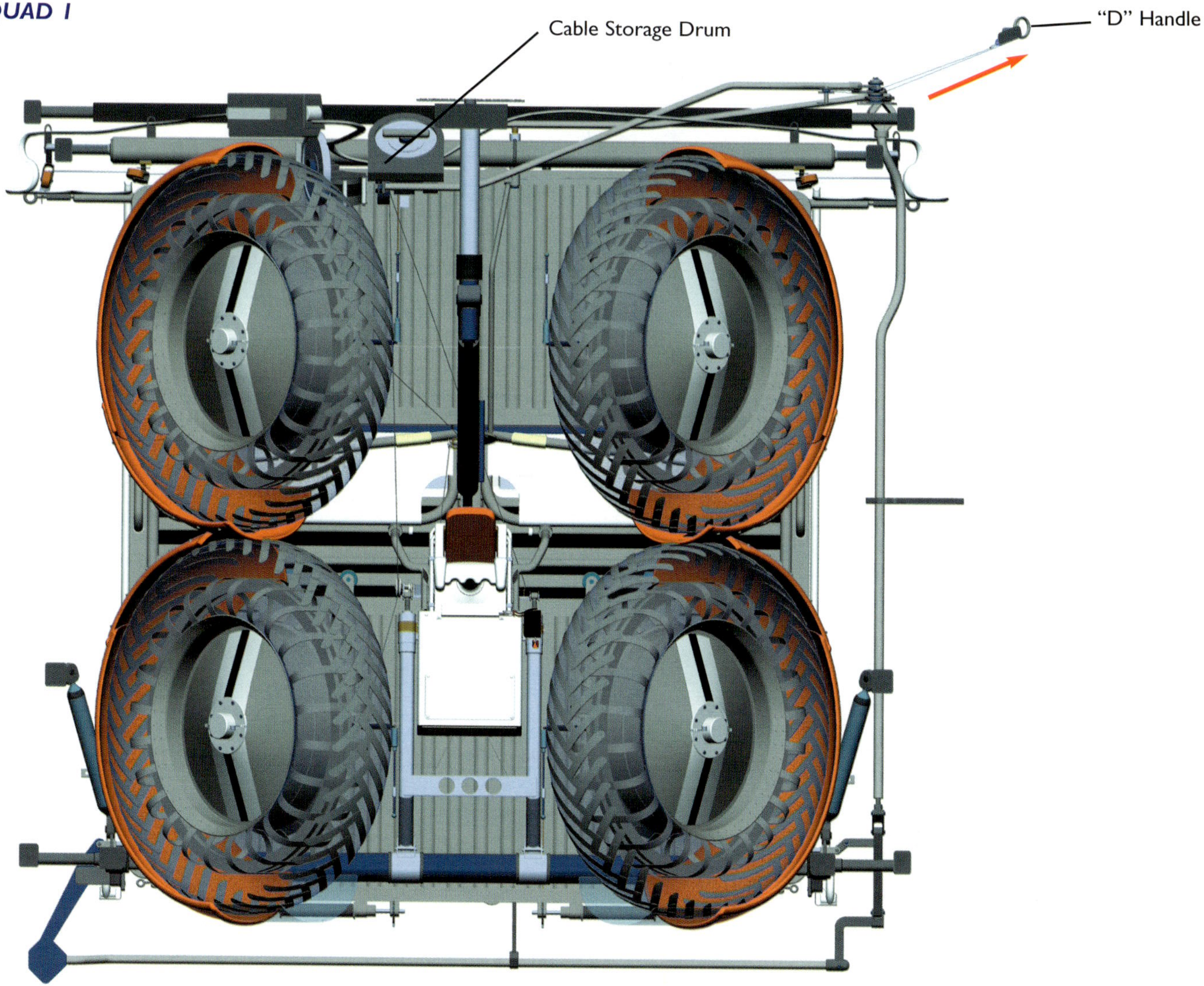

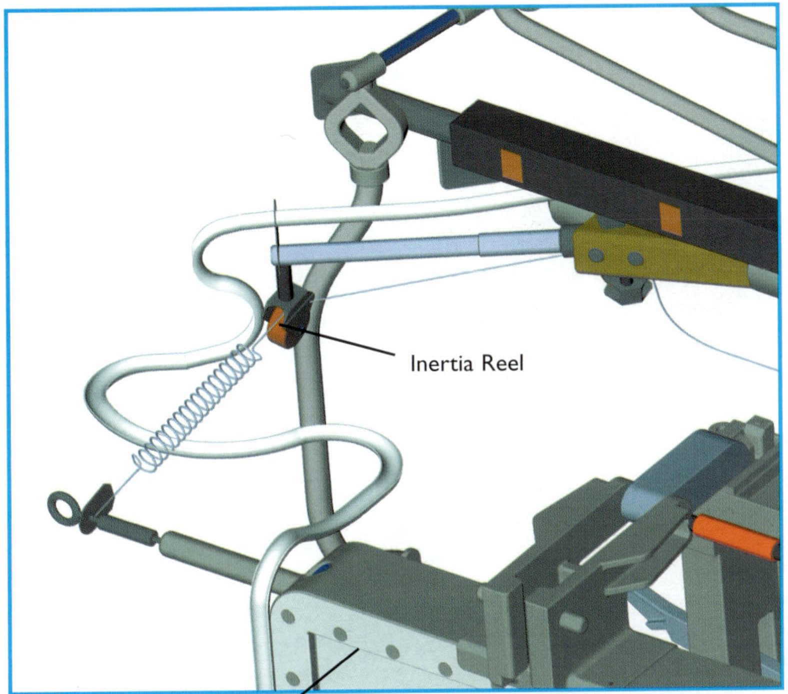

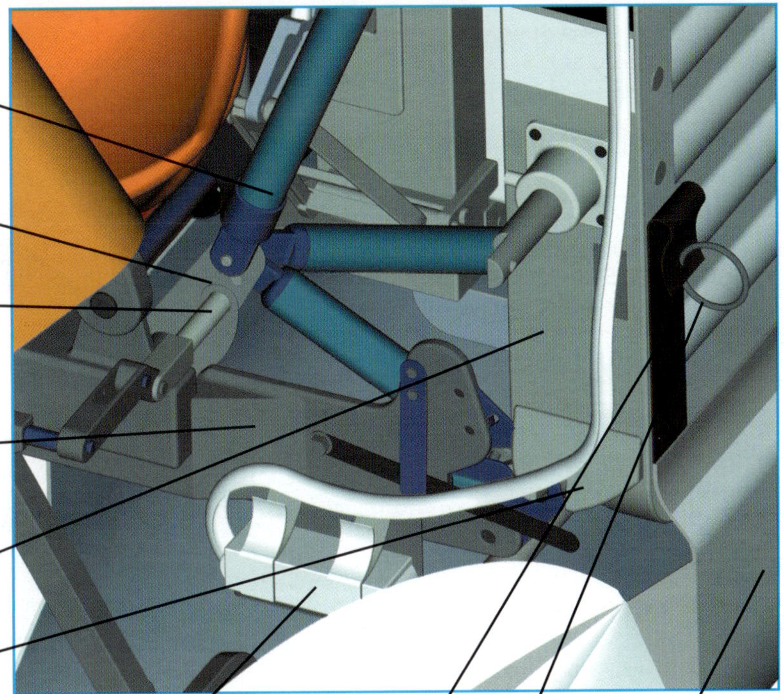

The LRV was designed to be deployed with minimal assistance from the astronauts. Deployment was started by astronauts pulling the D-handle located on the right side of the egress platform of the LM descent stage. Pulling this handle automatically caused a complex set of linkages to pull three release pins holding the LRV to the attach points in the LM / LRV support structure. These pins are labeled A, B and C. The release of these pins frees the LRV for deployment. The next step occurred when a spring-loaded push-off rod started the LRV rotating about lower support tube. The lower part of the LRV was rotated on two points formed by tripods attached to the LRV's center chassis.

Lunar Module Chronology

August 16, 1966
Apollo Lunar Module test model number 3 (TM-3) shipped to Cape Kennedy.

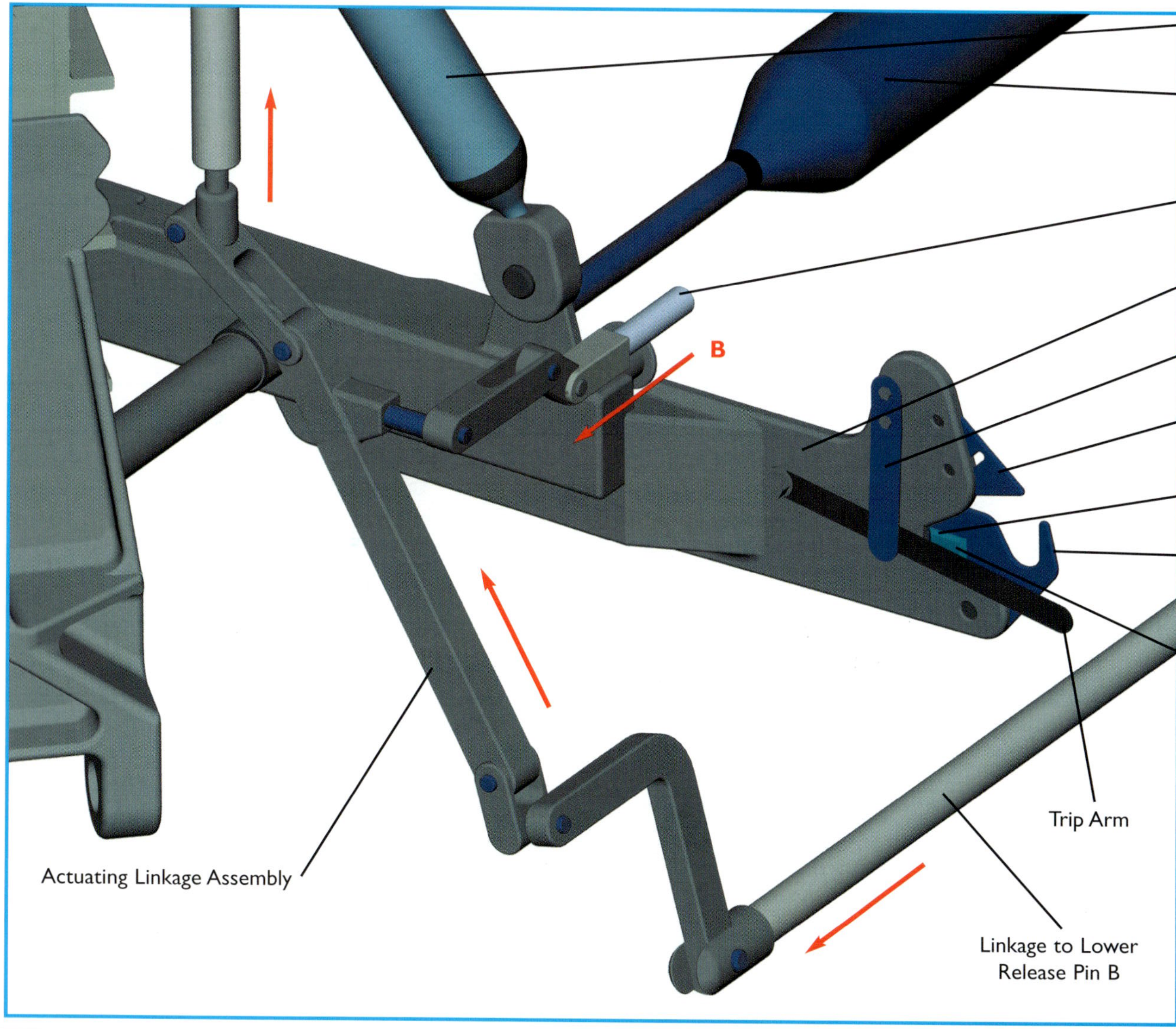

THE LRV MORE THAN DOUBLED THE EVA TIME OF THE ASTRONAUTS WHILE ON THE MOON

Lunar Module Chronology

September 30, 1966
Apollo Lunar Module test model TM-6 and test article LTA-10 shipped.

LRV SADDLE AND FORWARD CHASSIS RELEASE (LRV NOT SHOWN)

The push-off rod lowered the LRV to about 4 degrees from vertical; at this point the astronaut on the right side of the LRV, as viewed from its rear, pulled the deployment tape causing the cable storage drum to rotate. This action released cable – allowing gravity to rotate the LRV outboard. After the first 15 degrees of deployment motion, the two support arms picked up rotation of the LRV. This moves the rotational axis of the LRV to just outside of the LM storage bay in Quad 1.

In addition, two telescoping tubes attached to the saddle started to extend, pushing the LRV away from the LM. At 40 to 50 degrees of deployment motion a cable pulled pins that unlocked the forward and aft chassis sections so that they could unfold. To about 50 degrees of deployment the aft chassis section, which was under spring pressure, unfolded and latched into position. In addition, the wheel assemblies were released and torsion bars caused them to pivot and lock into position. Meanwhile the forward chassis section had been held at about a 45-degree position.

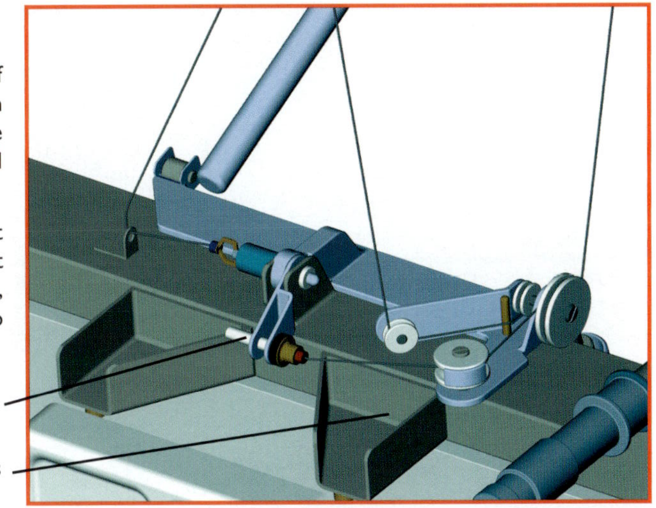

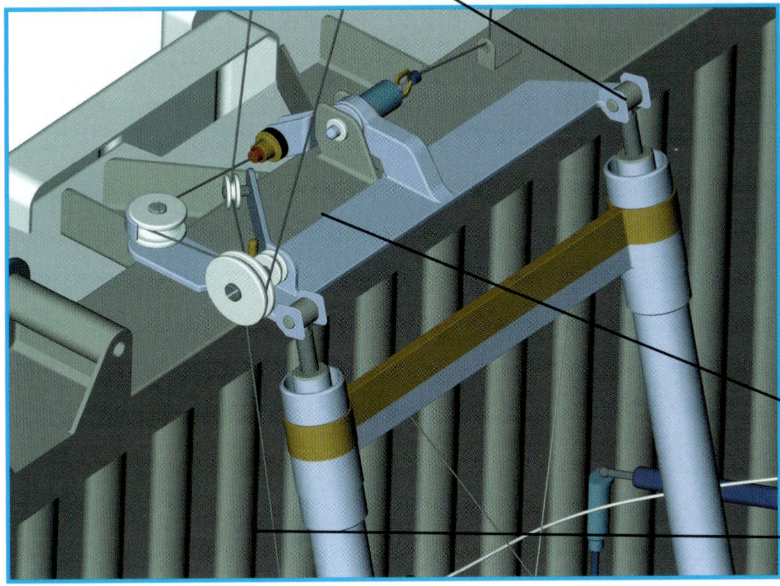

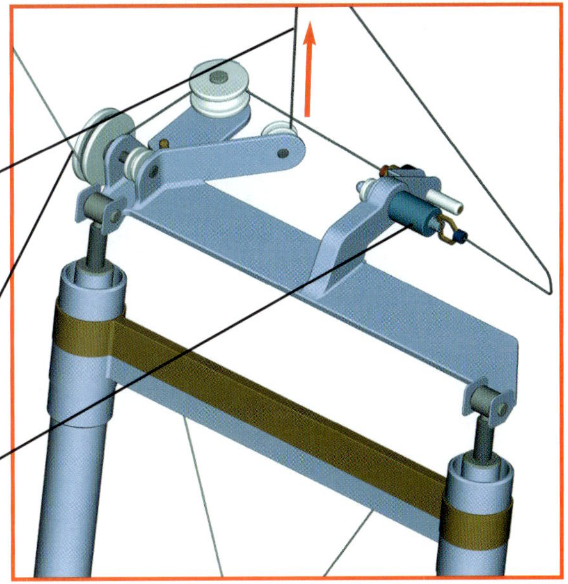

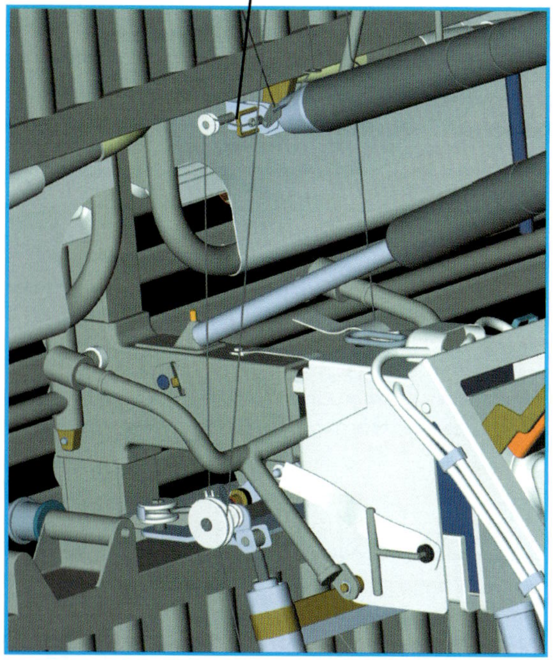
Aft Chassis Release
Pin Mechanism

Wheel Lock Strut
Aft Chassis

C

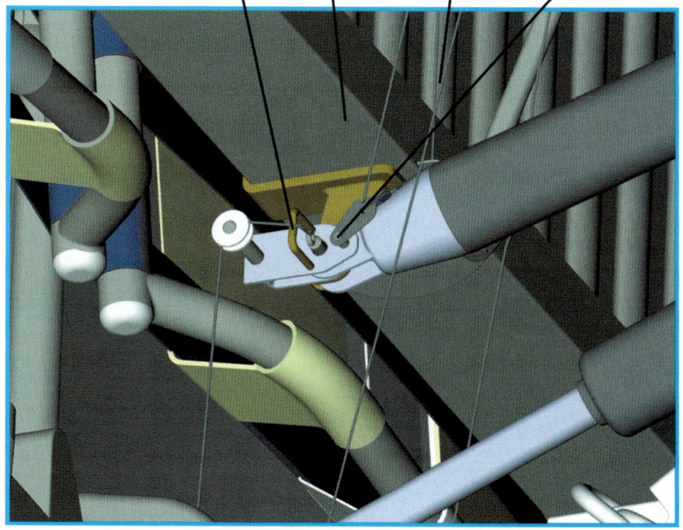
Wheel Lock Strut
Tether (Aft Chassis)
Contingency Cable
Aft Chassis
Release Pin
Contingency Cable
Release Pin

LRV DEPLOYMENT SEQUENCE

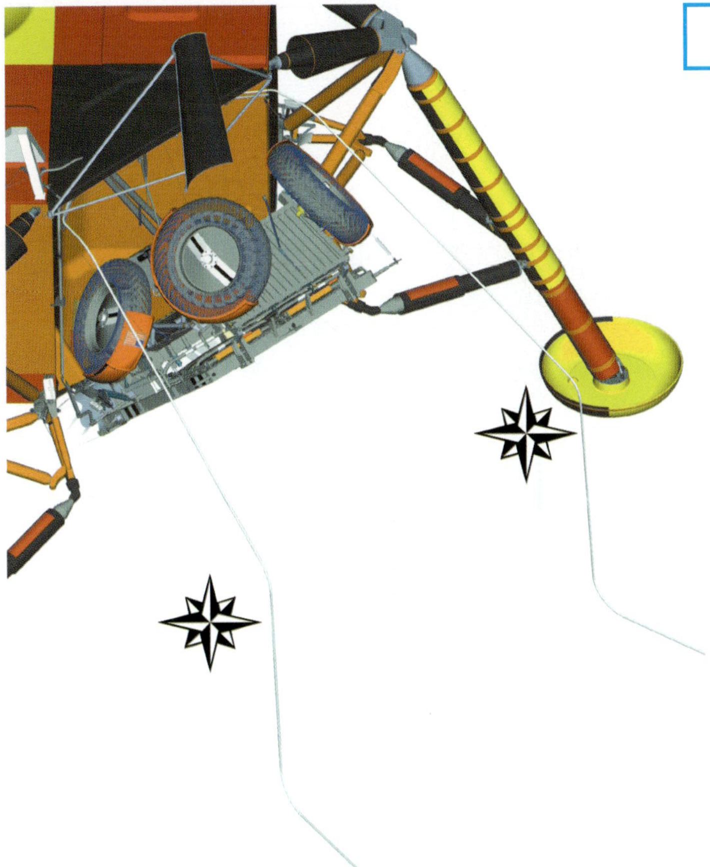

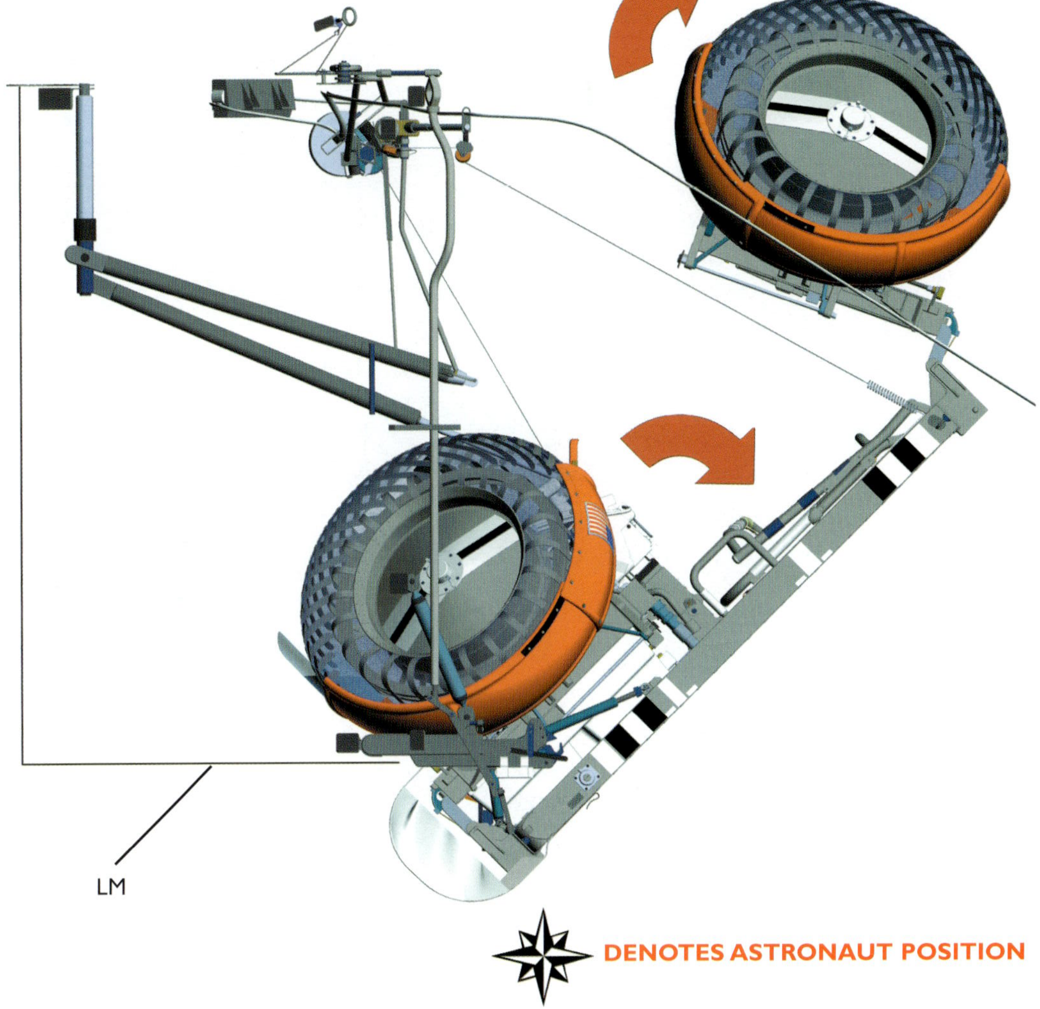

Lunar Module Chronology

March 20, 1967
Apollo 204 launch vehicle to launch the first unmanned Lunar Module.

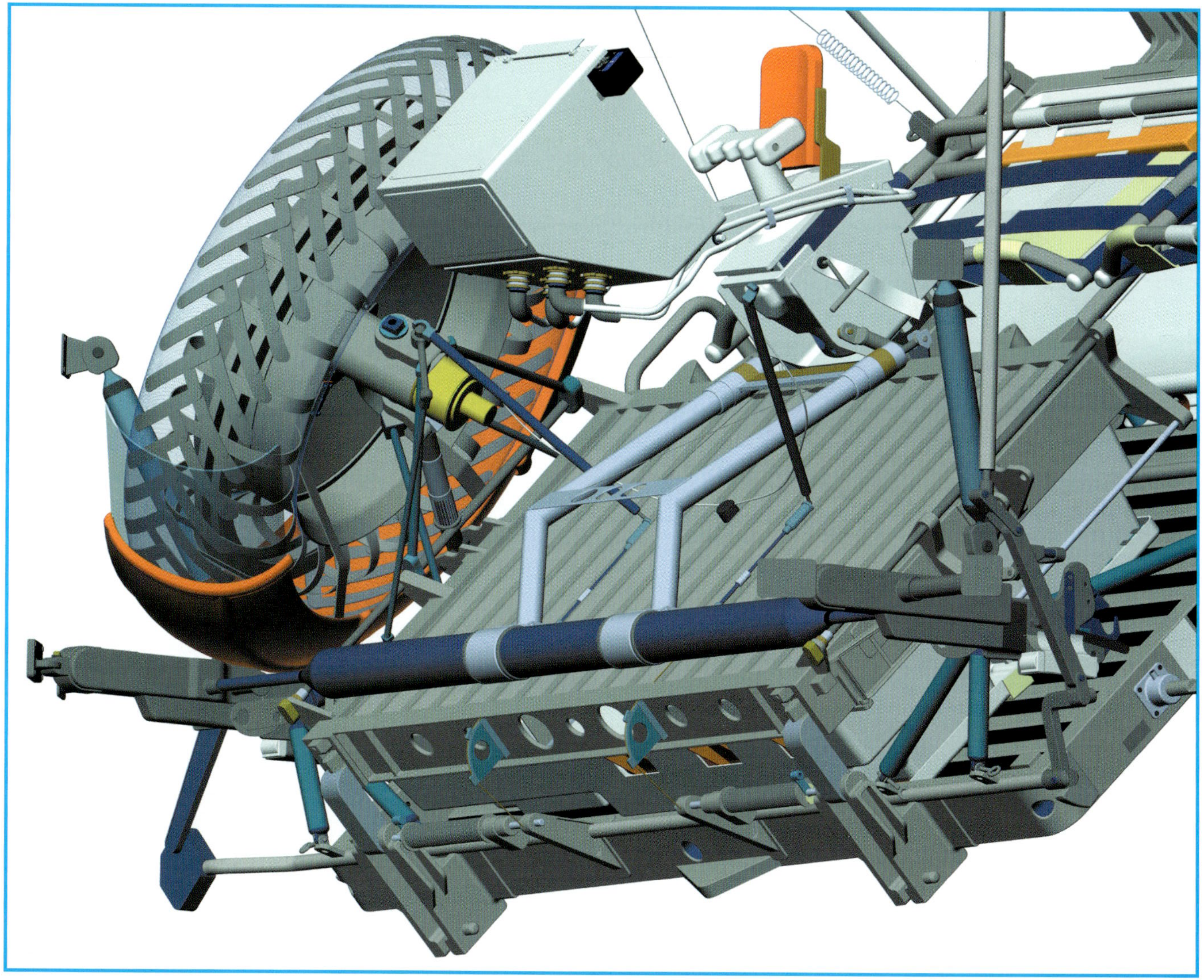

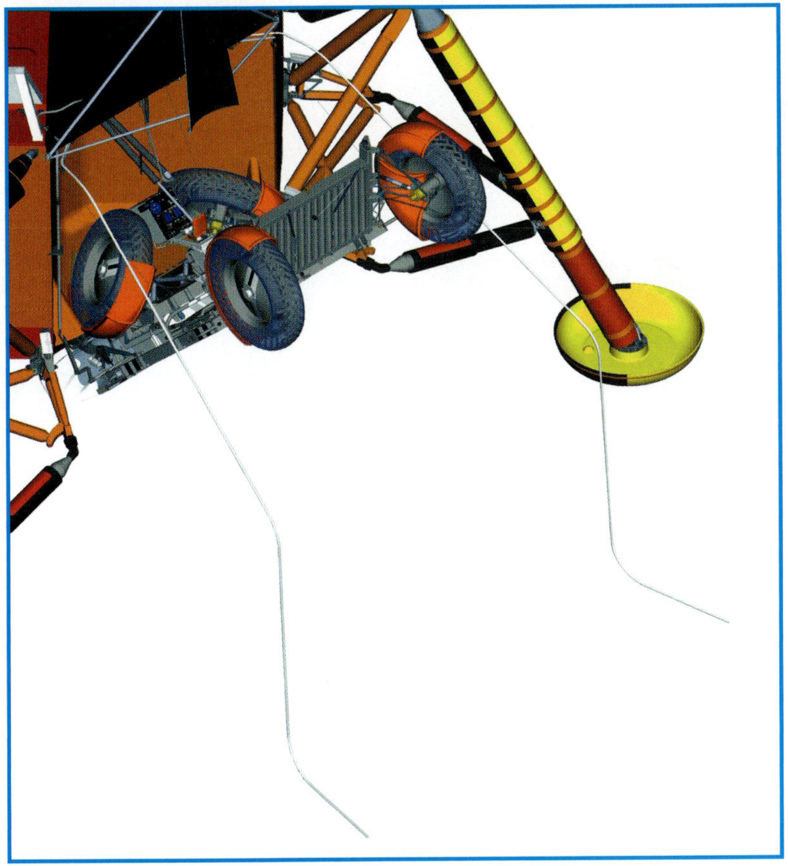

As the astronaut continued to pull the tape, the center and the aft chassis sections continued rotating until the rear wheel assemblies touched the lunar surface.

During this time, the forward chassis section was allowed to unfold and lock. Unfolding began when the center and aft sections had rotated to 73 degrees. At this point, the lobe cam on the LRV released latches on the support arms, which had served as the rotation axis. This allowed the telescoping tubes to extend further and hold the vehicle up and away from the LM.

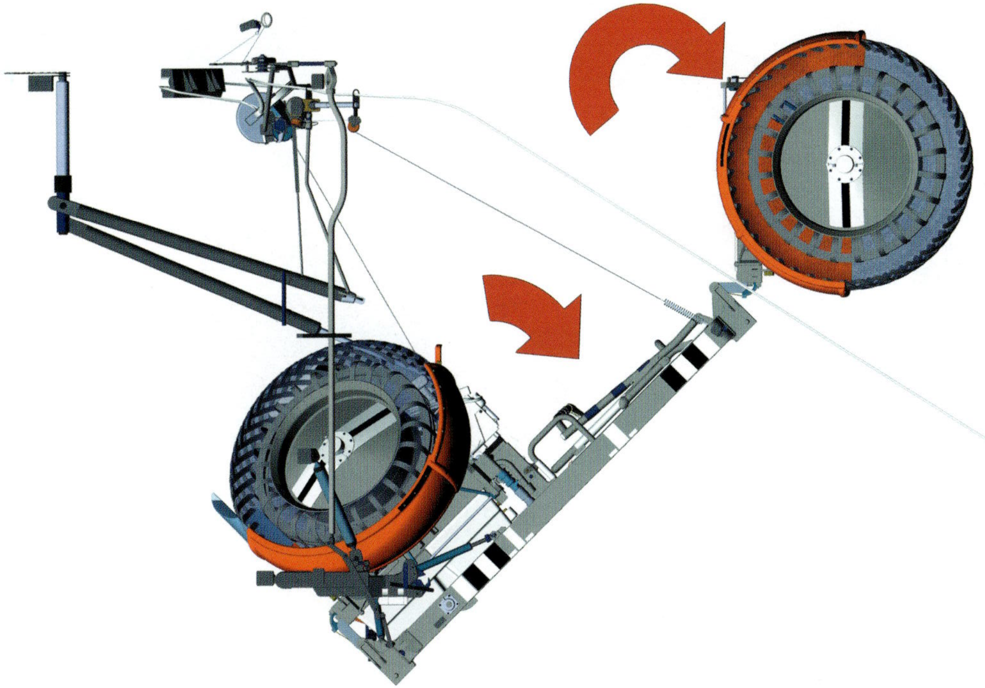

Lunar Module Chronology

September 7, 1967
LM-1 (*Apollo 5*) has serious schedule difficulties; all know problems resolved with the exception of propulsion system leaks.

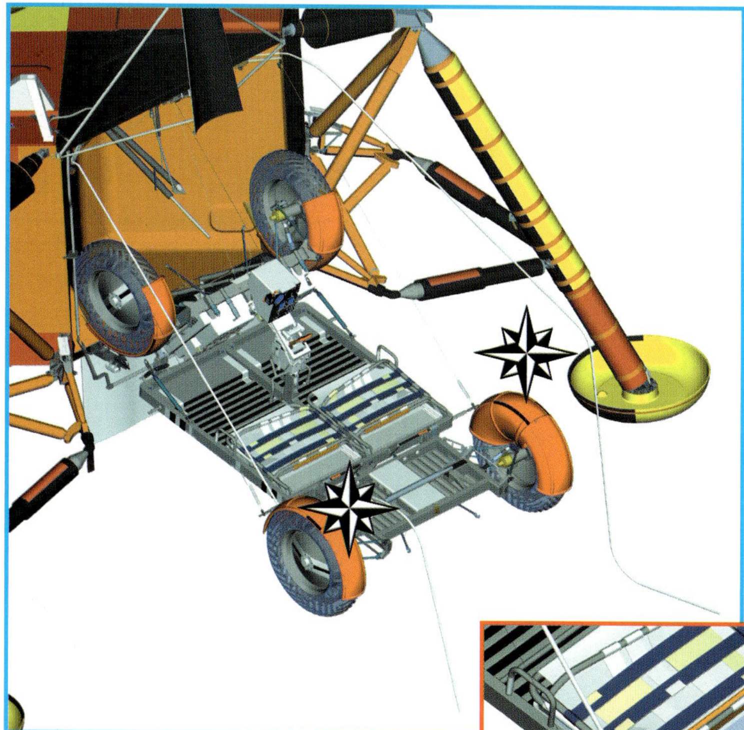

DEPLOYMENT TAPE IS USED TO SUPPORT THE WEIGHT OF THE LRV BY THE ASTRONAUT ON THE LEFT SIDE OF THE LRV

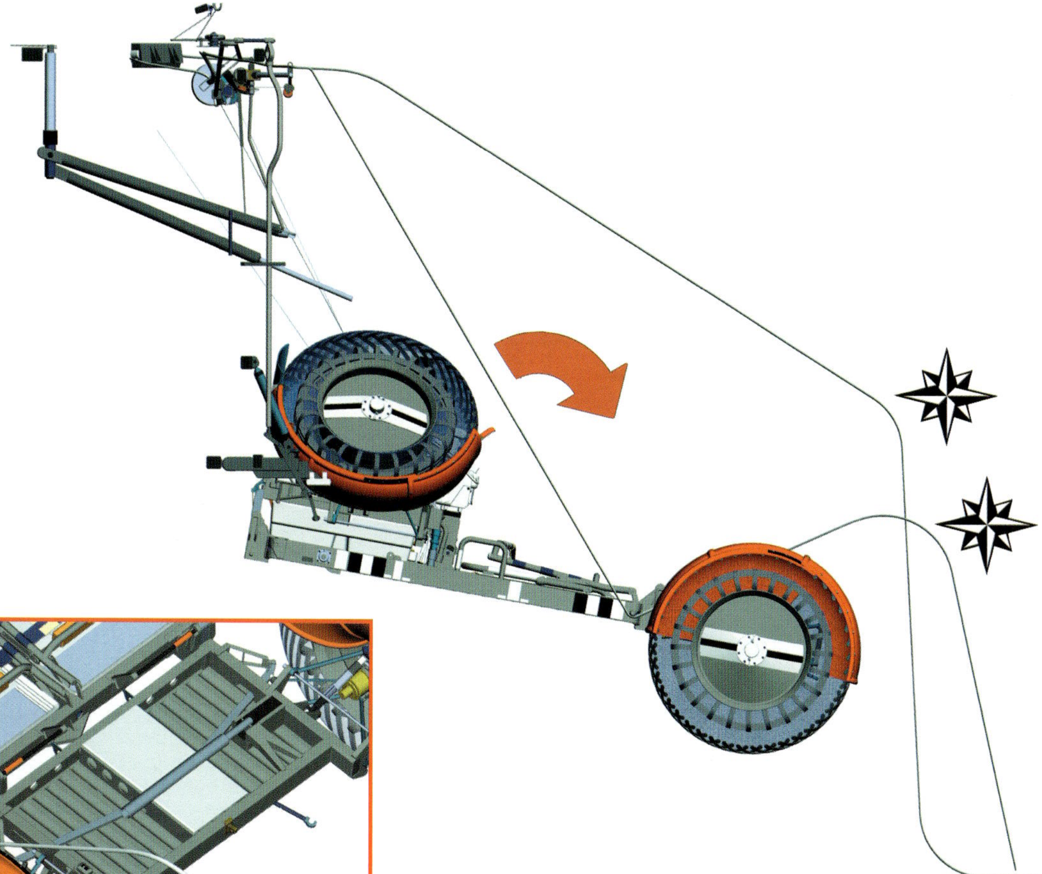

Lunar Module Chronology

October 30, 1967
Lunar Module delivery schedule revised with dates provided by Grumman.

Finally, an astronaut pulled the second operating tape, on the left side of the LRV, allowing the forward end of the unfolded LRV to be lowered to the lunar surface. The astronauts then disconnected the deployment hardware from the vehicle by pulling a series of pins.

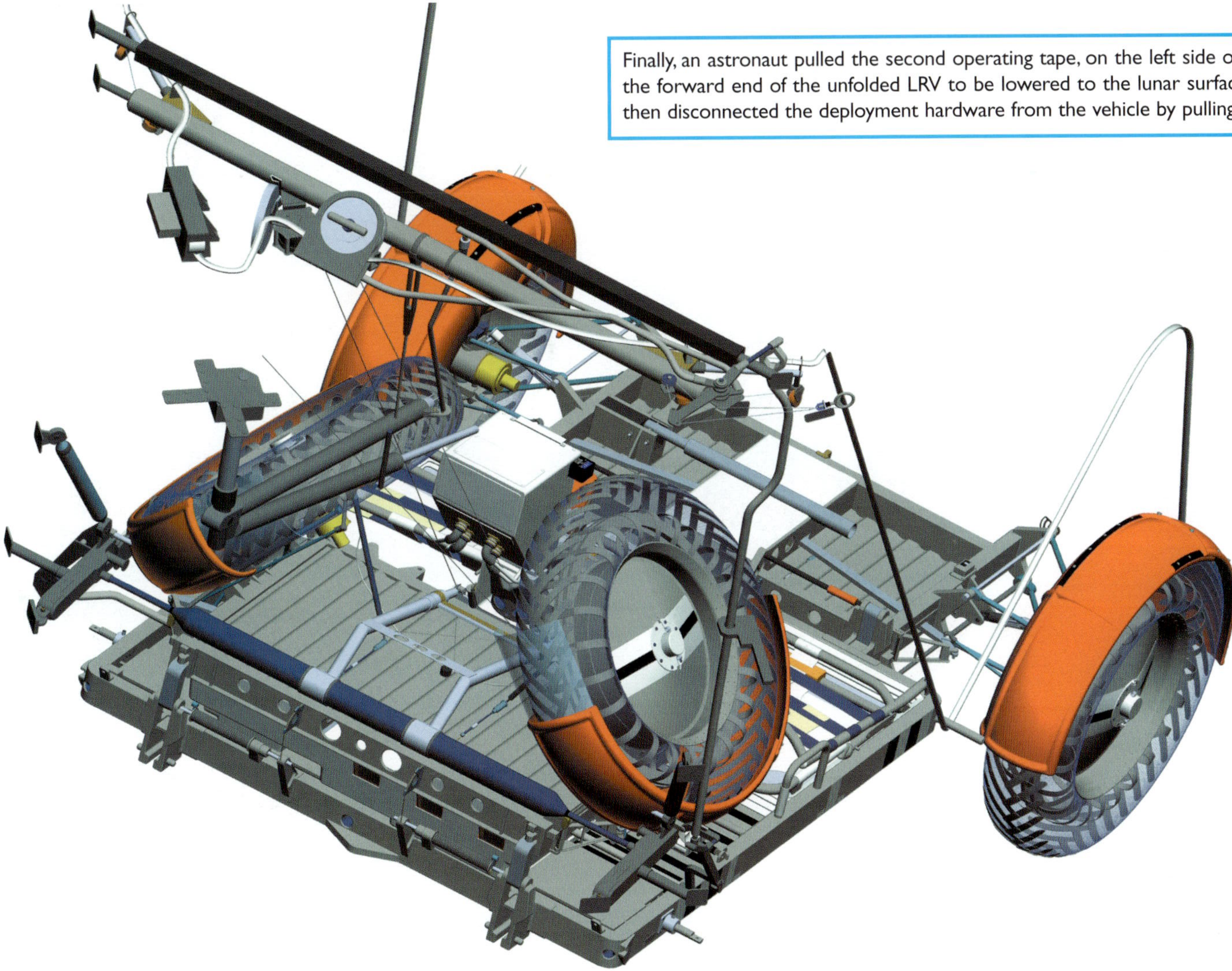

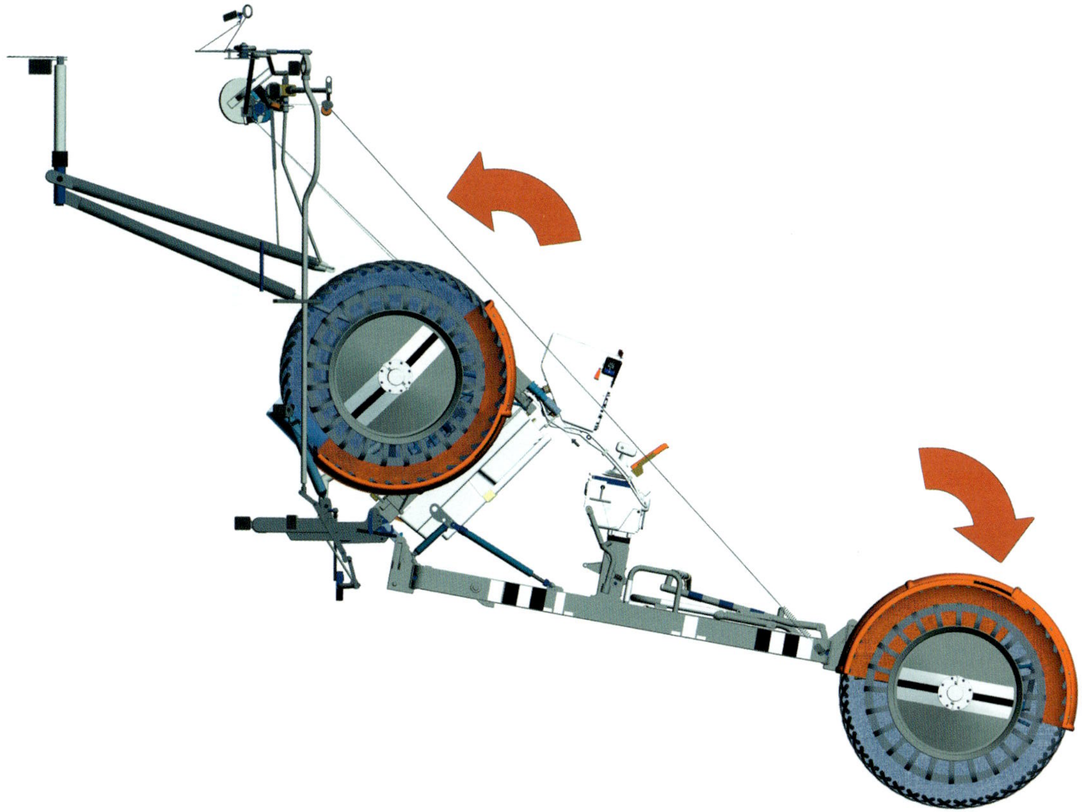

Lunar Module Chronology

December 17, 1967
Lunar Module 5 (*Apollo 11*) window shattered during a pressurization test.

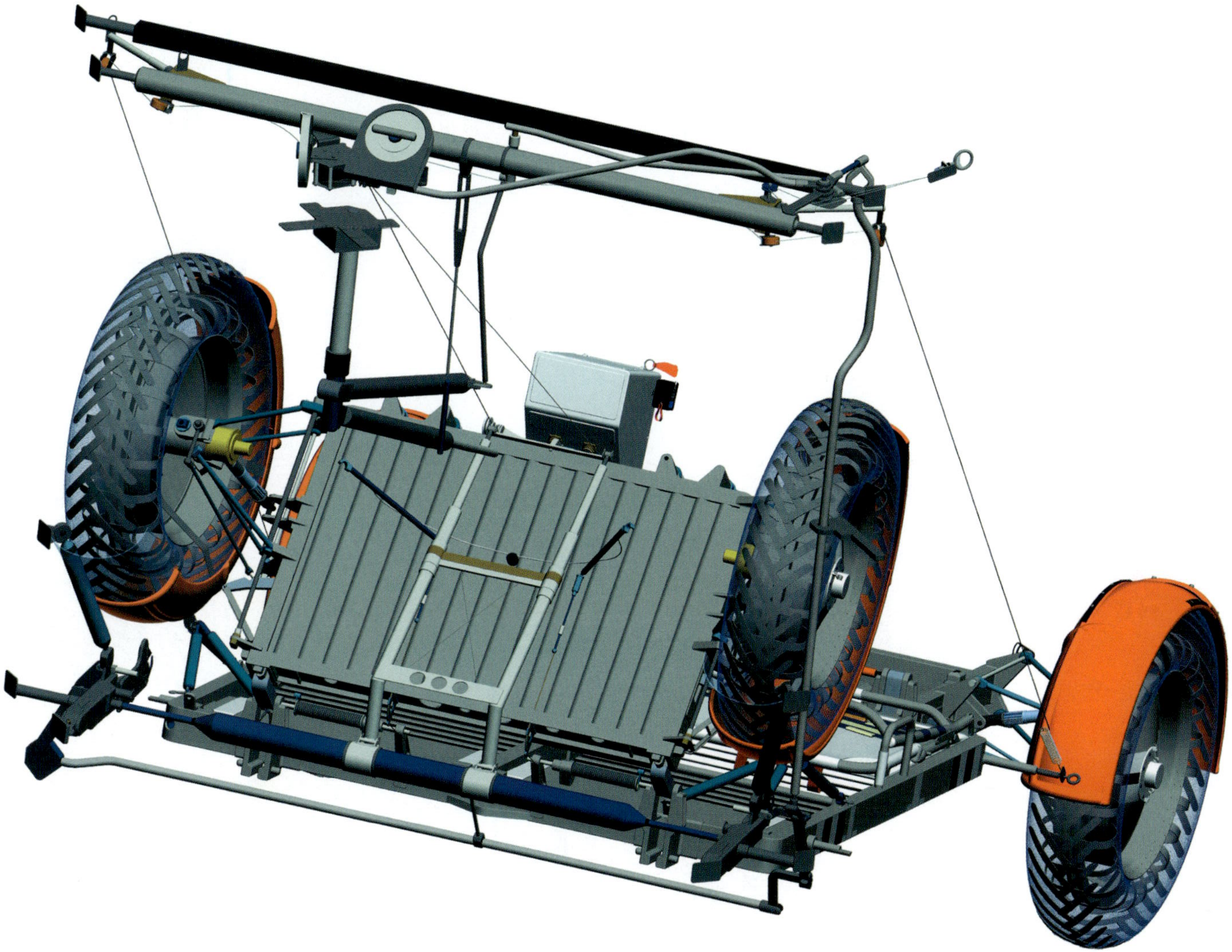

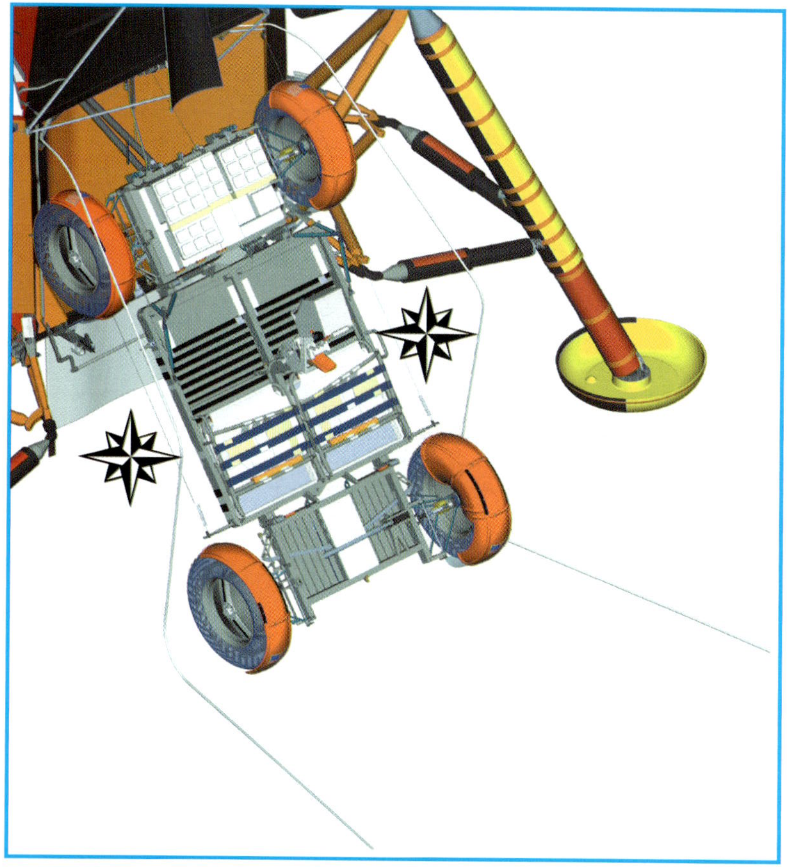

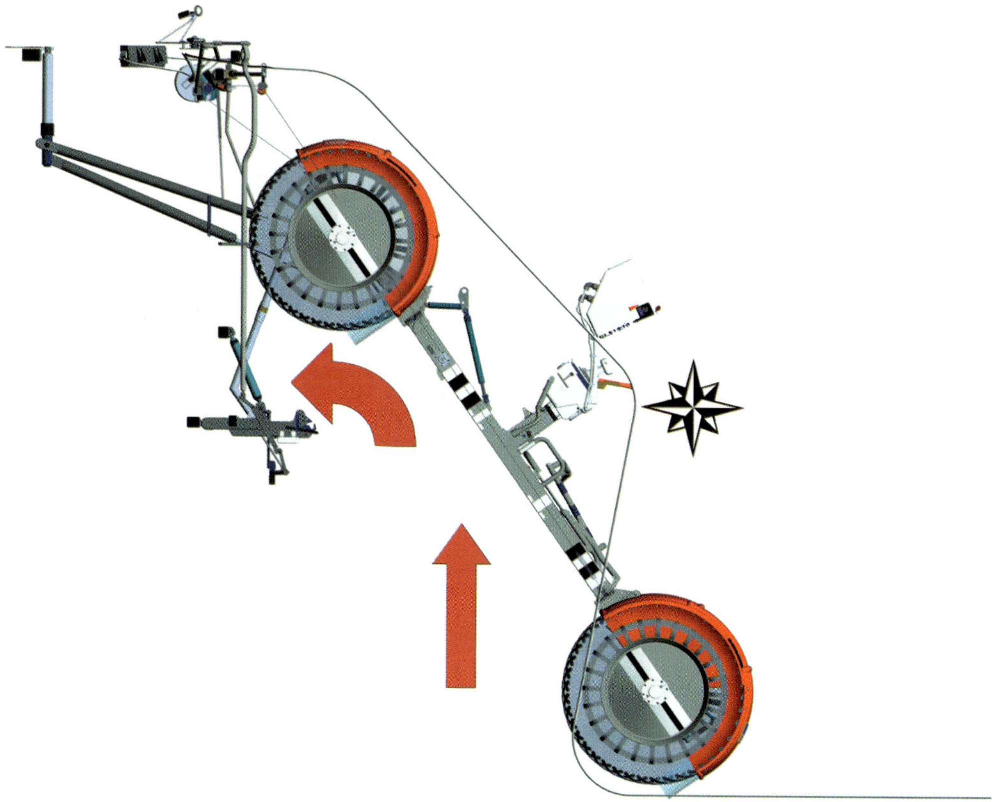

THE LRV WAS DEVELOPED IN A MERE 17 MONTHS BY THE BOEING CORPORATION

Lunar Module Chronology

December 22, 1967
First fire-in-the-hole test for the Lunar Module. Test firing duration was 680 milliseconds.

After the LRV was on the lunar surface, the astronauts visually inspected the vehicle and then deployment cables and other hardware were disconnected from the LRV. The crew then erected the seats and placed other crew station systems into operational order. One astronaut would board and power-up the vehicle, and then back it away from the LM to a designated area for payload loading, while the second crewman monitored and provided directions for the back-away operation. During this brief ride the driver verified that all LRV systems were functioning.

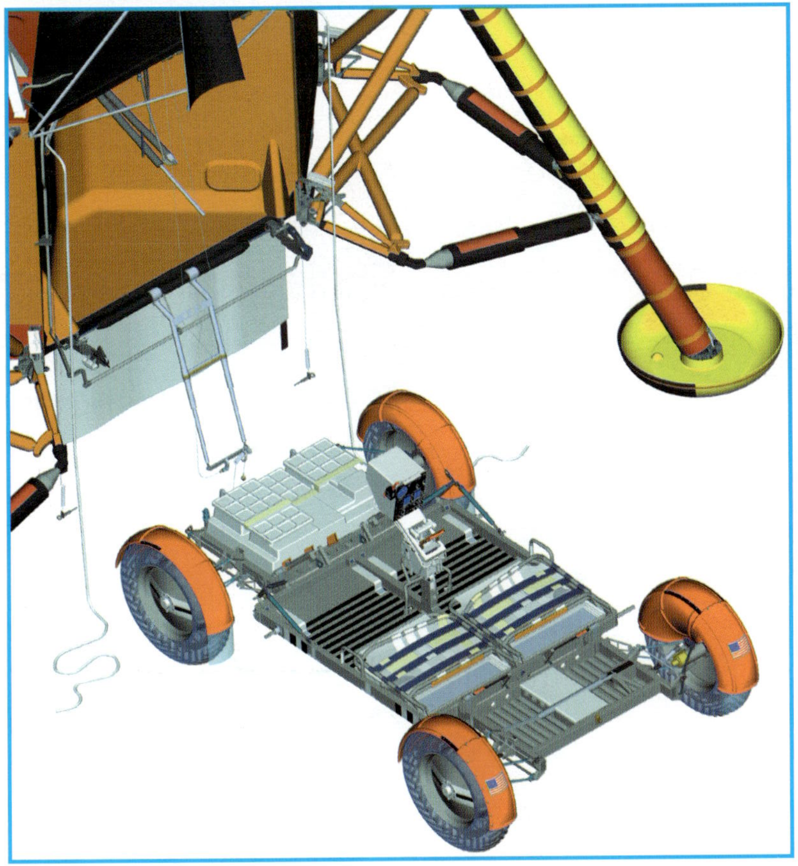

Lunar Module Chronology

January 22, 1968
Successful launch of *Apollo 5* carrying LM-1. Ascent stage orbit decayed on 1/24/68 and it subsequently was destroyed on Earth entry.

LRV CHASSIS DESIGN

Design and development of the LRV, by the Boeing Corporation, were under the direction of the NASA Marshall Space Flight Center, in Huntsville, Alabama. Boeing's major subcontractor was General Motors Delco Electronics Division's Santa Barbara Operation in California. The Delco electronics portion of the LRV work centered on the vehicle's mobility subsystem, which included the wheels, suspension, traction and drive systems.

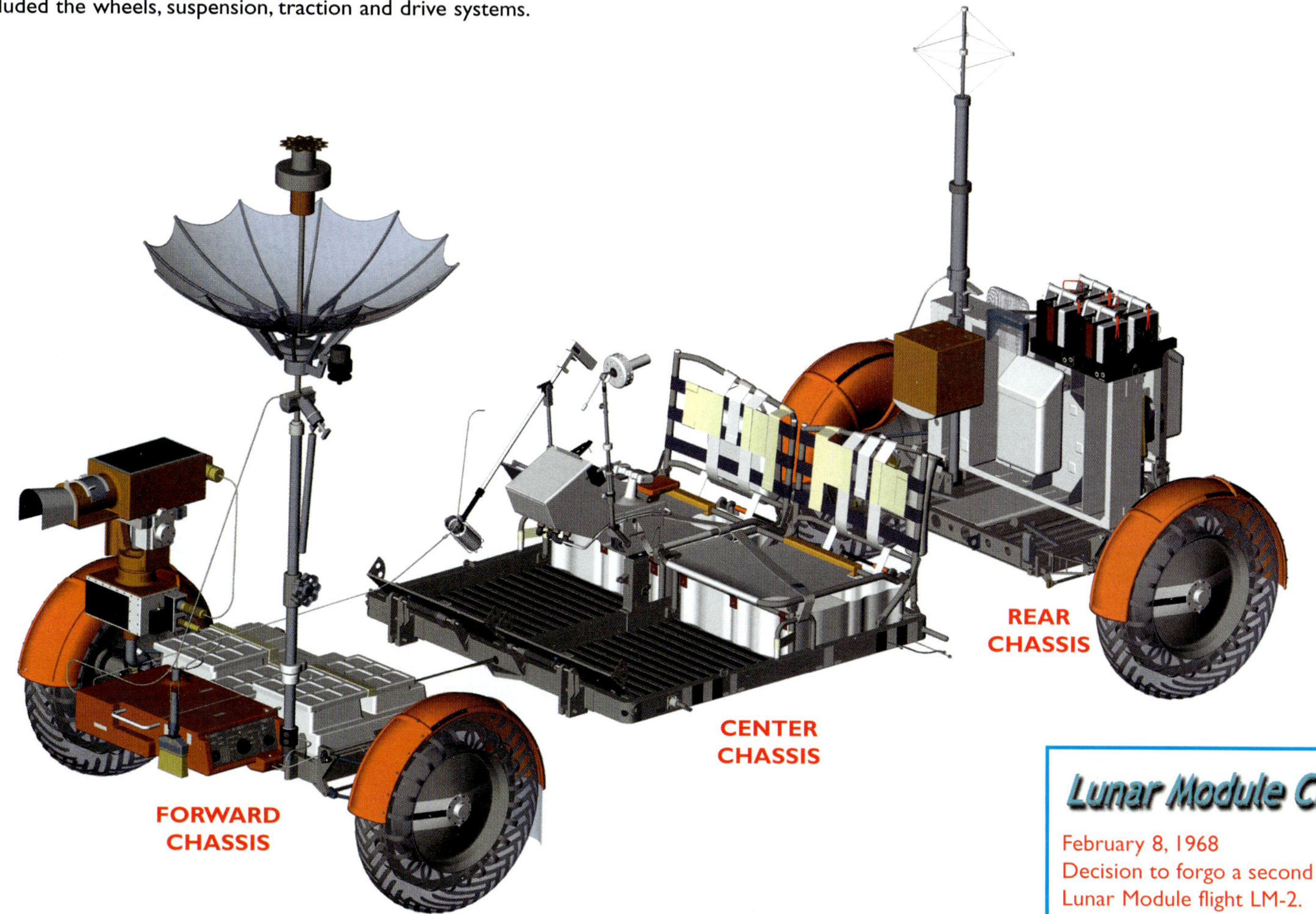

Lunar Module Chronology

February 8, 1968
Decision to forgo a second unmanned launch, Lunar Module flight LM-2.

FORWARD CHASSIS SECTION

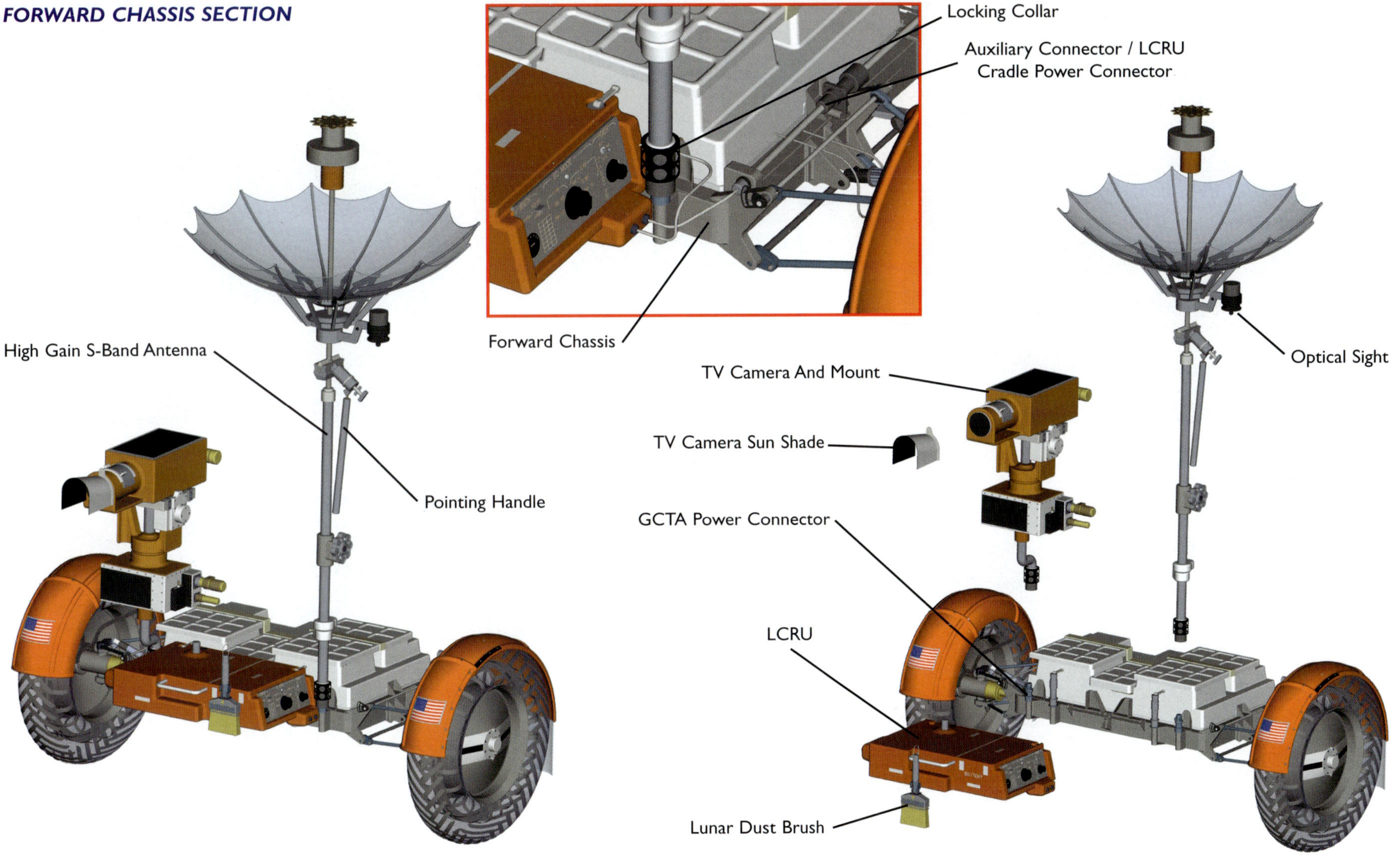

– Scott P. Sullivan –

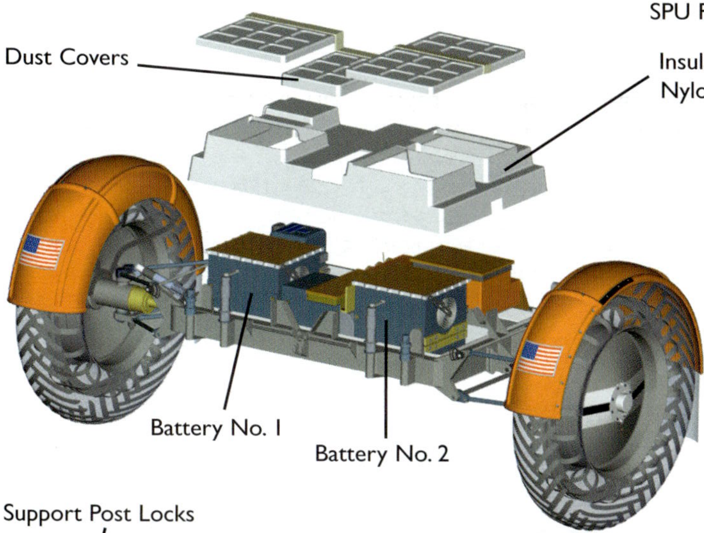
Dust Covers
Insulation Blanket 15 Layers of Aluminized Mylar and Nylon Netting with Beta Cloth on External Surfaces
Battery No. 1
Battery No. 2

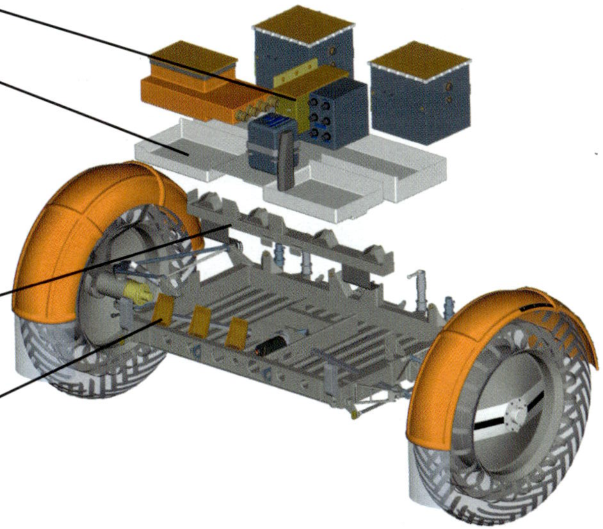
SPU Fusible Mass (2.25 lbs. of Wax) and Thermal Strip
Chassis Cross Member
Fiberglass Mounts

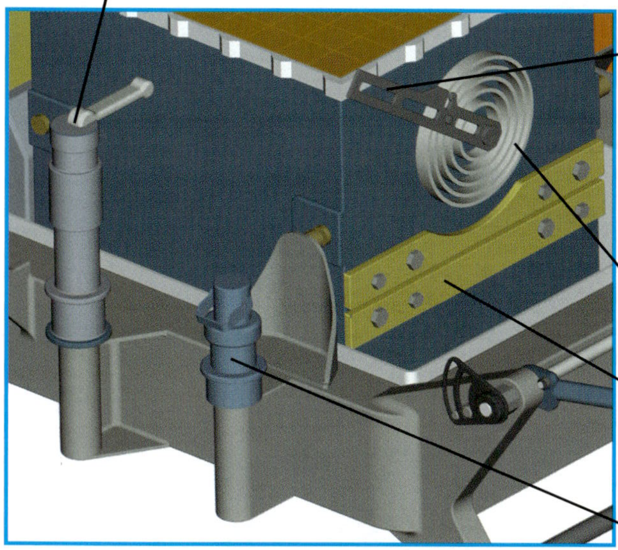
Support Post Locks
Dust Cover Raising and Lowering Connecting Link (Both Sides)
Actuator
Thermal Straps
LRCU Power Connector

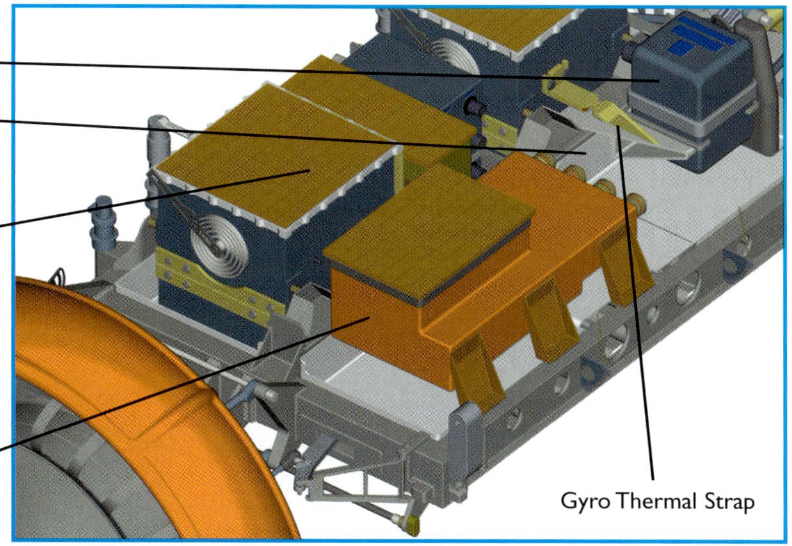
Gyro
Fiberglass Gyro Supports
Thermal Control Mirrors (Space Radiator)
Drive Control Electronics and Thermal Control Unit
Gyro Thermal Strap

SUSPENSION AND TRACTION DRIVE ASSEMBLIES

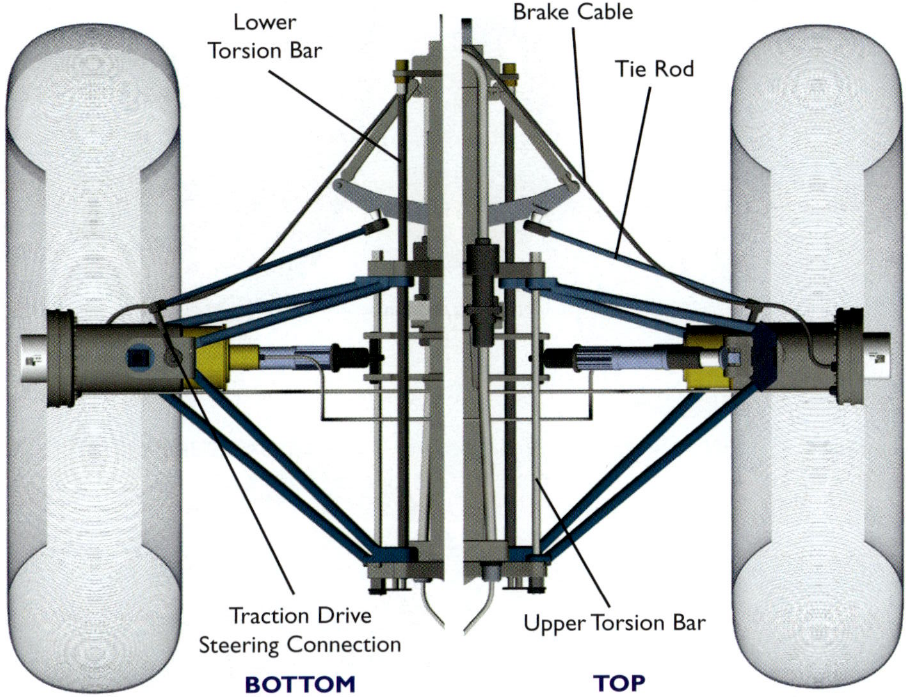

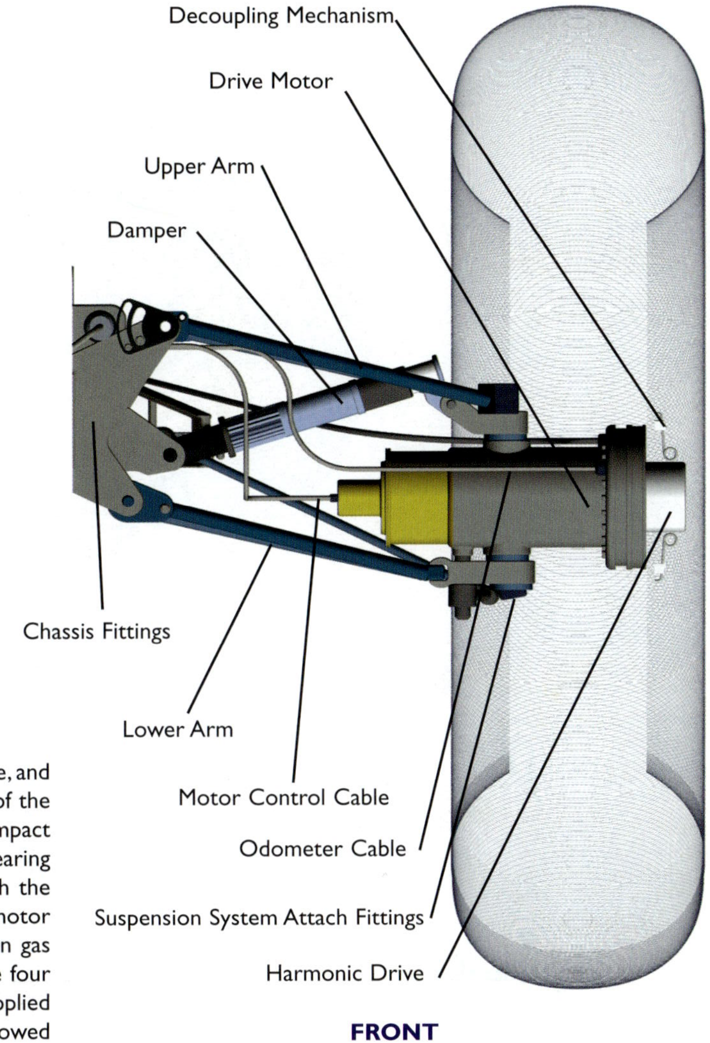

The mobility subsystem consisted of the chassis and equipment and controls required to propel, suspend, brake, and steer the Lunar Roving Vehicle. The wheels were an open design with a chevron tread covering 50 percent of the surface area. An inner frame or "bump stop" prevented excessive deflection of the wire mesh under high impact conditions. Each wheel could be uncoupled from the traction drive and allowed to "free wheel" on a bearing independent of the drive train. The uncoupling mechanism could also be used to re-engage the wheel with the traction drive. Its own separate traction drive assembly – consisting of a harmonic drive reduction unit, drive motor and brake assembly – drove each wheel. Each traction drive was hermetically sealed to maintain a nitrogen gas internal pressure of 7.5 psia. The gas aided in transferring heat to the outer casing of the drive motor. The four harmonic drive reduction units transmitted power the wheels. The 4¼-horsepower electric drive motors supplied input torque to the harmonic drives. The harmonic drive reducer motor speed rate was 80:1 and allowed continuous application of power to the wheels at all speeds without requiring gear shifting.

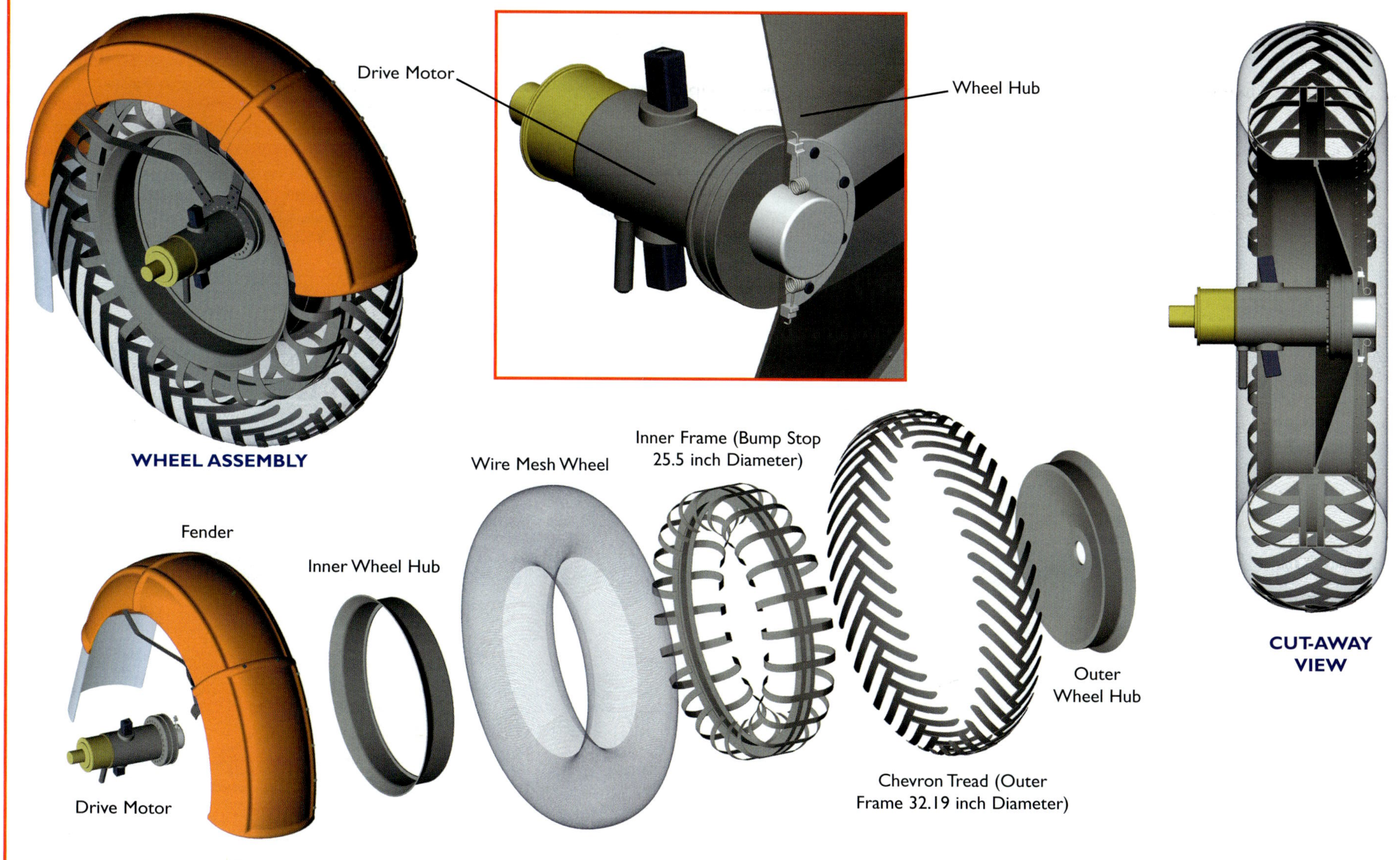
Drive Motor
Wheel Hub
WHEEL ASSEMBLY
Fender
Drive Motor
Inner Wheel Hub
Wire Mesh Wheel
Inner Frame (Bump Stop 25.5 inch Diameter)
Chevron Tread (Outer Frame 32.19 inch Diameter)
Outer Wheel Hub
CUT-AWAY VIEW

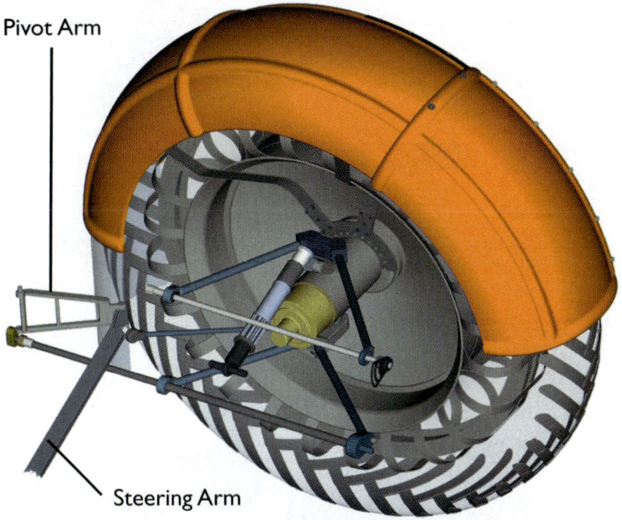

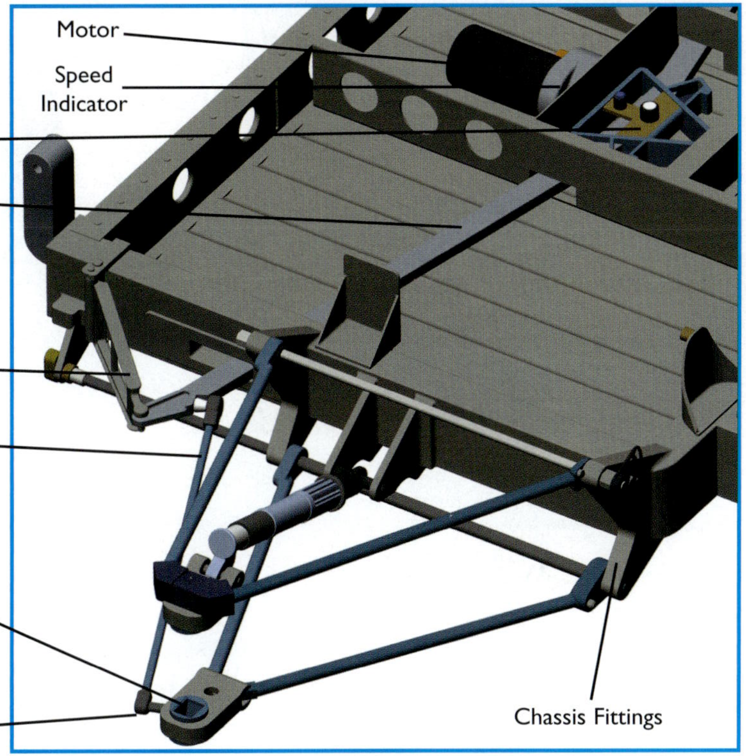

STEERING ASSEMBLY (WHEEL NOT SHOWN)

Each traction drive also contained an odometer pickup which transmits nine pulses to the navigation system signal processing unit for each wheel revolution.

The chassis was suspended from each wheel by a pair of triangular arms connected between the LRV's chassis and each traction drive. Loads were transmitted to the chassis through suspension arms to a separate torsion bar for each arm. Vertical travel of the wheel and its rate of travel were limited by a damper connected between the chassis and each upper suspension arm. The deflection of the suspension system and wire wheels allowed 14 inches of chassis ground clearance when the LRV is fully loaded and 17 inches when unloaded. The suspension system could be rotated 135 degrees to allow the LRV to be folded for stowage in the LM. Both the front and rear wheels could be steered. The LRV had a turning radius of 122 inches.

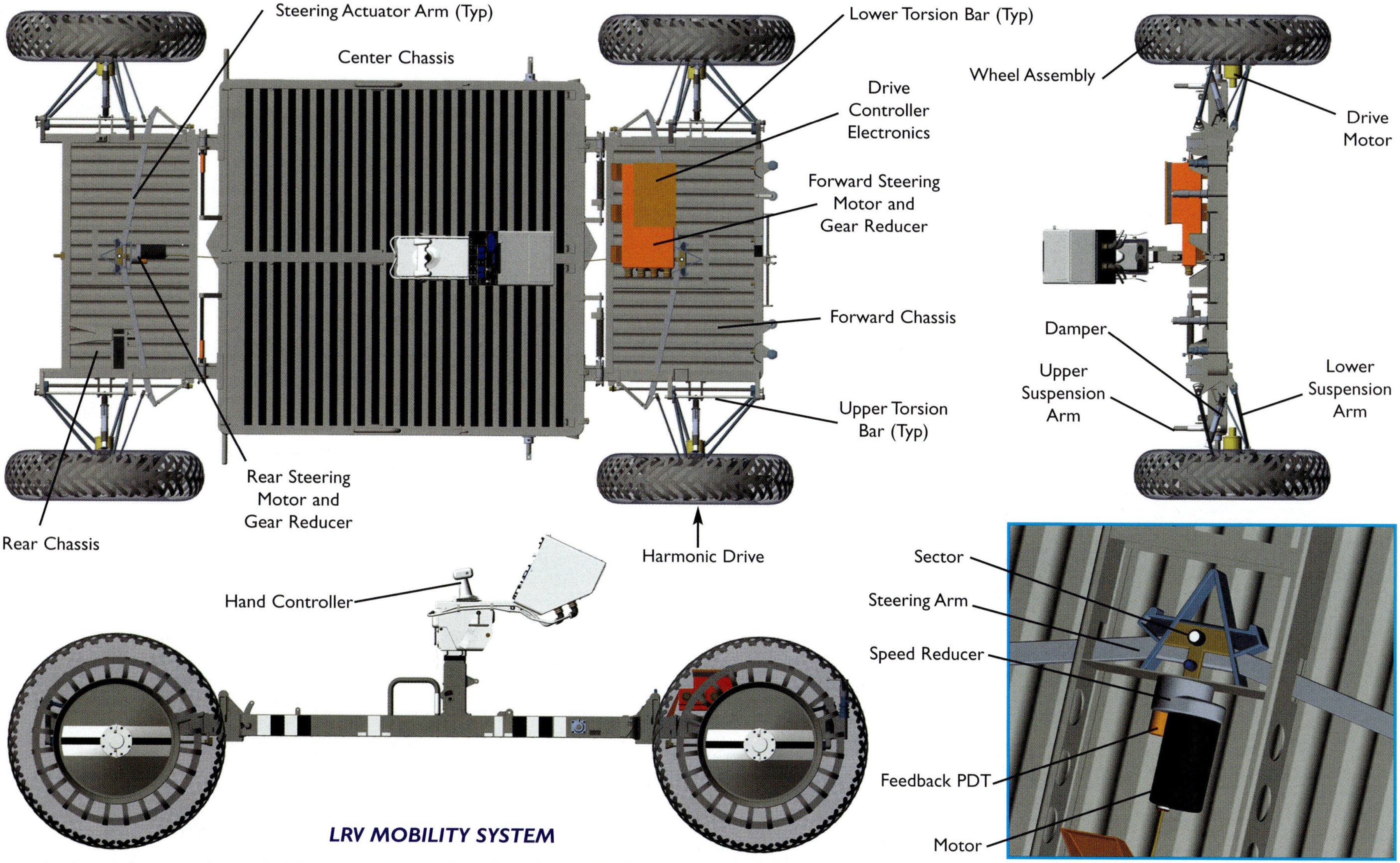

LRV MOBILITY SYSTEM

LRV NUMBER 1 IN ITS SUPPORT FIXTURE

THE LRV ON THE JOB

The front and rear steering assemblies were mechanically independent of each other. Provisions were made to steer with either set of wheels (or both sets). In the event of a steering malfunction, one set of wheels could be disconnected mechanically, and the mission could continue using the active steering system.

The hand controller provided speed, steering and braking commands. Tilting the controller forward of the neutral position proportionally increased the forward speed. Reverse power was applied when the controller was tilted backward past the neutral position and a reverse inhibit switch on the hand controller is thrown by the driver. Tilting the controller left or right of the neutral position initiated steering commands in the direction that the controller was moved. Braking was initiated when the controller was pulled backwards. At approximately three inches a spring-loaded catch engaged the handle to lock it in the park position. Forward and reverse were cut off when the braking action began.

The control and display console was divided into two main parts: navigation on the upper section of the panel, and controls for switching and monitoring of electrical loads on the lower section of the console. A sun shadow device determined the LRV heading with respect to the Sun. The heading could be compared with the gyro heading at regular intervals as a check against gyro shift.

The crew station equipment included the seats, foot rests, inboard and outboard handholds, arm rests, floor panels, seat belts, fenders and toe holds. The two seats were tubular aluminum frames spanned by nylon. They were folded flat onto the center chassis and unfolded somewhat like a lawn chair by the astronauts after the LRV was deployed from the LM. The seat backs were designed to support the astronaut's PLSS backpacks.

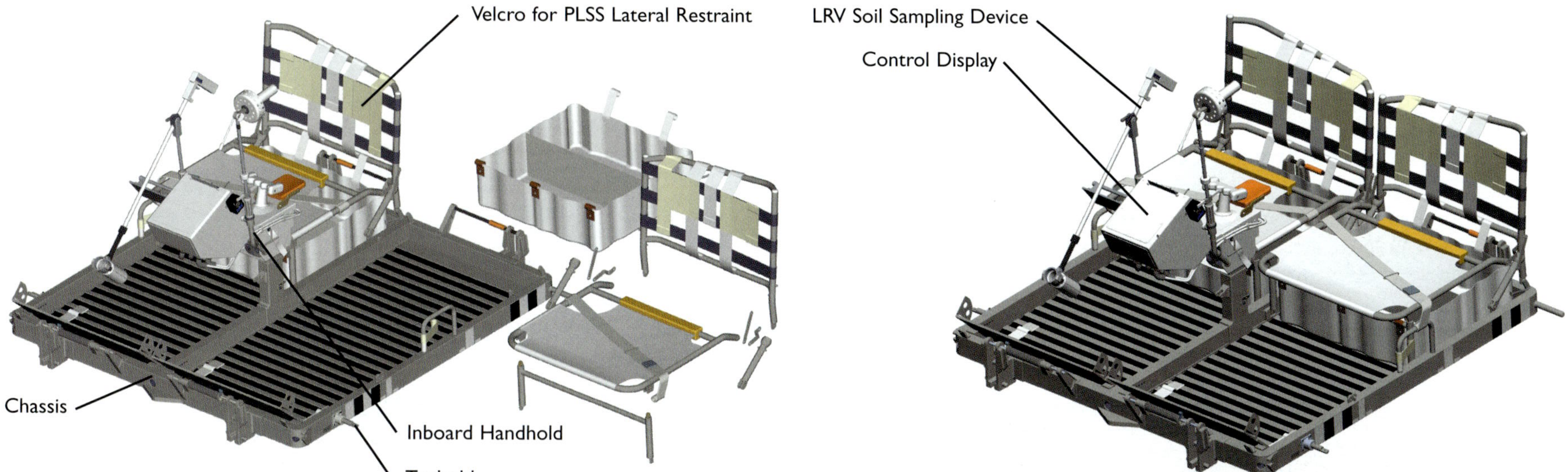

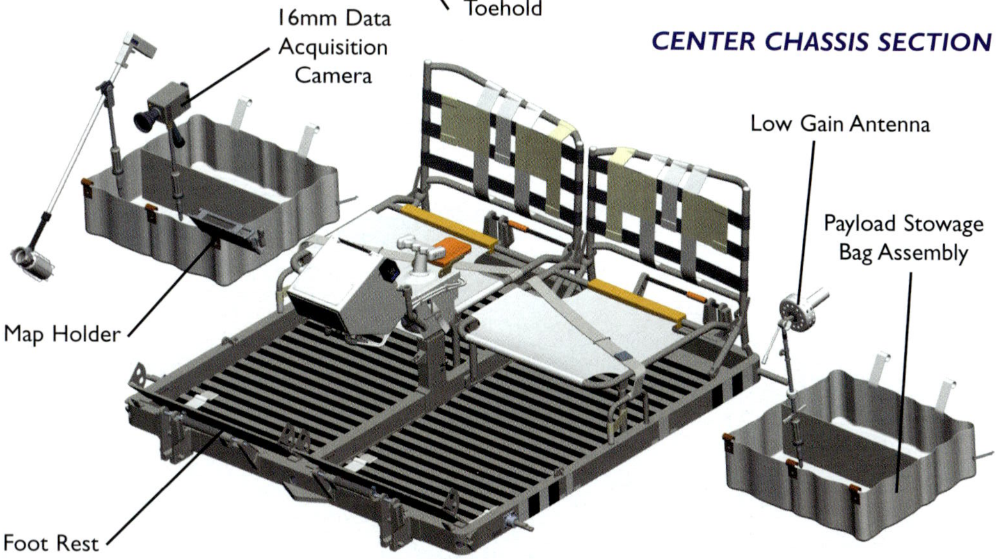

CENTER CHASSIS SECTION

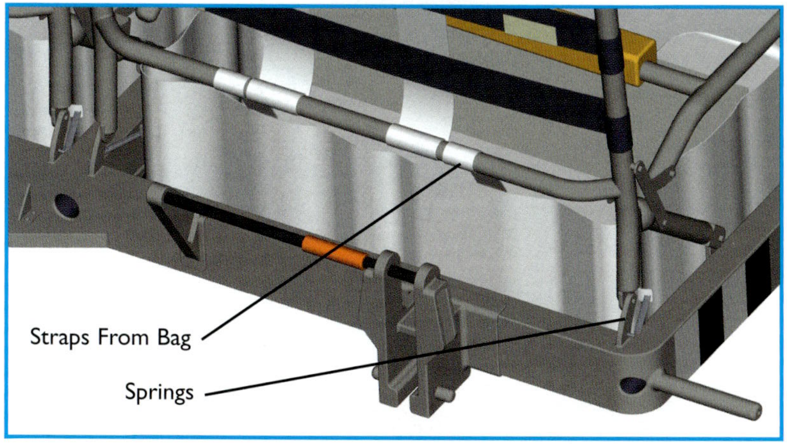

DETAIL OF STOWAGE BAG MOUNTING TO CREW SEAT FRAME

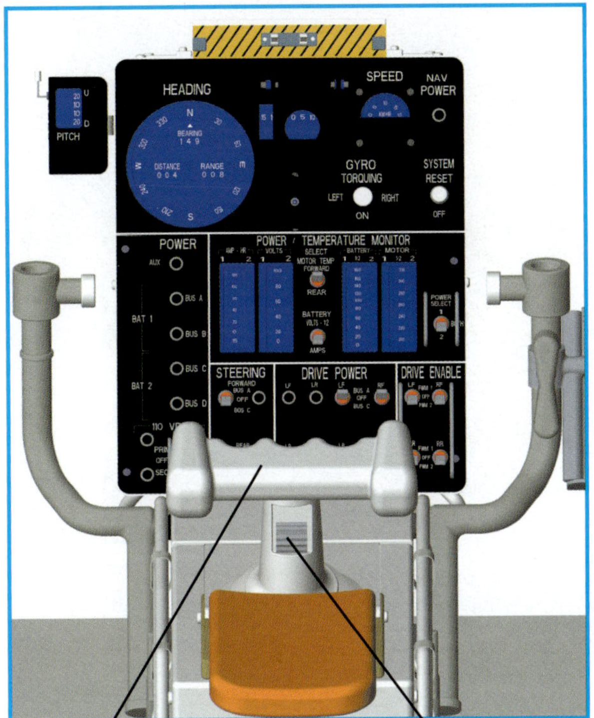

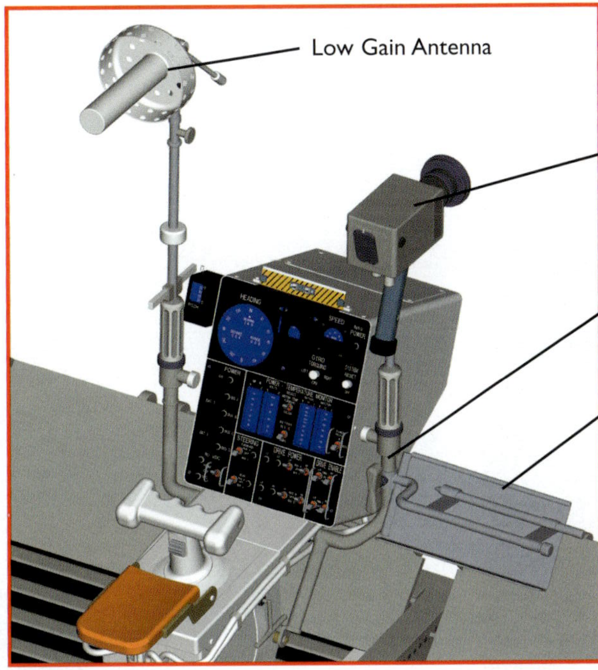

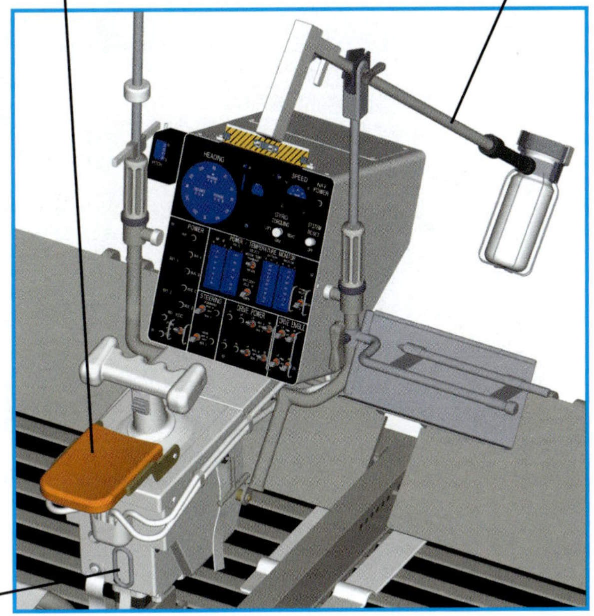

LRV DISPLAYS AND CONTROLS

The floor panels of the crew station area are made from beaded aluminum panels, which can support the full "lunar weight" of the astronauts when standing. Molded fiberglass fenders protect the vehicle and astronauts from lunar dust, which may be churned up by the wheels while traversing over the lunar surface. A toe hold on each side of the vehicle is used to aid the astronauts as they get out of the vehicle under one-sixth gravity conditions. Should it become necessary, the toe hold could be used as a tool to actuate a wheel decoupling mechanism.

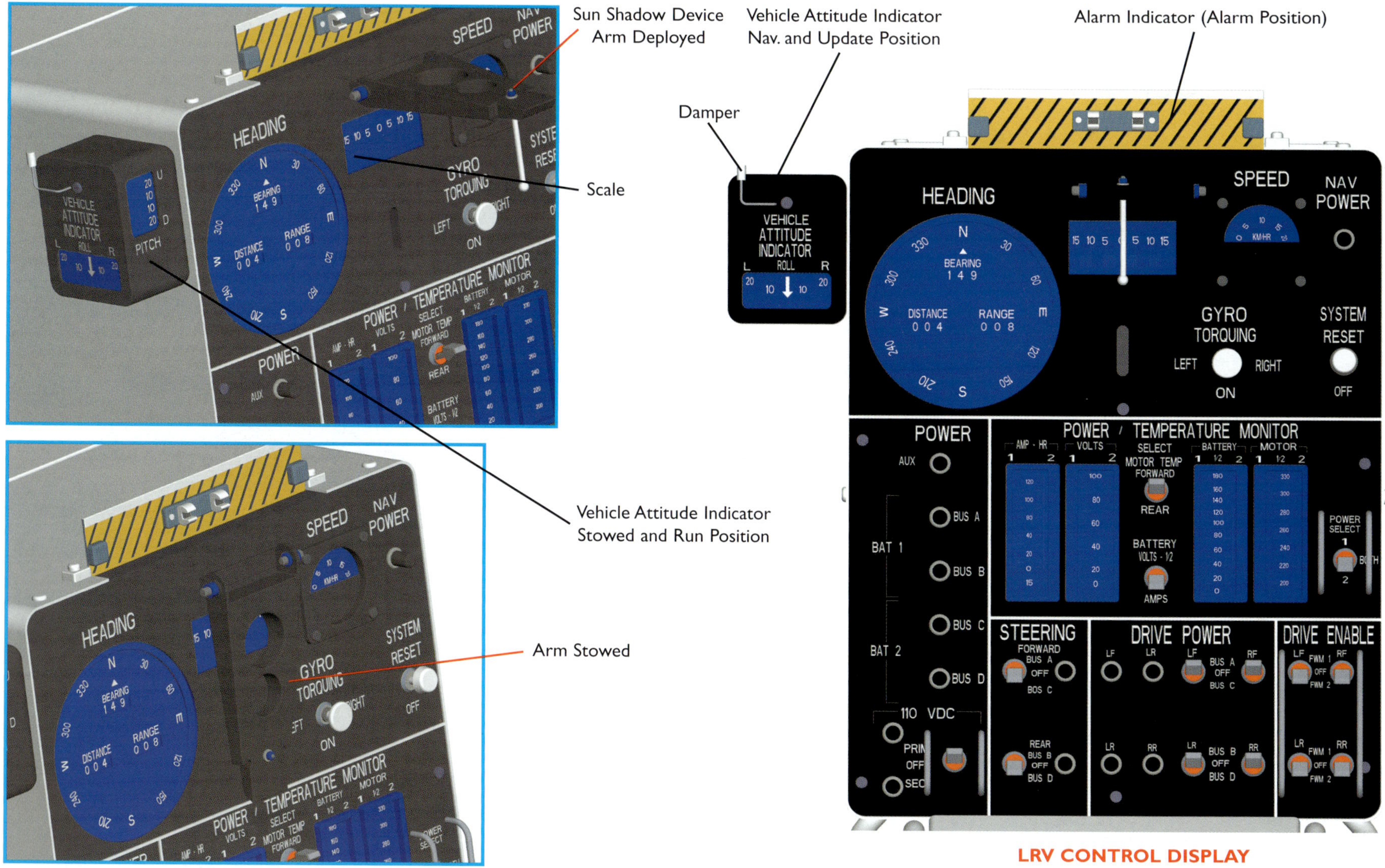

LRV CONTROL DISPLAY

AFT CHASSIS SECTION

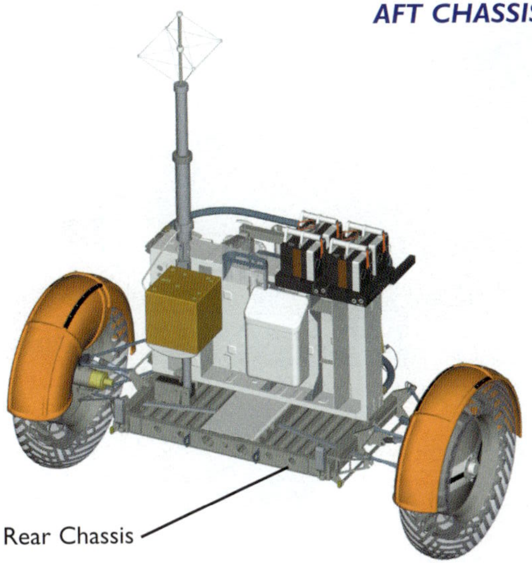

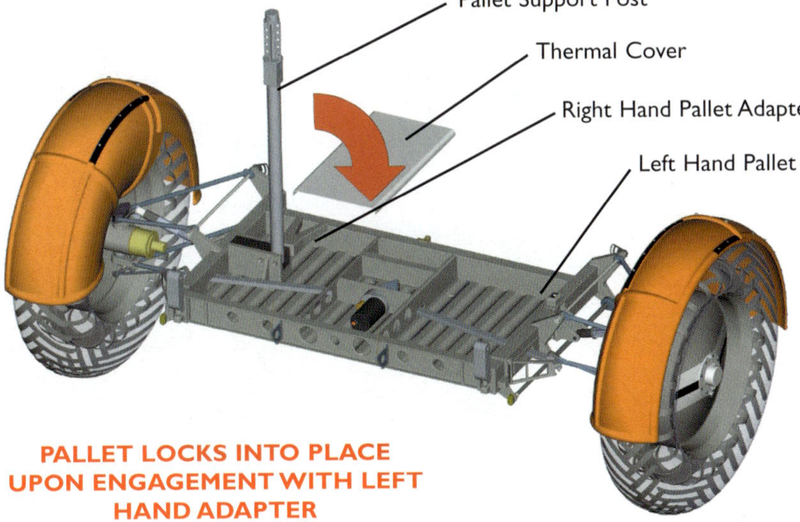

PALLET LOCKS INTO PLACE UPON ENGAGEMENT WITH LEFT HAND ADAPTER

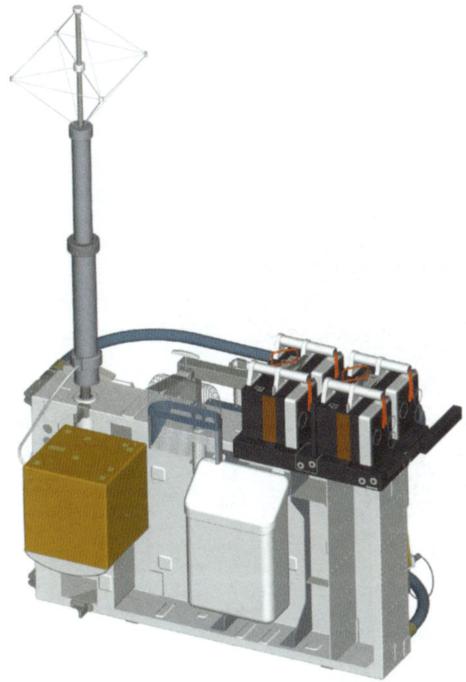

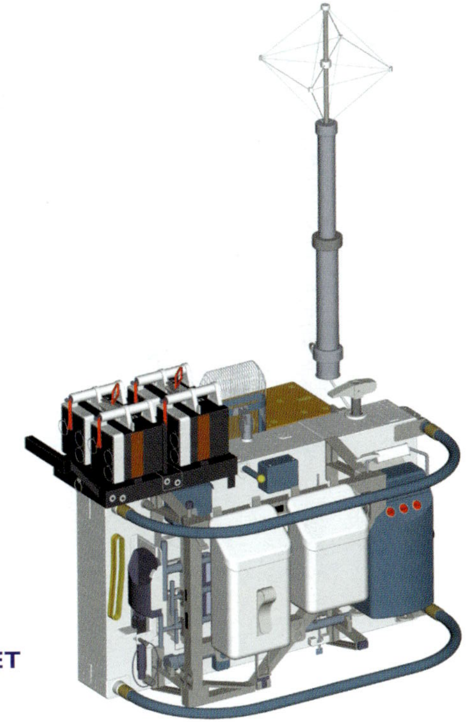

REAR PAYLOAD PALLET

Lunar Module Chronology

March 3 1969
Apollo 9 (LM-3 *Spider*) launches, marking the first manned flight of a Lunar Module, and the first flight of a complete Apollo spacecraft.

LRV AFT PALLET ASSEMBLY

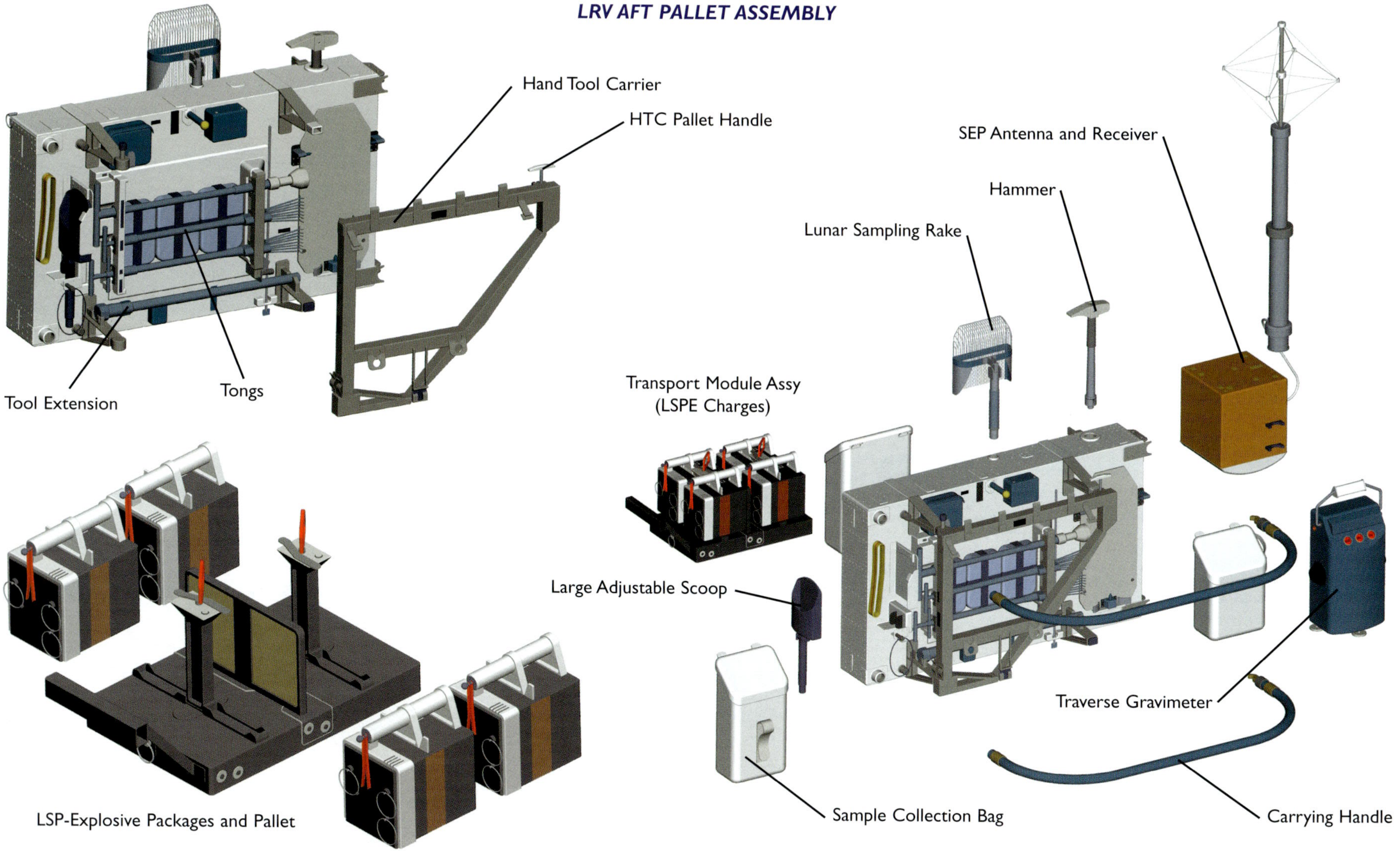

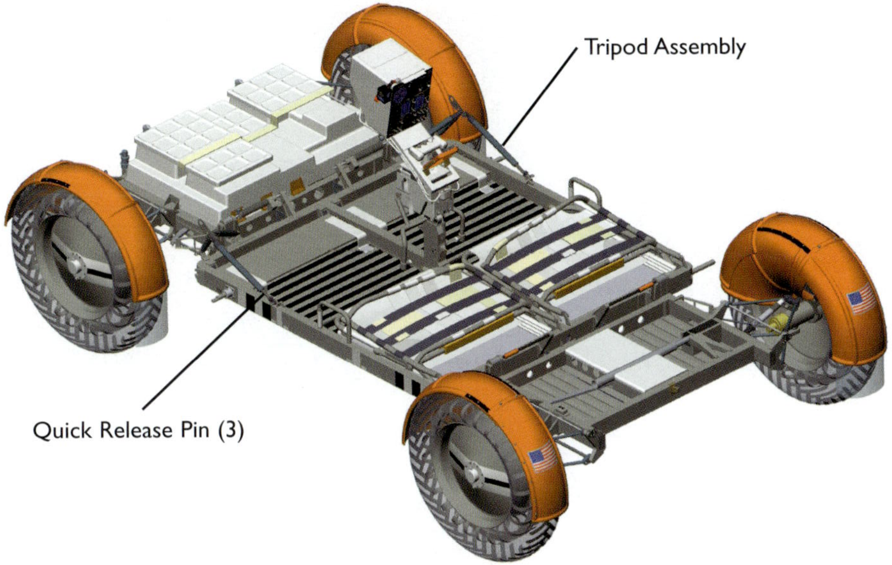

Both crewman unloaded the equipment from the LM and installed it on the LRV. A battery-powered Lunar Communications Relay Unit (LCRU) was mounted on the front of the forward chassis. An umbrella-like high-gain antenna for the TV transmissions and a television camera were also mounted on the forward chassis. The accessory staff and low-gain antenna for voice communications were mounted in the inboard handholds on either side of the control and display console. Finally, the tools and experiments were loaded on the pallet and mounted behind the seats.

THE LRV PRIOR TO EQUIPMENT ASSEMBLY AND ACTIVATION AND EQUIPMENT PALLET LOADING

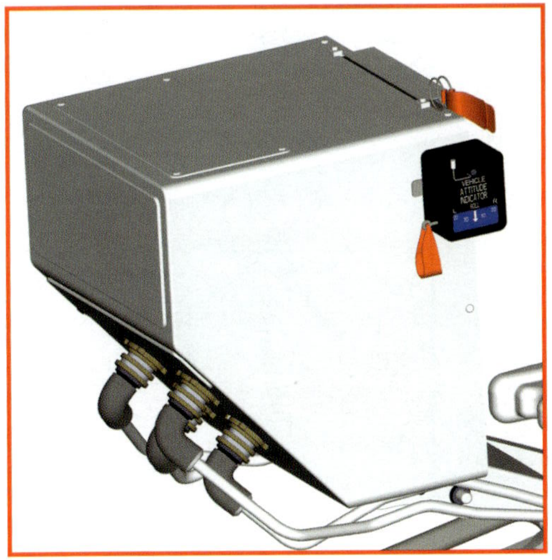

Lock Pin Warning Flag
(Lock Pin Removed and
Discarded After Deployment)

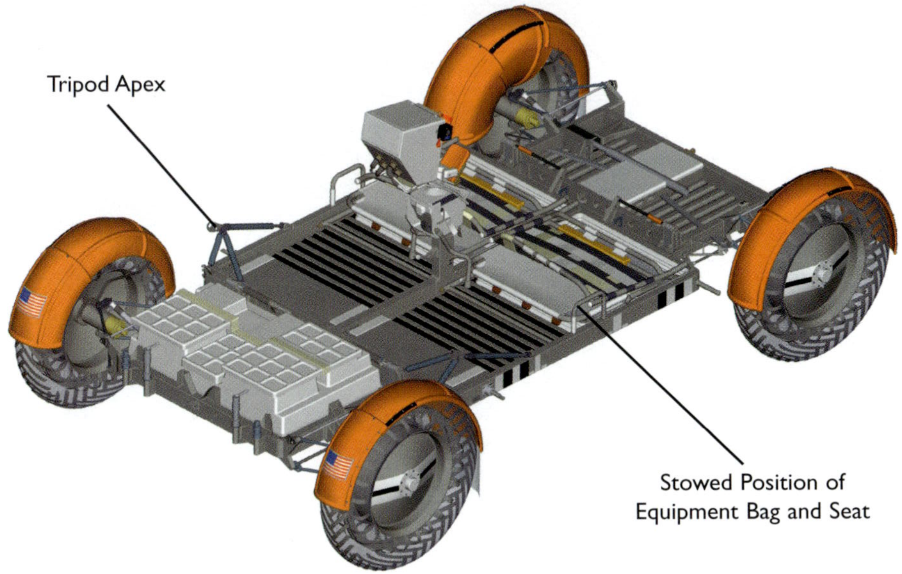

DEPLOYMENT HARDWARE AND STEERING RING LOCATIONS

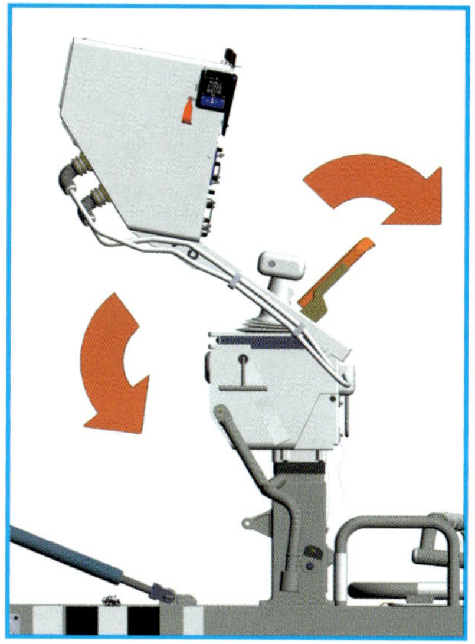

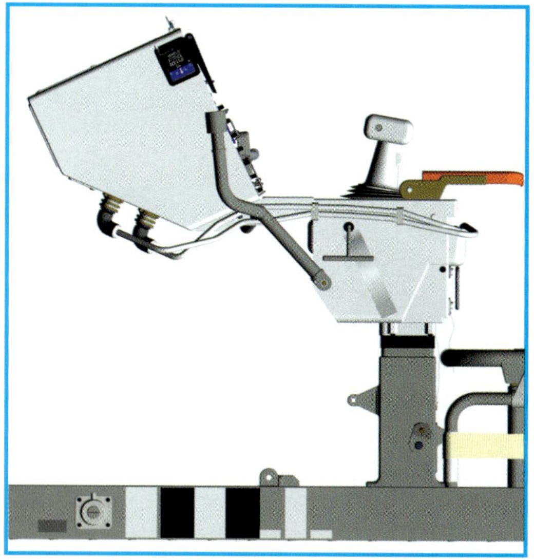

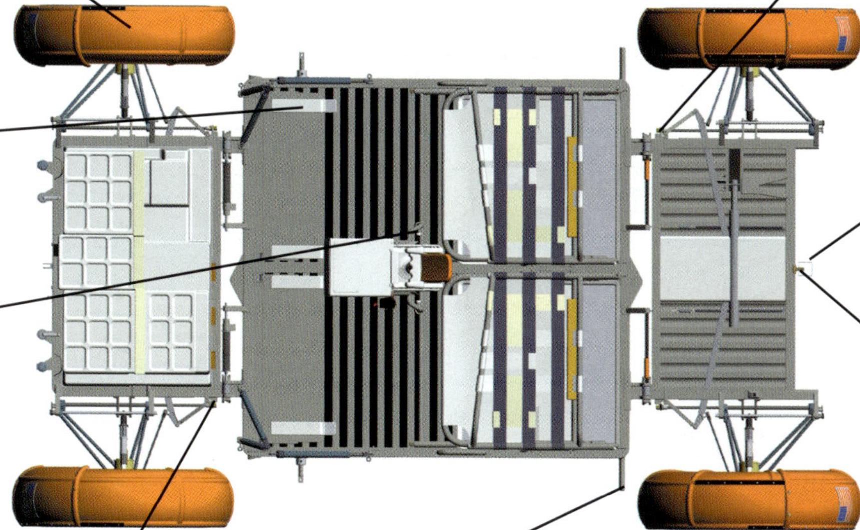

Lunar Module Chronology

May 18, 1969
Apollo 10 (LM-4 *Snoopy*) launches and conducts all phases of Apollo spacecraft operations with the exception of lunar landing.

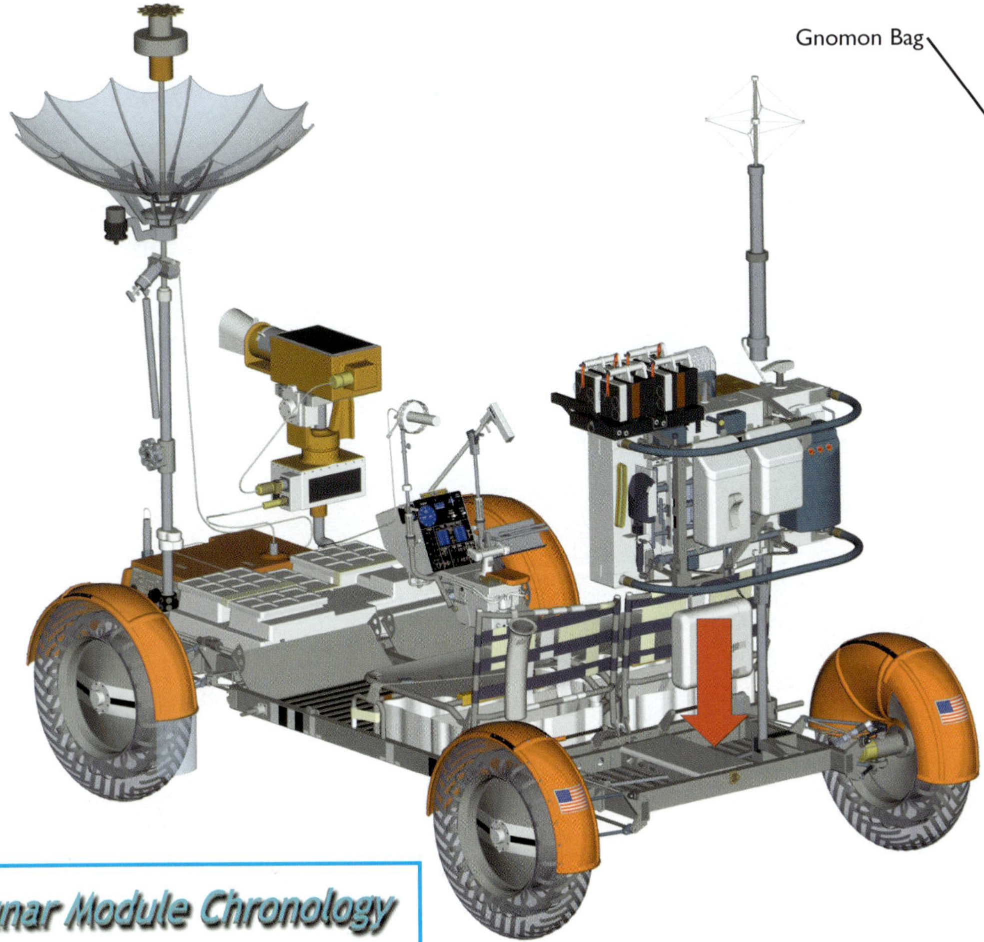

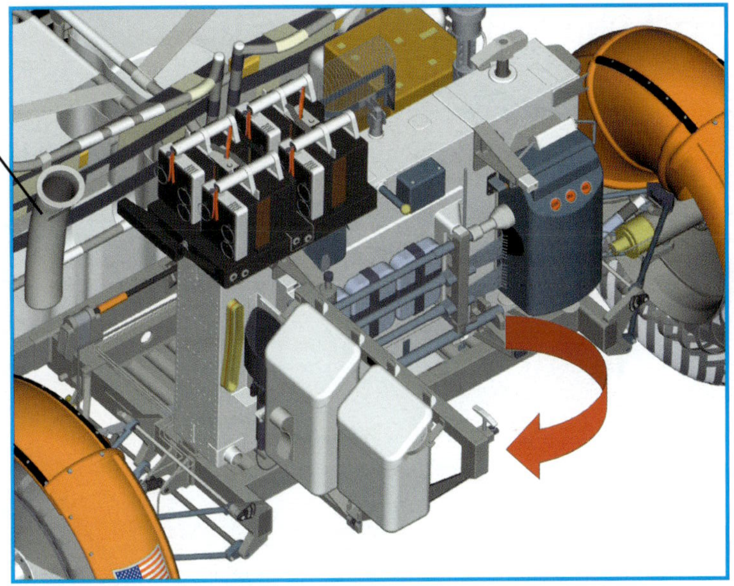

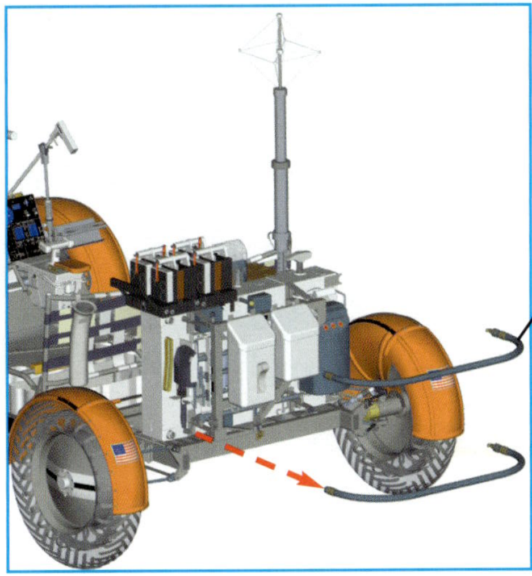

Lunar Module Chronology

July 20, 1969
Apollo 11 (LM-5 *Eagle*) touches down on the Moon in the Sea of Tranquillity, marking the first manned lunar landing.

THE APOLLO PERSONNEL LIFE SUPPORT SYSTEM (PLSS)

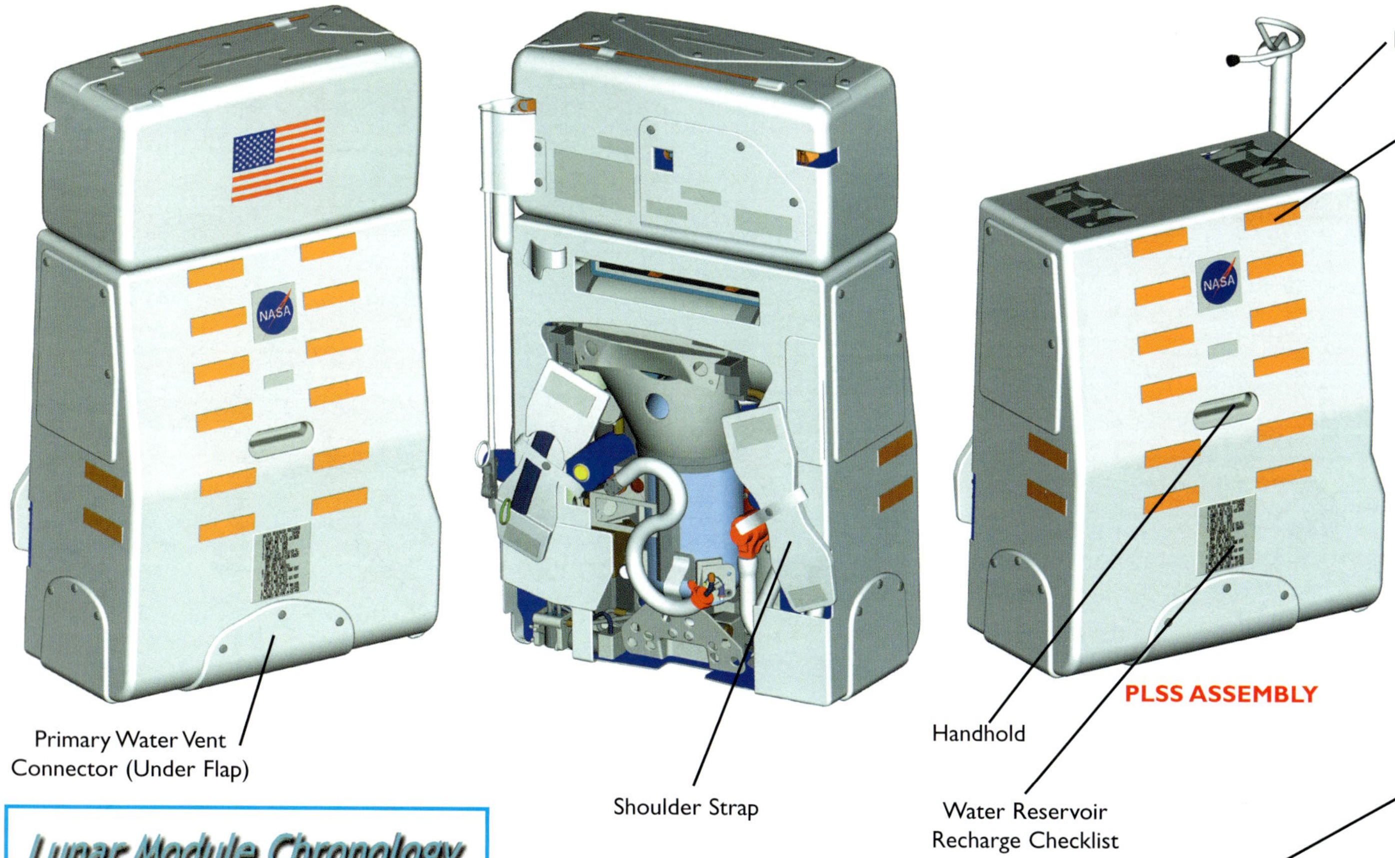

PLSS ASSEMBLY

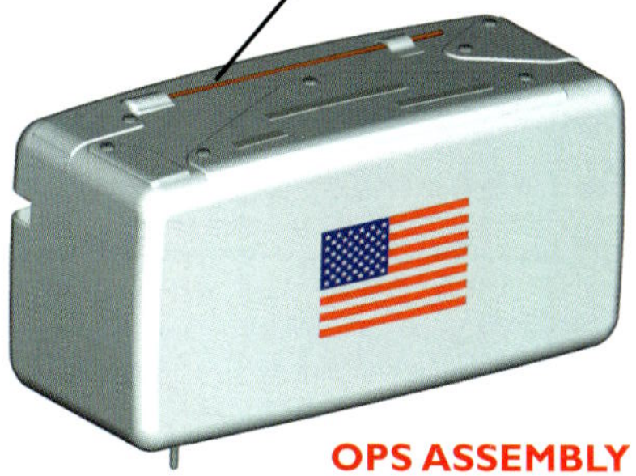

OPS ASSEMBLY

THE PLSS REMOTE CONTROL UNIT (RCU)

Lunar Module Chronology

November 19, 1969
Apollo 12 (LM-6 *Intrepid*) touches down on the Sea of Storms. This mission collected 75 pounds of lunar material.

The portable life support system (PLSS) provided an astronaut with a livable atmosphere inside his space suit during excursions on the lunar surface and in space. Worn on the back and connected to the suit's waist by umbilicals, it permitted up to 7 hours of extravehicular activity, depending on the astronaut's metabolic rate. The backpack supplied oxygen for breathing and suit ventilation, and refrigerated water and oxygen for body cooling. It pressurized the suit to 3.9 psi and removed contaminates from the oxygen, such as carbon dioxide, circulating through the suit. It also had a communication-telemetry set, controls to operate it, and devices to monitor its functions. For the lunar missions, the LM had two PLSS backpacks, stowed in the ascent stage. The LM carried enough supplies to refill each pack's oxygen tank and water reservoir, and replace its battery and two lithium hydroxide canisters twice. This would allow a total of three EVA trips.

The life support pack, with its controls, weighed 104 pounds; it was 26 inches high, 20.5 inches wide and 10.5 inches deep. A 16.8-volt silver-zinc battery powered it. A fiberglass hard cover protected the packs against micrometeoroids.

Five subsystems made up the PLSS system: primary oxygen supply, oxygen-ventilating circuit, water transport loop, feed water loop, and space suit communications system. For emergency use, an oxygen purge system (OPS) would supply an additional 30 to 90 minutes of oxygen according to the selected flow rate. The OPS is mounted on the pack but operates separately.

A thermal insulator made of fire-resistant Beta cloth and aluminized Kapton covered the pack and its hard cover shell to restrict heat leakage in or out, depending on the Moon's temperature. A similar insulator covered the oxygen purge system.

The oxygen purge system, connected to the suit by a separate umbilical, was designed for backup use in the event of emergencies such as a loss of suit pressure or depleted oxygen supply. However, an astronaut could use it independently as a life support chest pack during EVA transfer between the LM and the Command Module.

The system supplied either an open-loop purge flow or makeup flow directly to the suit. In both cases, it maintained suit pressure at 3.7 psi. In the full purge mode, it provided a 30-minute flow rate of 8.3 pounds of oxygen an hour, fulfilling breathing and cooling requirements, flushing out carbon dioxide, and defogging the helmet visor.

A lever attached to the pack's remote control unit operated the OPS, mounted separately on the top of the backpack. Its umbilical was attached to the suit connector that connected the suit to the LM environmental control subsystem when the astronaut was inside the LM. The purge unit weighed 35.1 pounds, was 18.4 inches long, 10 inches high, and 8 inches deep. Two spherical containers held a total of 5.7 pounds of oxygen stored at 6,950 psi. A battery-powered, temperature-controlled heater warmed the rapidly expanding oxygen to prevent subzero oxygen temperature at the space suit flow inlet.

Lunar Module Chronology

April 11, 1970
Apollo 13 (LM-7 *Aquarius*) became a "lifeboat" for the crew when an explosion inside the Service Module and subsequent damage to the CSM ended any chance of a lunar landing.

Lunar Module Chronology

February 5, 1971
Apollo 14 (LM-8 *Antares*) landed in the Fra Mauro Crater. The astronauts recorded 9 hours and 24 Minutes outside the LM.

Lunar Module Chronology

July 26, 1971
Apollo 15 (LM-10 *Falcon*) launched from Cape Kennedy. This was the first J-Series LM and the first use of the Lunar Roving Vehicle. The total EVA time on the lunar surface was 18 hours and 37 minutes.

OXYGEN PURGE SYSTEM (OPS)

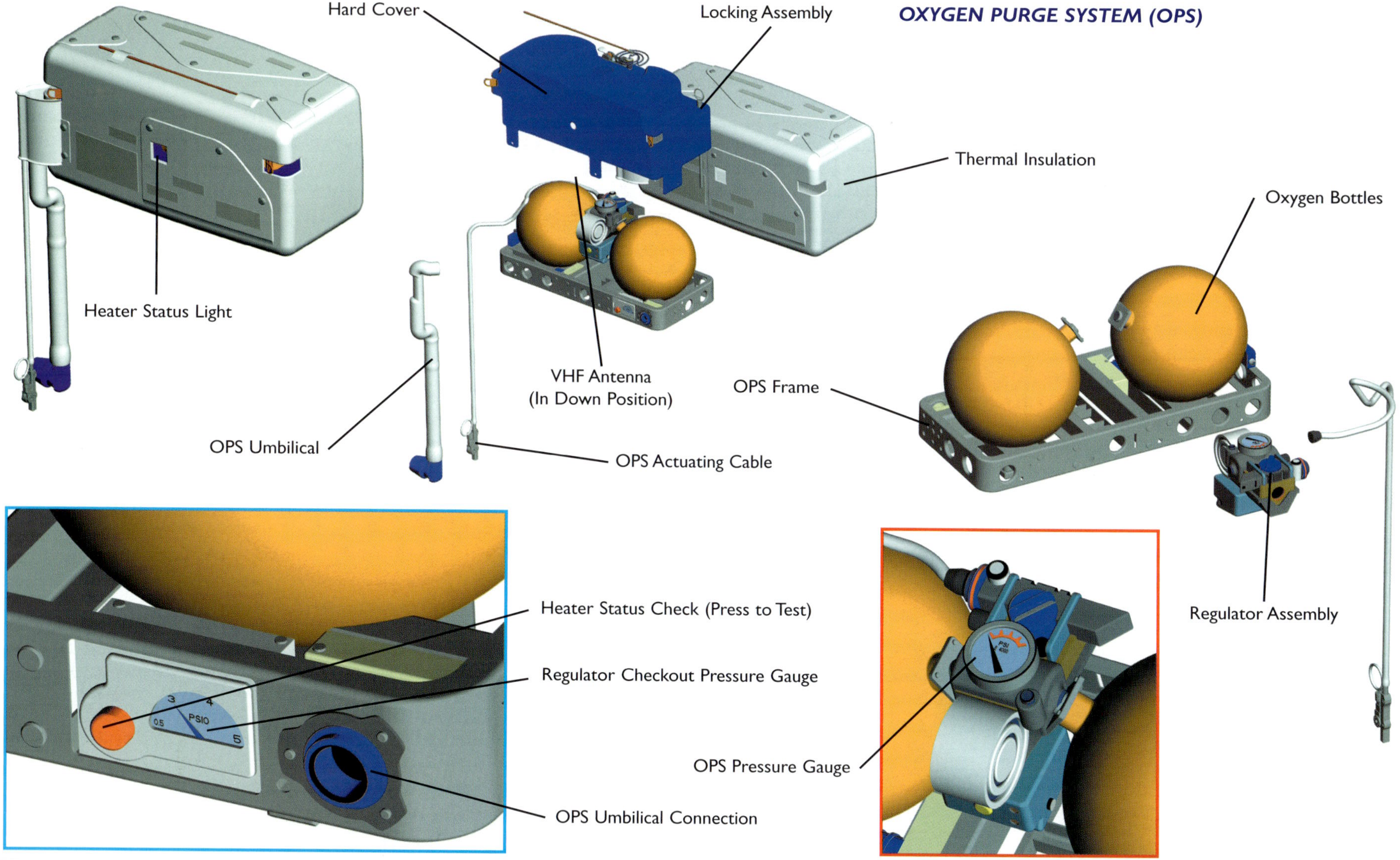

Remote Control Unit (RCU)

Oxygen Purge System

Personnel Life Support System

Lunar Module Chronology

April 16, 1972
Apollo 16 (LM-11 *Orion*) launched. This is the second J-mission; it totaled 20 hours and 14 minutes of EVA time.

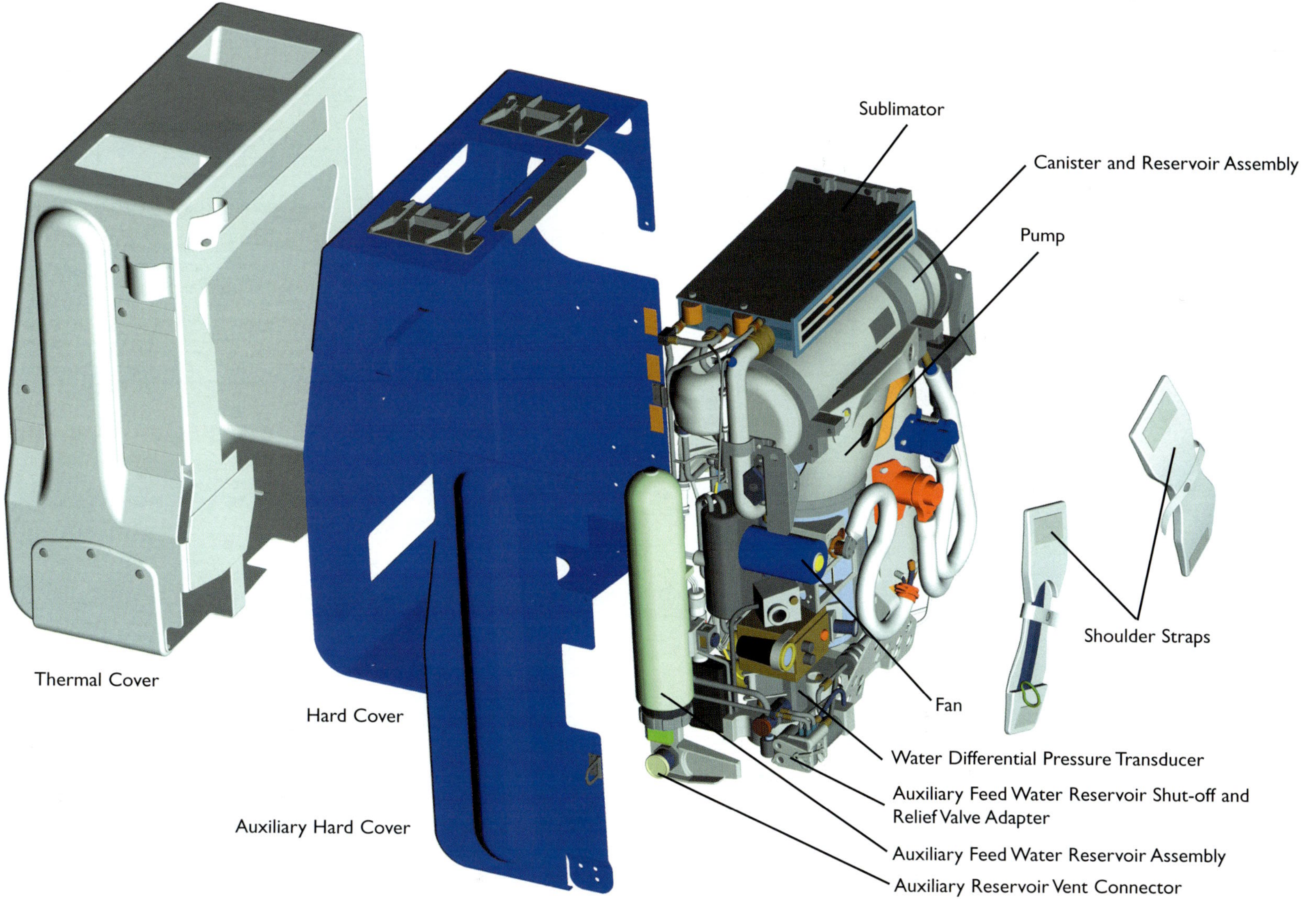
Sublimator
Canister and Reservoir Assembly
Pump
Shoulder Straps
Thermal Cover
Hard Cover
Fan
Water Differential Pressure Transducer
Auxiliary Feed Water Reservoir Shut-off and Relief Valve Adapter
Auxiliary Hard Cover
Auxiliary Feed Water Reservoir Assembly
Auxiliary Reservoir Vent Connector

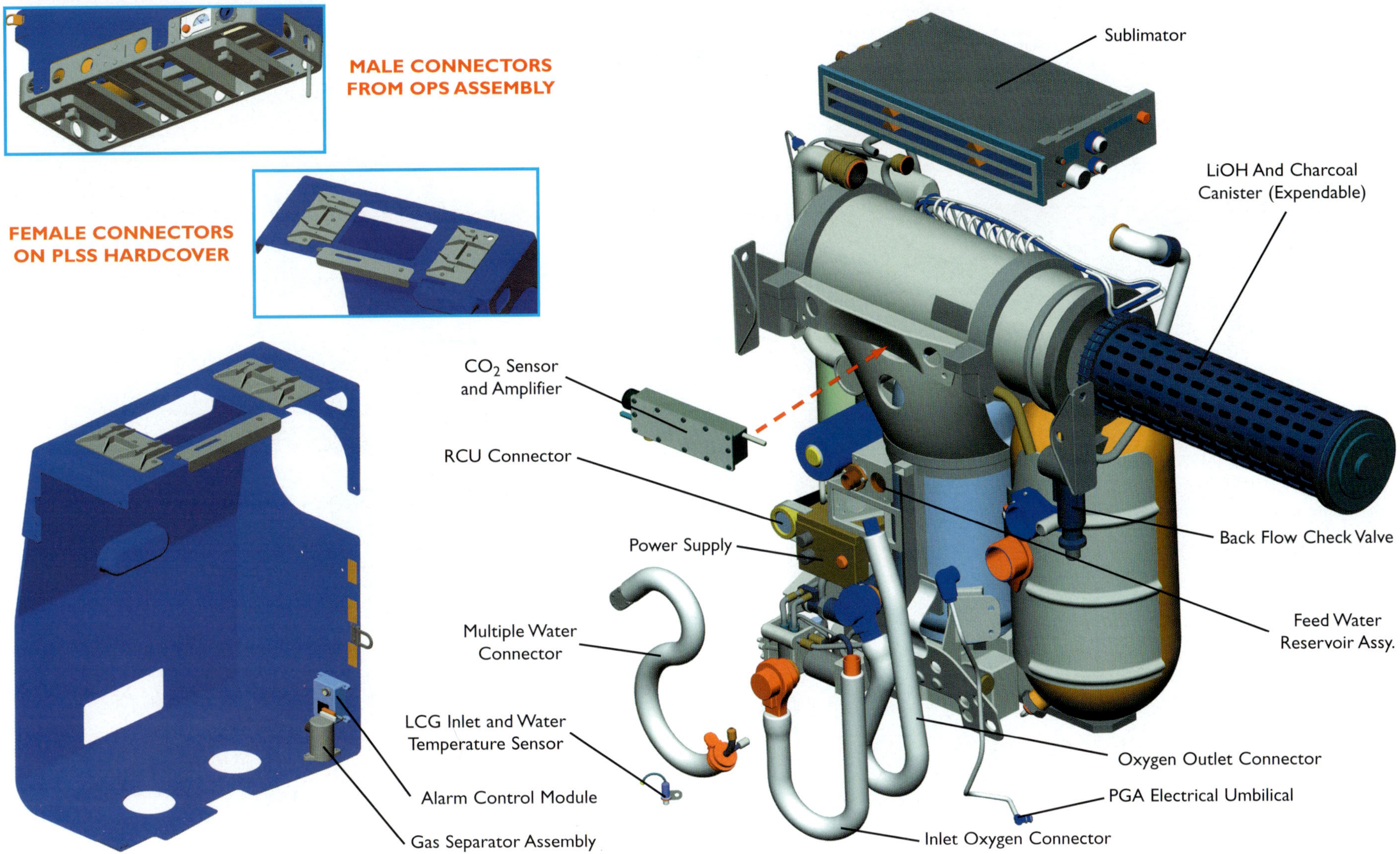
MALE CONNECTORS
FROM OPS ASSEMBLY
FEMALE CONNECTORS
ON PLSS HARDCOVER
Sublimator
LiOH And Charcoal
Canister (Expendable)
CO_2 Sensor
and Amplifier
RCU Connector
Power Supply
Back Flow Check Valve
Feed Water
Reservoir Assy.
Multiple Water
Connector
LCG Inlet and Water
Temperature Sensor
Oxygen Outlet Connector
PGA Electrical Umbilical
Alarm Control Module
Gas Separator Assembly
Inlet Oxygen Connector

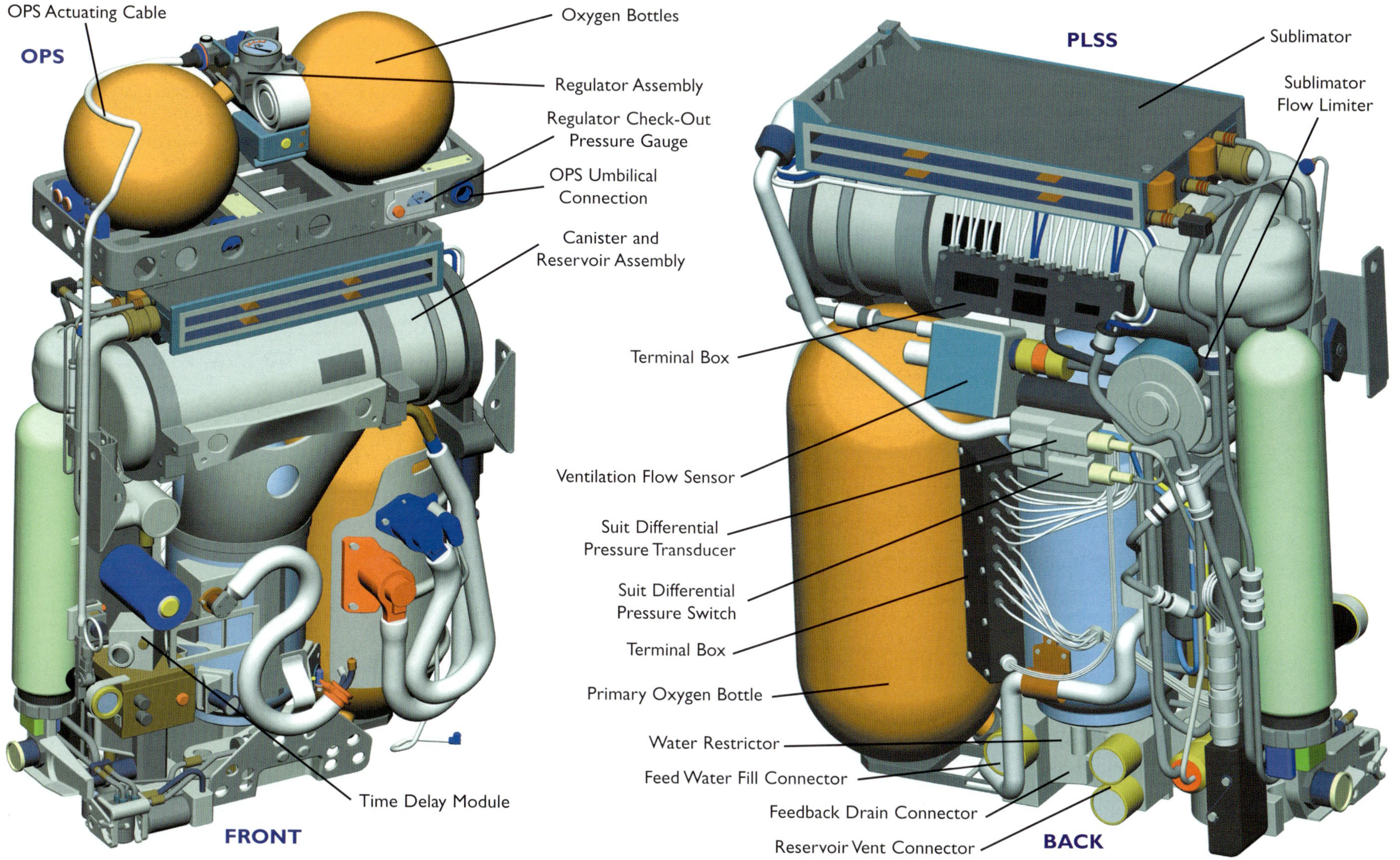
OPS Actuating Cable
OPS
Oxygen Bottles
Regulator Assembly
Regulator Check-Out Pressure Gauge
OPS Umbilical Connection
Canister and Reservoir Assembly
Time Delay Module
FRONT
PLSS
Sublimator
Sublimator Flow Limiter
Terminal Box
Ventilation Flow Sensor
Suit Differential Pressure Transducer
Suit Differential Pressure Switch
Terminal Box
Primary Oxygen Bottle
Water Restrictor
Feed Water Fill Connector
Feedback Drain Connector
Reservoir Vent Connector
BACK

THE PLSS REMOTE CONTROL UNIT (RCU)

The remote control unit (RCU), which was attached to the suit chest, had switches for the life support pack's water pump and oxygen fan, 4-position communication selector switch, a radio volume control, an oxygen quantity gauge, and warning indicators. The OPS actuator was attached to the RCU during EVA.

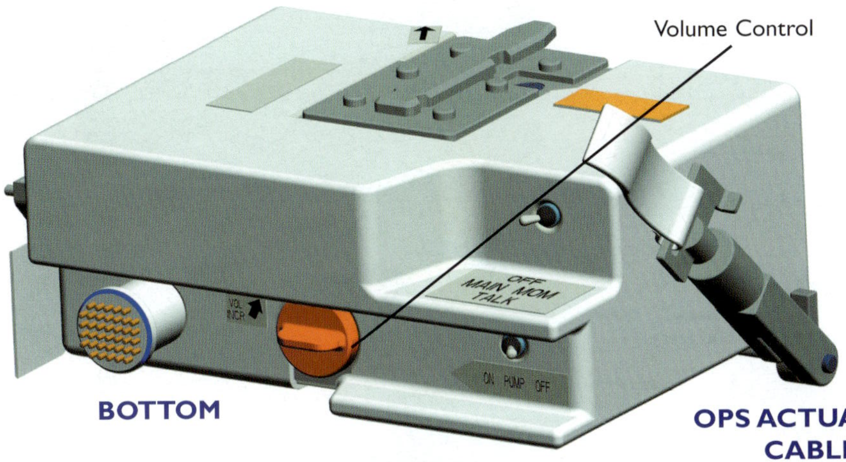

BOTTOM

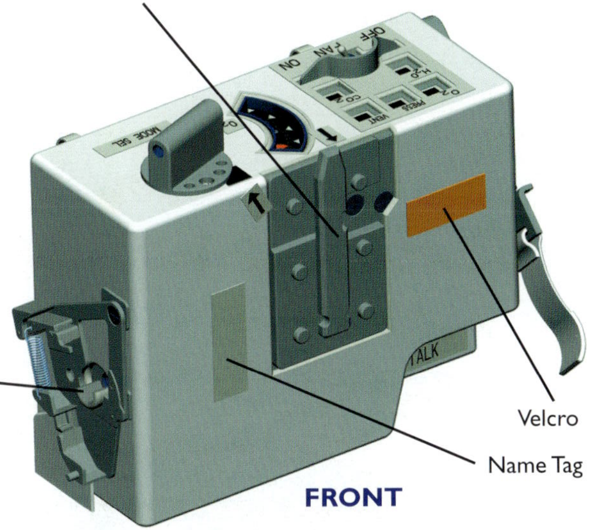

FRONT

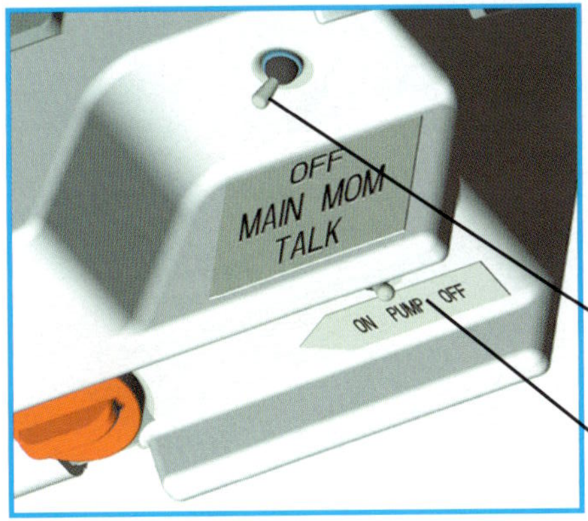

OPS ACTUATING CABLE

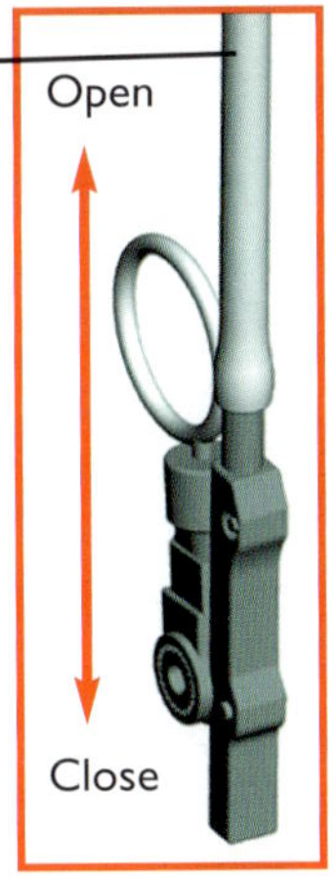

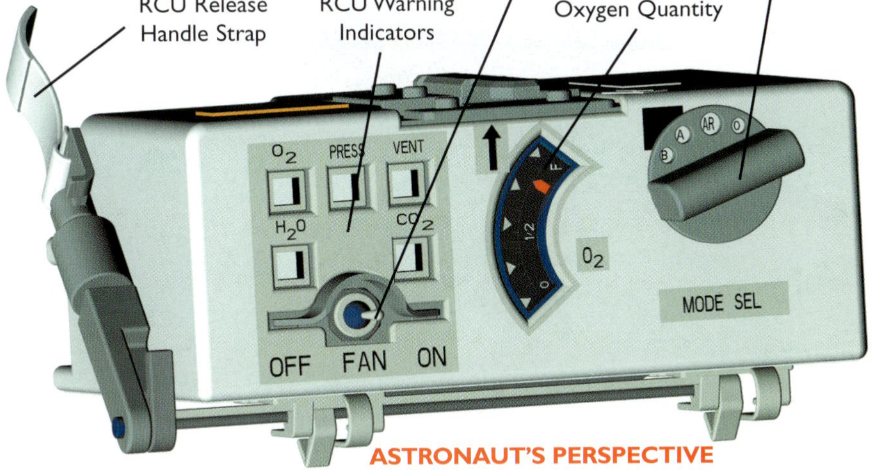

ASTRONAUT'S PERSPECTIVE

– *Scott P. Sullivan* –

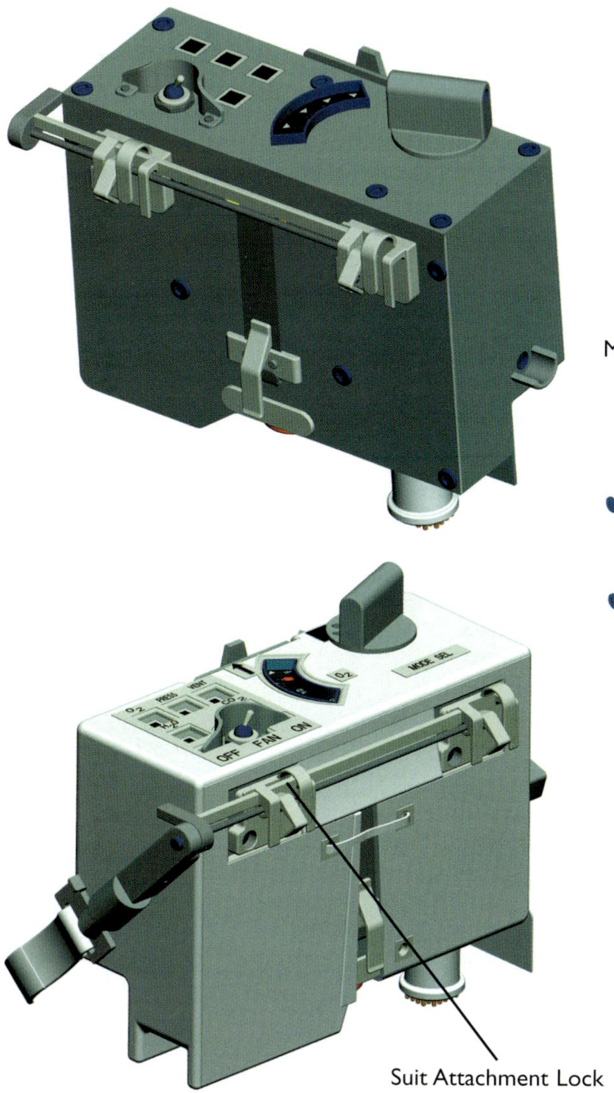

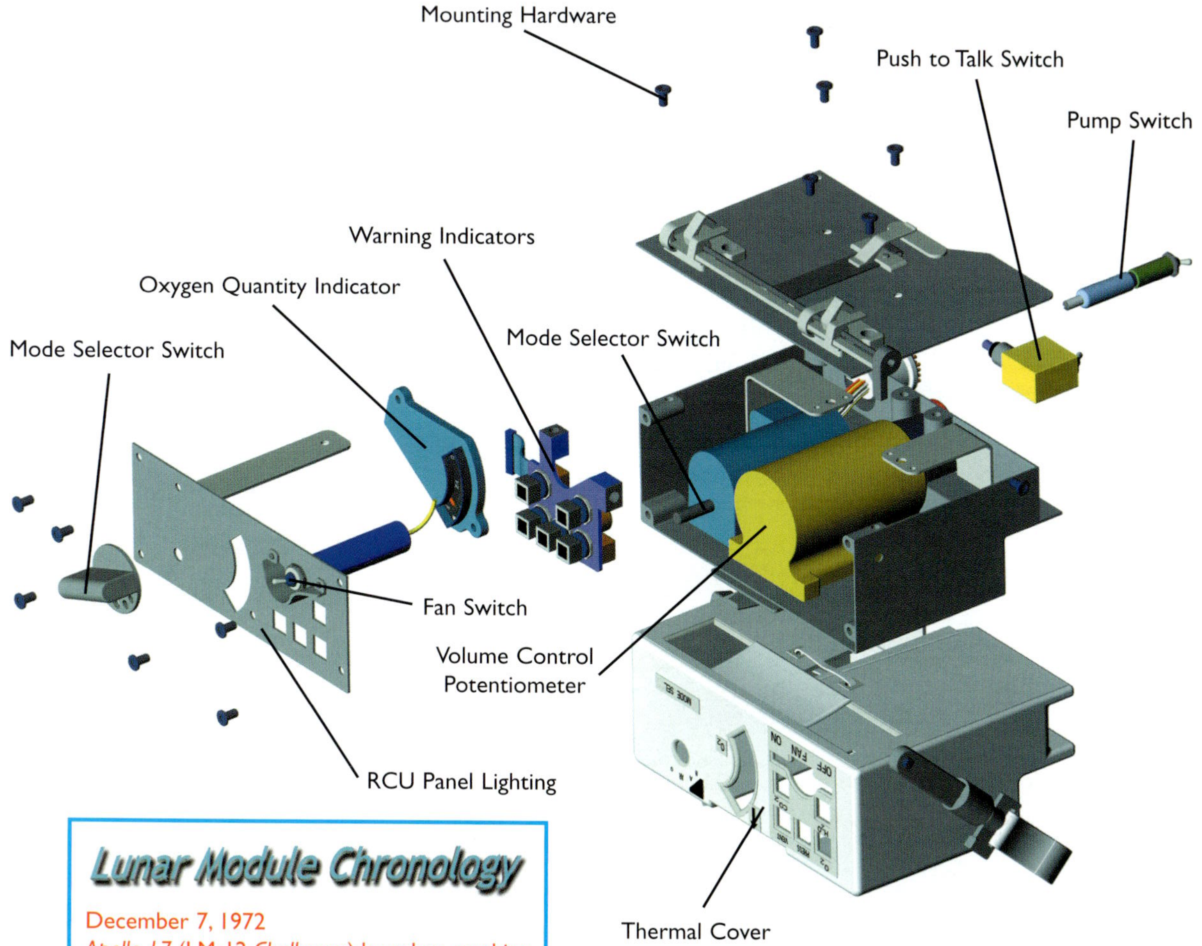

Lunar Module Chronology

December 7, 1972
Apollo 17 (LM-12 *Challenger*) launches, marking the final Apollo mission. On December 13, at 5:32 pm EST, the Apollo landings came to an end when the ascent stage lifted off from the Moon.

THE HISTORICAL ARTIFACT

These historic spacecraft are currently on display at the museums listed below.

1

2

3

4

5

1 LM 9 – (H-Series LM, replaced with J-Series LM on *Apollo 15*) Kennedy Space Center, Orlando Florida.

2 LM 2 – (unmanned backup) The National Air and Space Museum, Smithsonian Institution, Washington, DC.

3 LM 13 – (mission canceled) Cradle of Aviation Museum, Uniondale, New York.

4 Training model – Franklin Institute, Philadelphia, Pennsylvania.

5 LTA-1 – (Lunar Test Article) Cradle of Aviation Museum, Uniondale, New York.

– Scott P. Sullivan –

The NASA Mission Reports Series

Freedom 7 – ISBN 1-896522-80-7. America's first man in space, Alan B. Shepard, Jr., puts the US firmly in the race against the Soviets. *Includes a bonus CD-ROM.*

Friendship 7 – ISBN 1-896522-60-2. John Glenn pilots the US first manned orbital flight in 1962 and becomes a national hero. *Includes a bonus CD-ROM.*

Sigma 7 – ISBN 1-894959-01-9. Project Mercury Astronaut Walter M. Schirra, Jr., in his Sigma 7 spacecraft, orbited the Earth six times, completing another step in the grand U.S. program to send a man to the Moon and then return him safely to the Earth. *Includes a bonus CD-ROM.*

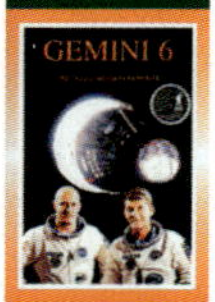

Gemini 6 – ISBN 1-896522-61-0. Piloted by Mercury astronaut Wally Schirra and Tom Stafford; the US achieves the first rendezvous in space with *Gemini 7*. *Includes a bonus CD-ROM.*

Gemini 7 – ISBN 1-896522-82-3. Frank Borman and James Lovell not only achieve the first rendezvous with *Gemini 6*, but go on to an endurance record of two weeks in space. *Includes a bonus CD-ROM.*

Gemini 12 – ISBN 1-894959-04-3. Space-walking had proven arduous and dangerous on previous Gemini flights, but after months of rigorous training, Buzz Aldrin proved that men could actually work comfortably in space. *Includes a bonus CD-ROM.*

Apollo 7 – ISBN 1-896522-64-5. Wally Schirra takes the helm again to prove that America's Apollo spacecraft can perform flawlessly for a full 8 days. *Includes a bonus CD-ROM.*

Apollo 8 – ISBN 1-896522-66-1. Borman, Lovell and Anders pilot the Apollo spacecraft out of Earth orbit to the Moon for the first time in human history. Genesis is read from Lunar orbit on Christmas Eve 1968. *Includes a bonus CD-ROM.*

Apollo 9 – ISBN 1-896522-51-3. This mission to test the full Apollo configuration, including the LM, had so many firsts it was almost a space program in its own right. *Includes a bonus CD-ROM.*

Apollo 10 – ISBN 1-896522-68-8. The penultimate full-up test of the CSM and LM in lunar orbit before the landing. The crew was probably the best ever to fly, and they had a blast doing it. *Includes a bonus CD-ROM.*

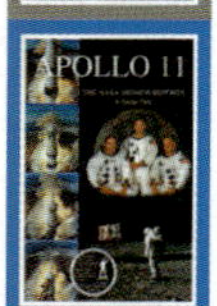

***Apollo 11**, Volume 1* – ISBN 1-896522-53-X. Neil, Buzz and Mike take that "One small step . . ." and succeed in completing the greatest scientific achievement in the history of mankind – the first manned landing on the Moon. *Includes a bonus CD-ROM of the EVA and over 1,200 color photos.*

***Apollo 11**, Volume 2* – ISBN 1-896522-49-1. The crew debrief of the mission, that was declassified specifically for this book. The story of the first moon landing in the crew's own words, including that pesky UFO sighting. *Includes a bonus CD-ROM with interactive lunar panoramas and exclusive interview with Buzz Aldrin.*

***Apollo 11**, Volume 3* – ISBN 1-896522-85-8. The crew performed many scientific tests on the Moon, and here is what they found. *Includes a bonus DVD with the landing video and EVA in a highly acclaimed format never seen before.*

Apollo 12 – ISBN 1-896522-54-8. "It may have been a small one for Neil, but it was a big one for me!" Pete Conrad and Al Bean prove that a landing can be bang on target. *Includes a bonus CD-ROM with the full EVA and exclusive interview with Dick Gordon.*

Apollo 13 – ISBN 1-896522-55-6. Possibly the US space program's finest hour. Find out what went wrong and how they saved this dramatic mission that almost resulted in the deaths of the crew. *Includes a bonus CD-ROM with an exclusive video interview with Captain Lovell.*

Apollo 14 – ISBN 1-896522-56-4. It was up to America's first astronaut, Alan Shepard, and his crew to get back to the Moon and to ... play a little golf! *Includes a bonus CD-ROM.*

***Apollo 15**, Volume 1* – ISBN 1-896522-57-2. Dave Scott and Jim Irwin land in the middle of a mountain range with their little Moon buggy to go for a spin, while Al Worden conducts many experiments in lunar orbit. *Includes a bonus CD-ROM.*

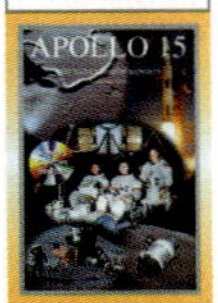

***Apollo 16**, Volume 1* – ISBN 1-896522-58-0. Captain John Young (The astronaut King) and Charlie Duke spend three days in the highlands at Descartes in a typically flawless Young fashion. *Includes a bonus CD-ROM with hours of EVA footage*

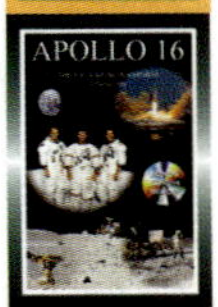

***Apollo 17**, Volume 1* – ISBN 1-896522-59-9. The last men on the Moon, Gene Cernan and Jack Schmitt (the first scientist to make the trip) almost make the whole thing look routine because they have so much fun. *Includes a bonus CD-ROM.*

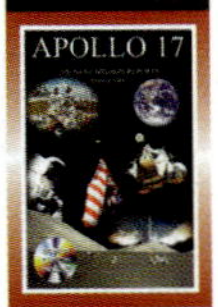

Space Shuttle STS 1-5 – ISBN 1-896522-69-6. The first amazing five missions of the Space Shuttle *Columbia*, with crew debriefings and mission reports. *Includes a bonus CD-ROM with hours of video and 100's of photos.*

***X-15:** The NASA Mission Reports* – ISBN 1-896522-65-3. Before and while the US space program was in full swing, there was another group of men becoming astronauts in a very promising precursor to the Shuttle program. Some wonderful film of rocket planes in full flight. *Includes a bonus CD-ROM.*

***Mars:** The NASA Mission Reports* – ISBN 1-896522-62-9. Every mission to Mars from *Mariner 4* to *Global Surveyor*. See how our knowledge of the red planet has grown exponentially since the 60's. *Includes a bonus CD-ROM with 1,000's of photos.*

***Mars:** The NASA Mission Reports, Volume 2* – ISBN 1-894595-05-1. This latest volume brings the exploration of Mars up to date, including the latest results from the amazingly successful *Spirit* and *Opportunity* rovers, as well as progress reports from *Mars Global Surveyor* and *Mars Odyssey*. *Includes a bonus CD-ROM.*

***Deep Space:** The NASA Mission Reports* – ISBN 1-894959-15-9. Missions to the outer solar system, Jupiter and beyond. Results from the past, *Voyagers 1* and 2; and the present, *Cassini-Huygens*; and a look at up-coming deep space missions. *Includes a feature packed bonus DVD.*

The Apogee Space Series

Rocket and Space Corporation Energia – ISBN 1-896522-81-5. For the first time in English, a pictorial history of the Russian Space program from the 1940's through Sealaunch.

The High Frontier – ISBN 1-896522-67-X. Gerard O'Neill writes about human colonies in space. This is the book that inspired the founders of the Space Frontier Foundation, among others. Includes 6 new contemporary chapters by leading visionaries. *Includes a bonus CD-ROM.*

Arrows to the Moon – ISBN 1-896522-83-1. Did you know that in 1958 the most advanced fighter aircraft in the world was not US or Russian? It was Canadian! When the Canadian government canceled the Avro Arrow program many of the aerospace engineers left to become an integral part of NASA.

The Unbroken Chain – ISBN 1-896522-84-X *by Guenter Wendt and Russell Still.* From the man who tucked all of the astronauts in their couches, shook their hands and closed the hatch from Alan Shepard through to the Shuttle. An amazing story! *Hardcover with a bonus CD-ROM.*

Creating Space – ISBN 1-896522-86-6 *by Mat Irvine.* The story of the Space age told through model kits. *Full color book with a foreword by Sir Arthur C. Clarke.*

Women Astronauts – ISBN 1-896522-87-4 *by Laura Woodmansee.* Every woman who has ever flown in space, how they did it, and more importantly, WHY they did it. *Full color book with a bonus CD-ROM, including exclusive interviews with current astronauts.*

On To Mars – ISBN 1-896522-90-4 *Edited by Dr. Robert Zubrin.* A collection of papers presented at the annual Mars Society Conferences, from the scientific issues to the moral and ethical questions posed by colonizing a new world. *Includes a bonus CD-ROM.*

The Conquest of Space – ISBN 1-896522-92-0 *by David Lasser.* The first scientifically accurate book written about space travel in the English language. This book is the one that inspired people like Sir Arthur C. Clarke to become interested in space. It still holds up today and is a wonderful reference for the novice.

Lost Spacecraft – ISBN 1-896522-88-2 *by Curt Newport.* The story of the search and recovery of Gus Grissom's *Liberty Bell 7* spacecraft by the man who did it. This was the subject of a major Discovery Channel program. *Includes a bonus CD-ROM.*

Virtual Apollo – ISBN 1-896522-94-7 *by Scott P. Sullivan.* If you ever wanted to know how to build a moonship, here it is – in awesome 3D color computer graphics.

Apollo EECOM – ISBN 1-896522-96-3 *by Sy Liebergot.* The amazing and heart-rending story of one of the key figures in US space history. The first man in Mission Control to acknowledge that *Apollo 13* really did have a problem. *Includes a bonus CD-ROM.*

A Vision of Future Space Transportation – ISBN 1-896522-93-9 *by Tim McElyea.* The official publication of SpaceDay 2003, by the man who produces the computer simulations of future spacecraft for NASA and the USAF. *Includes a bonus CD-ROM with animations, videos and 3D renderings.*

Space Trivia – ISBN 1-896522-98-X *by Bill Pogue*. The latest book from the *Skylab* astronaut who brought us the best seller *How Do You Go To The Bathroom In Space*. Entertains with 1000's of irreverent facts about how and why we go into space.

Interstellar Travel and Multi-Generational Spacecraft – ISBN 1-896522-99-8 *by Yoji Kondo and others from NASA's Goddard Space Flight Center*. If you really wish to travel to the stars, these guys are the ones who could "make it so!"

Dyna-Soar – ISBN 1-896522-95-5 *by Robert Godwin*. Called "A treasure trove . . ." by IEEE Spectrum. This is the complete story of America's first military space shuttle and how it could have flown in the early 1960's. *Includes a bonus DVD.*

The Rocket Team – ISBN 1-894595-00-0 *by Frederick Ordway III and Mitchell Sharpe*. This book tells the story of how a small group of amateur rocketeers led by Wernher von Braun became the rocket team, who did so much to make space flight a reality. *Hardcover with a bonus DVD featuring hours of rare videos.*

Women of Space – ISBN 1-894595-03-5 *by Laura Woodmansee*. "Cool Careers on the Final Frontier" – for girls, young women, and anyone else interested in learning about exciting careers in space exploration. *Includes a bonus CD-ROM.*

Columbia Accident Investigation Report – ISBN 1-894595-06-X. All of the details of the Gehman Report on the Space Shuttle Columbia's final flight, plus supplemental materials. *Includes a bonus CD-ROM featuring NASA movies, plus over 700 pages of additional material.*

The Simple Universe – ISBN 1-894595-11-6 *by Ian Brewster and Ken Shiwram*. If you have an interest in understanding the universe, this book is a good place to start because it takes a complex subject and makes it completely understandable.

New Moon Rising (*The Making of America's New Space Vision*) – ISBN 1-894595-12-4 *by Frank Sietzen and Keith Cowing*. The most comprehensive look available on what America's new space vision will do for human exploration of the Solar System – and how NASA will change as a result. *Hardcover with a bonus DVD.*

Moonrush (*How the Wealth of the Moon will improve life on Earth*) – ISBN 1-894595-10-8 *by Dennis Wingo*. This book examines how the exploration of space, specifically a commercial base on the Moon and Mars would transform our economies on the Earth. The Moon holds great promise for a high tech manufacturing future.

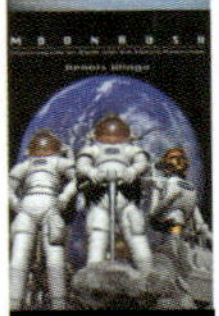

Rocket Science – ISBN 1-894595-09-4 *by Alfred Zaehringer*. From the man who actually originally coined the term "rocket science" comes an informative and comprehensive discussion of the history, state of the art, and future of space propulsion, presented in a completely understandable way.

How NASA Learned to Fly in Space – ISBN 1-894595-07-8 *by David M. Harland*. Apollo claimed the glory, but it was Project Gemini that "stretched the envelope" of space flight to make going to the Moon feasible. Read about how NASA learned to fly in space at a time when the agency was young and lean.

Virtual LM – ISBN 1-894595-14-0 *by Scott P. Sullivan*. The follow-up to *Virtual Apollo*, this book details the design and construction of the Apollo Lunar Module in marvelous full-color graphics. *Includes a bonus CD-ROM.*

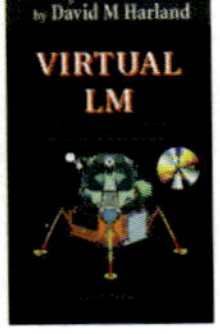

Apogee Books Space Series

#	Title	ISBN	Bonus	US$	UK£	CN$	
1	Apollo 8	1-896522-66-1	CDROM	$18.95	£13.95	$25.95	______
2	Apollo 9	1-896522-51-3	CDROM	$16.95	£12.95	$22.95	______
3	Friendship 7	1-896522-60-2	CDROM	$18.95	£13.95	$25.95	______
4	Apollo 10	1-896522-52-1	CDROM	$18.95	£13.95	$25.95	______
5	Apollo 11 Vol 1	1-896522-53-X	CDROM	$18.95	£13.95	$25.95	______
6	Apollo 11 Vol 2	1-896522-49-1	CDROM	$15.95	£10.95	$20.95	______
7	Apollo 12	1-896522-54-8	CDROM	$18.95	£13.95	$25.95	______
8	Gemini 6	1-896522-61-0	CDROM	$18.95	£13.95	$25.95	______
9	Apollo 13	1-896522-55-6	CDROM	$18.95	£13.95	$25.95	______
10	Mars	1-896522-62-9	CDROM	$23.95	£18.95	$31.95	______
11	Apollo 7	1-896522-64-5	CDROM	$18.95	£13.95	$25.95	______
12	High Frontier	1-896522-67-X	CDROM	$21.95	£17.95	$28.95	______
13	X-15	1-896522-65-3	CDROM	$23.95	£18.95	$31.95	______
14	Apollo 14	1-896522-56-4	CDROM	$18.95	£15.95	$25.95	______
15	Freedom 7	1-896522-80-7	CDROM	$18.95	£15.95	$25.95	______
16	Space Shuttle STS 1-5	1-896522-69-6	CDROM	$23.95	£18.95	$31.95	______
17	Rocket Corp. Energia	1-896522-81-5		$21.95	£16.95	$28.95	______
18	Apollo 15 - Vol 1	1-896522-57-2	CDROM	$19.95	£15.95	$27.95	______
19	Arrows To The Moon	1-896522-83-1		$21.95	£17.95	$28.95	______
20	The Unbroken Chain	1-896522-84-X	CDROM	$29.95	£24.95	$39.95	______
21	Gemini 7	1-896522-80-7	CDROM	$19.95	£15.95	$26.95	______
22	Apollo 11 Vol 3	1-896522-85-8	DVD*	$27.95	£19.95	$37.95	______
23	Apollo 16 Vol 1	1-896522-58-0	CDROM	$19.95	£15.95	$27.95	______
24	Creating Space	1-896522-86-6		$30.95	£24.95	$39.95	______
25	Women Astronauts	1-896522-87-4	CDROM	$23.95	£18.95	$31.95	______
26	On To Mars	1-896522-90-4	CDROM	$21.95	£16.95	$29.95	______
27	Conquest of Space	1-896522-92-0		$23.95	£19.95	$32.95	______
28	Lost Spacecraft	1-896522-88-2		$30.95	£24.95	$39.95	______
29	Apollo 17 Vol 1	1-896522-59-9	CDROM	$19.95	£15.95	$27.95	______
30	Virtual Apollo	1-896522-94-7		$19.95	£14.95	$26.95	______
31	Apollo EECOM	1-896522-96-3		$29.95	£23.95	$37.95	______
32	Visions of Future Space	1-896522-93-9	CDROM	$27.95	£21.95	$35.95	______
33	Space Trivia	1-896522-98-X		$19.95	£14.95	$26.95	______
34	Interstellar Spacecraft	1-896522-99-8		$24.95	£18.95	$30.95	______
35	Dyna-Soar	1-896522-95-5	DVD*	$32.95	£23.95	$42.95	______
36	The Rocket Team	1-894959-00-0	DVD*	$34.95	£24.95	$44.95	______
37	Sigma 7	1-894959-01-9	CDROM	$19.95	£15.95	$27.95	______
38	Women Of Space	1-894959-03-5	CDROM	$22.95	£17.95	$30.95	______
39	Columbia Accident Rpt	1-894959-06-X	CDROM	$25.95	£19.95	$33.95	______
40	Gemini 12	1-894959-04-3	CDROM	$19.95	£15.95	$27.95	______
41	The Simple Universe	1-894959-11-6		$21.95	£16.95	$29.95	______
42	New Moon Rising	1-894959-12-4	DVD*	$33.95	£23.95	$44.95	______
43	Moonrush	1-894959-10-8		$24.95	£17.95	$30.95	______
44	Mars Volume 2	1-894959-05-1	DVD*	$28.95	£20.95	$38.95	______
45	Rocket Science	1-894959-09-4		$20.95	£15.95	$28.95	______
46	How NASA Learned	1-894959-07-8		$25.95	£18.95	$35.95	______
47	Virtual LM	1-894959-14-0	CDROM	$29.95	£22.95	$42.95	______
48	Deep Space	1-894959-15-9	DVD*	$TBA	£TBA	$TBA	______

* NTSC Region 0

BONUS *VIRTUAL LM* CD-ROM

The included bonus CD-ROM contains information to complement the renderings in *Virtual LM*. Manuals and checklists that describe the LM and detail its operation are included. Additionally, detailed photographs of the LM mission simulator cockpit, and the LTA-1 engineering test article show what the real spacecraft were like.

Spacecraft Documents:
- Apollo Operations Handbook, Volume I
- Apollo Operations Handbook, Volume II
- Apollo 14 Lunar Module Activation Checklist
- Apollo 14 Lunar Module Timeline
- Apollo 15 Lunar Module Cue Cards
- Apollo 15 Lunar Module Data Cards
- Apollo 16 Contingency Checklist

Photographs:
- LM Mission Simulator: Cockpit Interior
- LM Test Article-1: Descent Stage
- LM Test Article-1: Ascent Stage

VIRTUAL APOLLO

Virtual LM is the successor to the highly acclaimed *Virtual Apollo*, and has been crafted with similarly high-quality, full-color renderings, done in exacting detail, and providing an array of views of the Apollo Lunar Module, complete with details of design and fabrication.

The Apollo Lunar Module was a true space ship. Its sole purpose was to land men on the Moon, facilitate and protect them while they were there, and then return them to space, where they would rendezvous and head for home. And during every mission it succeeded in full.

The Lunar Module was the essential complement to the Command and Service Modules detailed in *Virtual Apollo*. Together these two books tell the story, in brilliant pictures, of the spacecraft that took men to the Moon and returned them safely to Earth.

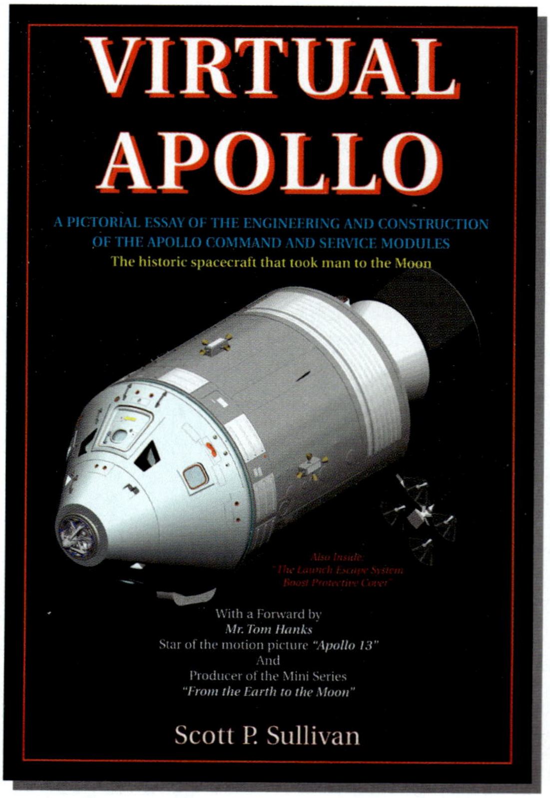